U0184325

乾隆朝《泰陵图》（出自《中国美术分类全集 · 中国建筑艺术全集 · 清代陵墓建筑》）

Drawing of the Tomb of Peace (*Tailing tu*), drawn in Qianlong Dynasty. (*Complete Series of Chinese Architectural Art, the authoritative volume on Qing-dynasty funerary architecture*)

西陵全图，光绪十三年至宣统元年（1887—1909 年）。故宫博物院藏

Site plan of the West Qing Tomb, finished in Guangxu Emperor's 13rd regnal year to Xuantong Emperor's 1st regnal year (1887-1909). In the collection of the Palace Museum

清西陵鸟瞰图（冯建逵绘）
Bird's-eye view of the Western Qing Tombs (Drawing by FENG Jiankui)

泰陵前区组群正立面渲染图
Front elevation rendering of the building complex of the Tomb of Peace's front area group

泰陵龙凤门起组群正立面渲染图

Front elevation rendering from the Dragon and Phoenix Gate of the building complex of the Tomb of Peace

泰陵神道碑亭起组群正立面渲染图
Front elevation rendering from the Pavilion for the Stela of the building complex of the Tomb of Peace

泰陵隆恩门起组群正立面渲染图
Front elevation rendering from the Gate of Monumental Grace of the building complex of the Tomb of Peace

泰陵琉璃花门起组群正立面渲染图
Front elevation rendering from the Gate with Glazed Roof Tiles of the building complex of the Tomb of Peace

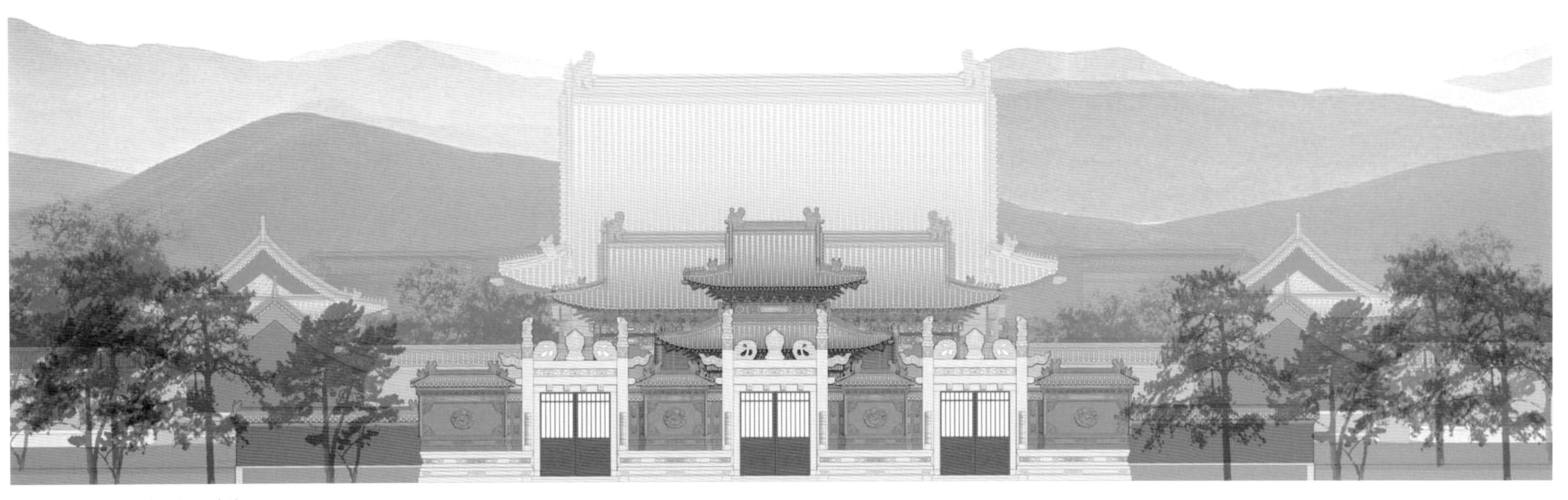

慕陵龙凤门起组群正立面渲染图
Front elevation rendering from the Dragon and Phoenix Gate of the building complex of Tomb of Veneration

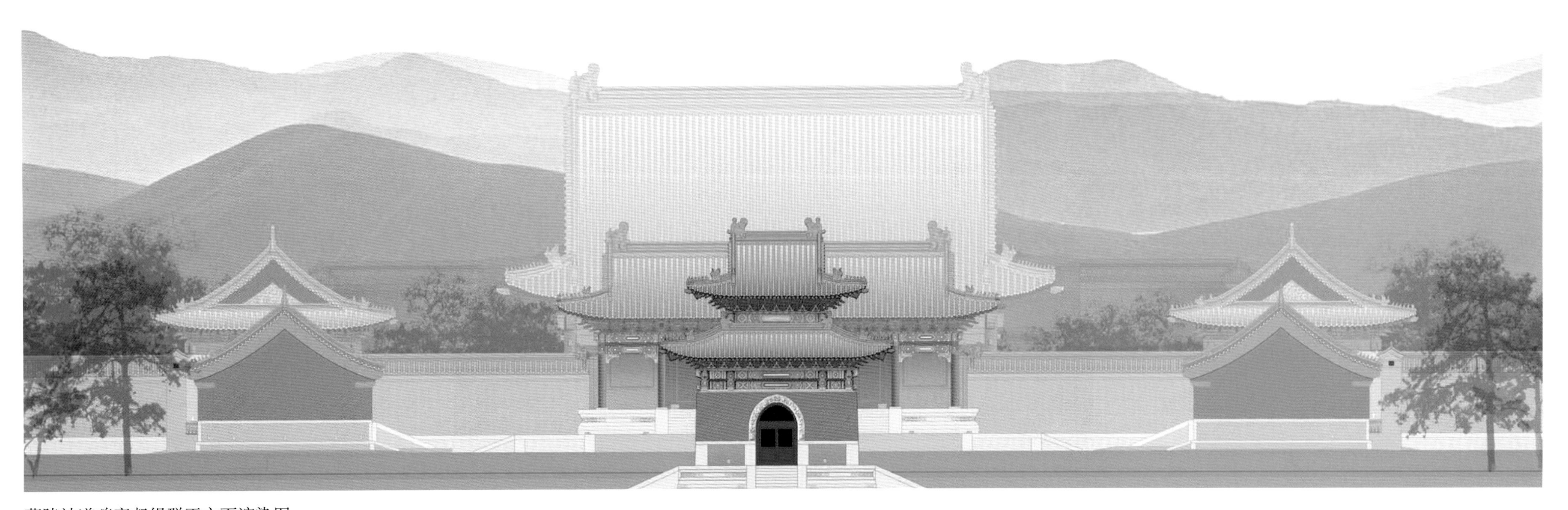

慕陵神道碑亭起组群正立面渲染图
Front elevation rendering from the Pavilion for the Stela of the building complex of Tomb of Veneration

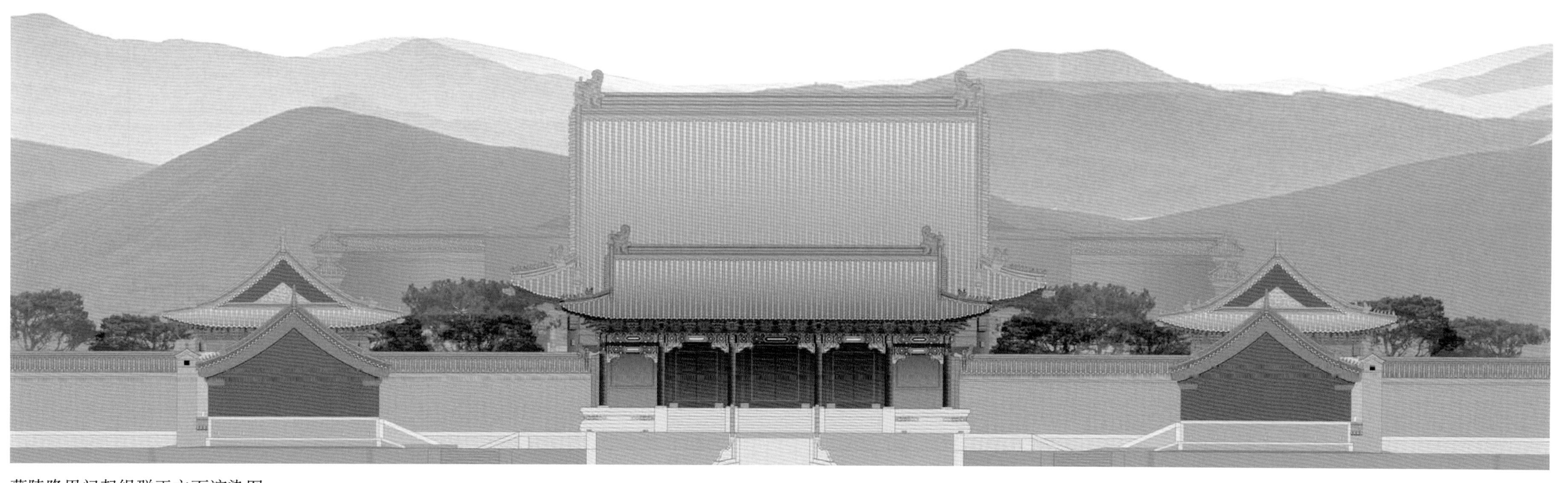

慕陵隆恩门起组群正立面渲染图
Front elevation rendering from the Gate of Monumental Grace of the building complex of Tomb of Veneration

0 2.5 5 10m

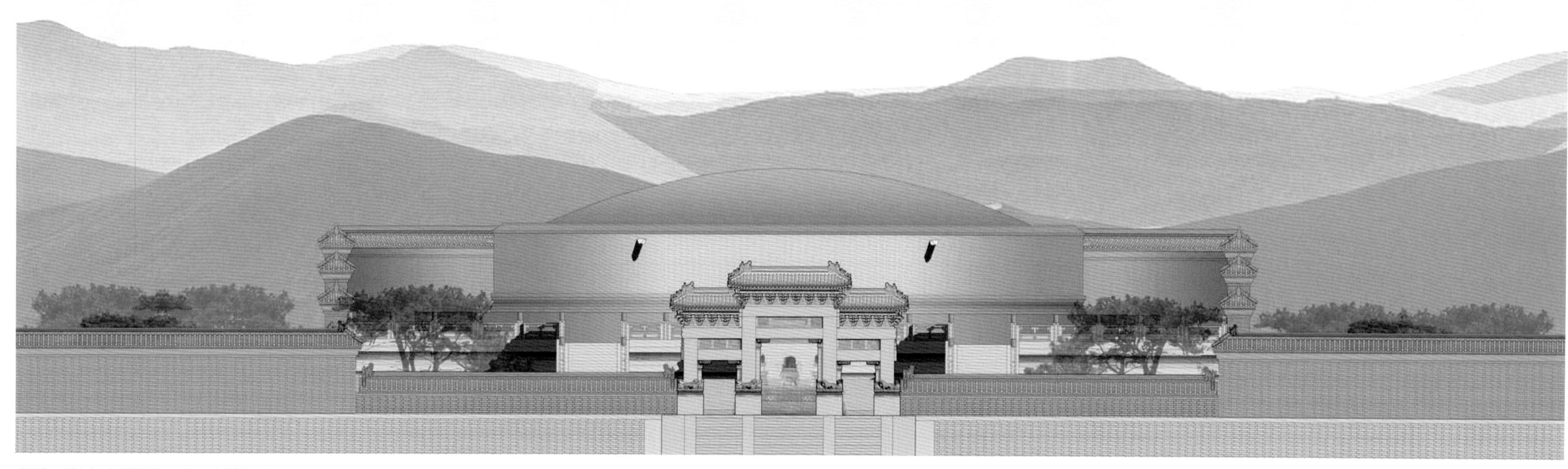

慕陵石牌坊起组群正立面渲染图
Front elevation rendering from the Marble Memorial Gateway of the building complex of Tomb of Veneration

0 2.5 5 10m

永福寺组群正立面渲染图
Front elevation rendering of the building complex of Temple of Perpetual Happiness

永福寺大雄宝殿组群正立面渲染图
Front elevation rendering of the building complex of the Daxiong Hall (the main hall in a Buddhist temple), Temple of Perpetual Happiness

永福寺牌坊起组群正立面渲染图

Front elevation rendering from the Memorial Gateway of the building complex of Temple of Perpetual Happiness

0 1 2 4m

永福寺组群侧立面渲染图

Side elevation rendering of the building complex of Temple of Perpetual Happiness

0 1 2 4m

国家出版基金项目

「十二五」国家重点图书出版规划项目

中国古建筑测绘大系·陵寝建筑

清西陵

天津大学建筑学院 易县清西陵文物管理处 合作编写
王其亨 主编
朱蕾 陈书砚 王其亨 编著

中国建筑工业出版社

Traditional Chinese Architecture Surveying and Mapping Series:
Tomb Architecture

WESTERN QING TOMBS

Compiled by School of Architecture, Tianjin University & Western Qing Tombs
Cultural Relics Management Office, Yixian
Chief Edited by WANG Qiheng
Edited by ZHU Lei, CHEN Shuyan, WANG Qiheng

China Architecture & Building Press

Contents

目录

Introduction

Located west of Lianggezhuang, Yi county, Hebei province, the Western Qing Tombs, one of China's listed UNESCO World Heritage Sites, is the second large-scale imperial Tomb complex built by the Qing dynasty south of the Shanhai Pass. The Western Qing Tombs cover a total area of 178.35 km sq., bordering Lianggezhuang village to the east and Twin Pagoda (*shuangta*) Peak of Yunmeng Mountain to the west, reaching to the northern side of Qifeng Mountain to the north and Flame Memorial Gateway to the south. The important protection area is 1942 hectares, and the development control area (buffer area) is 4758 hectares.

Construction of the Western Qing Tombs spanned one hundred and eighty-six years, beginning in with building at Tomb of Peace (Tailing, the tomb of emperor Yongzheng) in 1730, the eighth reign year of emperor Yongzheng, and lasting until in the completion of Tomb of Serendipity (Chongling, the tomb of emperor Guangxu) in 1915. There would eventually be built fourteen tombs: the tombs of four emperors—Tomb of Peace, Tomb of Good Fortune (Changling, the tomb of emperor Jiaqing), Tomb of Veneration (Muling, the tomb of emperor Daoguang), and Tomb of Serendipity; three tombs for empresses—East Tomb of Peace, West Tomb of Good Fortune, and East Tomb of Veneration; three tombs for imperial concubines affiliated with Tomb of Peace, Tomb of Good Fortune, and Tomb of Serendipity respectively; one tomb for princes and another for princesses. In these tombs a total of seventy-nine people were buried including four emperors, nine empresses, fifty-seven imperial concubines, two princes, five sons of emperors, and two princesses.

The Tomb complex was surrounded by geomantic (*fengshui*) walls and consisted of the Temple of Perpetual Happiness (Yongfu Si), a structure located adjacent to the tombs. Additionally, there was a traveling palace or "temporary palace" in nearby Lianggezhuang and supporting facilities like Spirit Ways (*shendao*), water engineering systems, camps (*yingfang*), government offices (*yashu*), and guard houses (*xunfang*). In the late Qing dynasty, the Xin(cheng)-Yi(xian) railway line was built that stopped at a local station near the Temple of Perpetual Happiness. At the periphery were built subordinate tombs for four princes of the first rank (*qinwang*), including Chun Du, Yu Dao, Yu Zhuang, and Guo Yi, as well as for the prince of the second rank

导言

世界文化遗产地清西陵是清王朝开辟的关内第二处规模宏大的皇家陵区，位于今河北省易县梁格庄以西，北至奇峰岭，南抵火焰牌楼，西达云蒙山双塔站顶峰，其中，重点保护范围面积1842hm^2，建设控制地带（即缓冲区）面积4758hm^2。

雍正八年（1730年）泰陵鼎建，1915年崇陵全工告竣，历时186年，共建成14座陵寝。包括泰陵、昌陵、慕陵和崇陵4座帝陵，泰东陵、昌西陵和慕东陵3座后陵，以及分别隶从于泰陵、昌陵和崇陵的3座妃园寝；另建王爷园寝2座、皇子园寝1座、公主园寝1座。葬皇帝4位、皇后9位、嫔妃57位、亲王2位、皇子6位、公主2位，共80人。

陵寝周围配建风水围墙、永福寺、梁格庄行宫以及从属各陵的神道、水系工程、营房、衙署和汛房，并于清末增建新易铁路及车站。外围建淳度亲王、裕悼亲王、裕庄亲王、果毅亲王、果恭郡王和固伦端顺公主等陪葬园寝。陵域遍植古松，茂

图1 清西陵风水地势全图（日本东京大学东洋文化研究所藏，引自国家图书馆「清代样式雷建筑图档展」）

图2-1 泰陵前区鸟瞰（出自《中国美术分类全集·中国建筑艺术全集·清代陵墓建筑》）

图2-2 泰陵龙凤门北神道上行进中所见，建筑外部空间的序列组织出神入化，造诣极高（出自《风水理论研究》）

Fig.1 Terrain of the Western Qing Tombs. (In the collection of the Institute of Oriental Culture at the University of Tokyo; here from National Library of China, "Exhibition of the Qing-dynasty Yangshi Lei Archives")

Fig.2-1 Bird's-eye view of the front area of Tomb of Peace. (*Complete Series of Chinese Architectural Art, the authoritative volume on Qing-dynasty funerary architecture*)

Fig.2-2 Different views along the spirit way north of the Dragon and Phoenix Gate at Tomb of Peace. (*Fengshui lilun yanjiu*)

(*junwang*) Guo Gong, and the princess of the first rank (Gulun *gongzhu*) Duan Shun. Planted with luxuriant old pines, the site conveyed solemnity and reverence (Fig.1).

The Western Qing Tombs show the evolution of the burial scheme in the Ming and Qing dynasties, specifically the origin of the Zhaomu System. The building of the second imperial Tomb impelled emperor Qianlong (temple name Gaozong; born Hongli) to initiate this system to regulate the location of burial sites either on the east or the west side of the entire burial complex, alternating with each successive ruler, to coordinate the two (east and west) clusters and their further development at policy level. During the construction, attention was to be paid to balance, cultural tradition, and innovation. At the same time, all stages of architecture and landscape planning, in addition to detail design work, were informed by a practice that emphasized adjustment to local topography and surface characteristics and timely modification (if necessary) to further improve the system.

The placement of architecture to fit perfectly with the natural topography (mountains and rivers) at the Western Qing Tombs is the ultimate embodiment of the traditional Chinese ecological concept. When selecting the site for his tomb, emperor Yongzheng was looking for a place where "all *(long, xue, sha, shui*; *fengshui* terms for topographic elements) should be beautiful, all (*xingshi, liqi; fengshui* terms for energy manifested in the physical environment) should be propitious", a place where architecture could be built in harmony with its surroundings. In fact, to the west of the Tomb complex lies Yongning Mountain; at the southern end winds the Yi (shui) River; to the east and south-west stand the Nine Dragons (Jiulong) and Nine phoenixes (Jiufeng) mountains, while Yuanbao Mountain and the East and West Huagai mountain ranges create marvelous scenery in front (south) of the cemetery. Natural mountains and rivers surround, protect, and guard the tombs that have been nestled into the landscape. Thanks to the perfect choice of location, the skillful site layout that embodies the required geomantic principles, and the strict control of visibility and size of individual monuments, a variety of spatial contrasts were achieved, including the juxtapositions of sparsity and density, openness and closure, macro- and microscopic views, and monotony and change. In addition, the adoption of *guobai* (i.e. leaving a sufficient distance between two buildings so that, if someone stands in the rear building and looks forward, he can see the whole front building in addition to a thin strip of sky formed by the inner borderline of the rear building), *jiajing* (i.e. using barriers such as trees or buildings at both sides to emphasize a distant-view scenery) and *kuangjing* (i.e. using for example door and window to selectively present a scenery) strengthened the artistic effect and the sensual experience of space, making the Western Qing Tombs a masterpiece of design creativity and originality "in harmony with the landform of mountains and rivers" (Fig.2).

郁葱胜，庄严敬肃（图1）。

清西陵是明清陵制演变过程的实物载体，东西陵昭穆制度的滥觞。第二皇家陵区的辟建促使清高宗弘历颁定东西陵昭穆制，政策层面上长久协调二陵的关系及发展；在次第修建各建置过程中，注重平衡、因承和创新，在规划、建筑、景观和细部设计的各级尺度都显出因地制宜、因时而化的革新，进一步完善陵寝制度。

清西陵建置与山川胜势完美匹配，是中国传统景观生态理念的极致展现。雍正帝选勘陵址时严苛地追求『龙穴砂水无美不收，形势理气诸吉咸备』的山水格局，注重并强调建筑与环境和谐统备的意象。陵区在气势磅礴的永宁山映衬下，易水河萦绕南端，九龙山和九凤山迤逦两旁，前有元宝山和东西雄峙的华盖山构成恢弘对景。天然造就的山川形势对镶嵌于其中的各陵寝形成了拱卫、环抱、朝揖之势（图2-1）。在几近完美的选址条件下，陵区建置经营巧妙结合风水地势，通过对谒陵流线、单体建筑尺度及视距的精准控制，实现空间疏朗与紧凑、开敞与封闭、宏观与微观、凝滞与激变的对比；使用风水过白夹景与框景强化空间艺术感受效果，堪称『配合山川之胜势』规划设计藻思的成功杰作（图2-2）。

清西陵是清代皇家工程高超技术水平的集中反映。泰陵工程与工部《工程做法》颁布同期，建置遵循官式做法规范，成为研究《清工部工程做法》的重要实物参照；清西陵工程档案中的格子本使用了在二维介质上记录三维空间数据的平格方法，与现代数字高程模型（DEM）技术并无二致；清末民初的崇陵工程大胆尝试新材料、新技术，不但引进水泥而且灵活地使用火车、摄影、电灯、电报和电话等技术手段。

清西陵是营建资料保留最完整的皇家陵寝之一，是世界记忆遗产——『样式雷』建筑图档的实物见证。清西陵作为绵延近二百年的重大国家级工程项目，其勘察设计图和工程档案是样式雷图档的重要组成。这些珍贵的历史文献和建筑遗存相互印证，不仅是揭橥中国古代建筑规划与设计理念和方法的重要基础，而且是世界遗产名录和世界记忆工程两个项目对人类文明瑰宝进行全面保护的重要例证（图3）。

清西陵的建置记录着清王朝由盛至衰的演变过程。泰陵和昌陵完整宏伟的陵寝规模反映王朝盛期的辉煌，慕陵裁撤建筑与崇陵缩减规模真实地标记帝国由盛而衰的历史轨迹，堪称清中期至清末民初重大历史事件的史实资料。

现代遗产保护工作与清代陵寝护卫传统相衔接，使得清西陵各建置大部分保存完好，还留下号称『华北之最』的万余株古松，成为现存格局最完整的清代陵寝组群，是中国两千年来陵寝建筑艺术辉煌壮丽的一页。

The Western Qing Tombs reflect the superb technical level used in government-sponsored Qing-dynasty construction. The building of Tomb of Peace coincides with the compilation and publication of *Engineering manual of the Ministry of Works* (*Gongbu gongcheng zuo-fa*; 1734). With structures implementing the building standards specified in the text, Tomb of Peace is key physical reference for the study of this technical manual. Additionally, the survey drawings among the archival documents of the Western Qing Tombs have reference lines forming grids (of squares) to show locations accurately; heights are recorded at the intersections of reference lines (without further definition of the surface), which turns these drawings into 3d graphic representations like the digital elevation models (DEM) used today. At Tomb of Serendipity (the tomb of emperor Guangxu), built from the late-Qing to 1910s, the planners and builders experimented with new materials such as cement and new technologies such as railway, camera, electric light, telegraph, and telephone.

The Western Qing Tombs are not only the best-documented imperial Tomb complex (regarding construction), but also the physical evidence of one of China's documentary heritages ("Qing Dynasty Yangshi Lei Archives") that is included in the Memory of the World Register since 2007. Since the Western Qing Tombs were a significant national engineering project for almost two centuries, the design and survey drawings and related project documents form an integral part of the Yangshi Lei Archives. The historical documents and the architectural remains are mutually reinforced each other. They are not only the key to understand and study the planning and design theories and methods of traditional Chinese architecture, but—inscribed on the World Heritage List and the Memory of the World Register—they also bear testimony to the importance the high culture of late imperial China has had for the history of human civilization (Fig.3).

The Western Qing Tombs witnessed the rise and fall of the last of China's dynasties—the Qing. The completeness and grand scale of Tomb of Peace and Tomb of Serendipity reflect the glory of this dynasty at its peak; while the simplified layout of Tomb of Veneration and the downsized scale of Tomb of Serendipity mark the turning point in the dynastic development from prosperity to decline. The Western Qing Tombs are thus a true mirror of the historical events that took place between the mid Qing to the late Qing and 1910s.

The combination of modern heritage protection in our contemporary age and the Qing tradition of tomb guarding established the Western Qing Tombs as the best-preserved complex among all the extant Qing tombs, including ten-thousand-year-old pine trees known as "North China's best", and as a brilliant episode in the long and colorful history (2000+ years) of imperial Chinese funerary art and architecture.

1. Construction History

Construction of the Western Qing Tombs was initiated in 1730, the eighth reign year of emperor Yongzheng, and completed in 1915. For environmental and security reasons, a strict boundary system was adopted for the relatively large tomb area with markers indicating the property lines in the form of a geomantic wall a fire-barrier belt, red, white or green boundary stones, and government mountains. Originally, Tomb of Peace was surrounded by a 20-km-long geomantic wall that started at the Great Red Gate, erected in-between Jiulong and Jiufeng mountains which are like eastern and western watchtowers or que. The wall ran eastwards through the Jiulong Mountain to Sanfeng Ridge, westward through Jiufeng Mountain to Taining Temple. Following the gradual expansion of the site, the boundaries were correspondingly enlarged (Fig.4).

Emperor Yongzheng (temple name: Shizong; born: Yinzhen) decided to build a tomb for himself in 1726, his fourth reign year. Three years later, in 1729, as no proper site had been found near Tomb of Filial Piety and Tomb of Admiration—two of the later East Qing Tombs—and because the originally selected site at Jiufeng Mountain in Zunhua, Hebei province, had been disapproved due to its bad soil, emperor Yongzheng chose Taiping Valley in Yi County located southwest of the capital (Beijing). Construction of his Tomb began in the next year and was completed in 1737, the second reign year of his fourth son and successor—emperor Qianlong—who named Yongzheng's Tomb "Peace". In the same year, the tomb of Yongzheng's concubines was completed, and construction at East Tomb of Peace began where Qianlong's mother, empress Xiaosheng, was buried in 1777, the forty-second year of her son's reign.

Emperor Jiaqing (temple name: Renzong; born: Yongyan) initiated the construction of his tomb in 1799, his fourth reign year. His Tomb was built west of Tomb of Peace in accordance with the Zhaomu System, and completed in 1803, together with the tomb for his concubines. Emperor Jiaqing died in 1820, his twenty-fifth reign year, and his son and successor emperor Daoguang named his father's tomb 'Tomb of Good Fortune'. West Tomb of Good Fortune, the Tomb for the second empress consort of emperor Jiajing, was built over the course of two years, from 1851, the first reign year of emperor Xianfeng, to 1852.

Emperor Daoguang (temple name: Xuanzong; born: Minning) began to build his tomb in Baohua Valley at the East Qing Tombs according to the Zhaomu System in 1821, his first reign year. It was completed after six years but soon afterwards demolished because water entered the Underground Palace (*digong*). A new tomb was built in Longquan Valley at

一、建置沿革

清西陵始建于雍正八年（1730年），完工于1915年。陵区幅员十分广阔，出于生态与安防等多方考虑，使用一套严谨的边界系统，包括风水墙，火道及红、白、青桩，官山。泰陵初营，修风水墙20多公里，自九龙山、九凤山东西阙立之间的大红门起，向东经九龙山至三峰岭止，向西经九凤山至泰宁寺止。随陵区陆续扩建，风水墙，火道及红、白、青桩，官山界定范围相应扩大（图4）。

雍正四年（1726年）清世宗陵寝备建。因孝、景二陵附近无上吉佳壤，又因土质废遵化九凤朝阳山吉地，雍正七年（1729年）末最终相地于京师西南的易县太平峪。雍正八年（1730年）陵寝工程正式动工，乾隆二年（1737年）完工，命名为『泰陵』，其妃园寝同期完工。

乾隆二年（1737年）『泰东陵』动工，乾隆四十二年（1777年）孝圣宪皇后入葬。

嘉庆四年（1799年）依昭穆次序，清仁宗陵寝兴建于泰陵之西。嘉庆八年（1803年）完工，其妃园寝同期完工。嘉庆二十五年（1820年）嘉庆帝崩，道光帝钦定陵名为『昌』。咸丰元年（1851年）『昌西陵』兴建，咸丰二年（1852年）完工。

道光元年（1821年）清宣宗陵寝依制在清东陵宝华峪兴工，约六年完工，因地宫渗水全部拆除。道光十一年（1831年）于清西陵龙泉峪冉次动工，道光十五年（1835年）告竣。其妃园寝同期兴建、拆迁，咸丰五年（1855年）为葬皇太后而增崇为『慕东陵』。

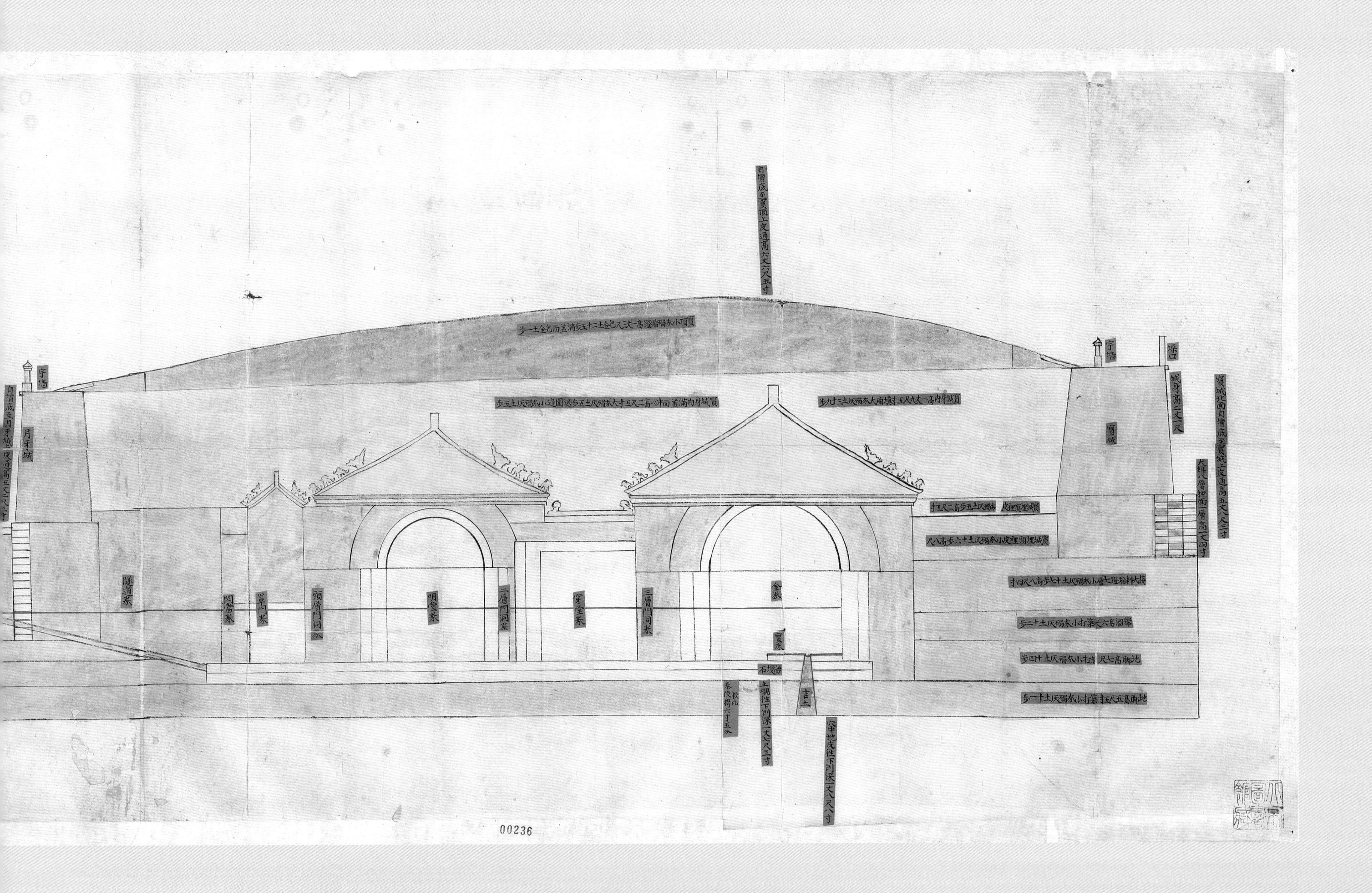
宇墙
月牙城
寶城
垛口
隧道券
閃當券
罩門券
明堂券
穿堂券
金券
吉土
00236

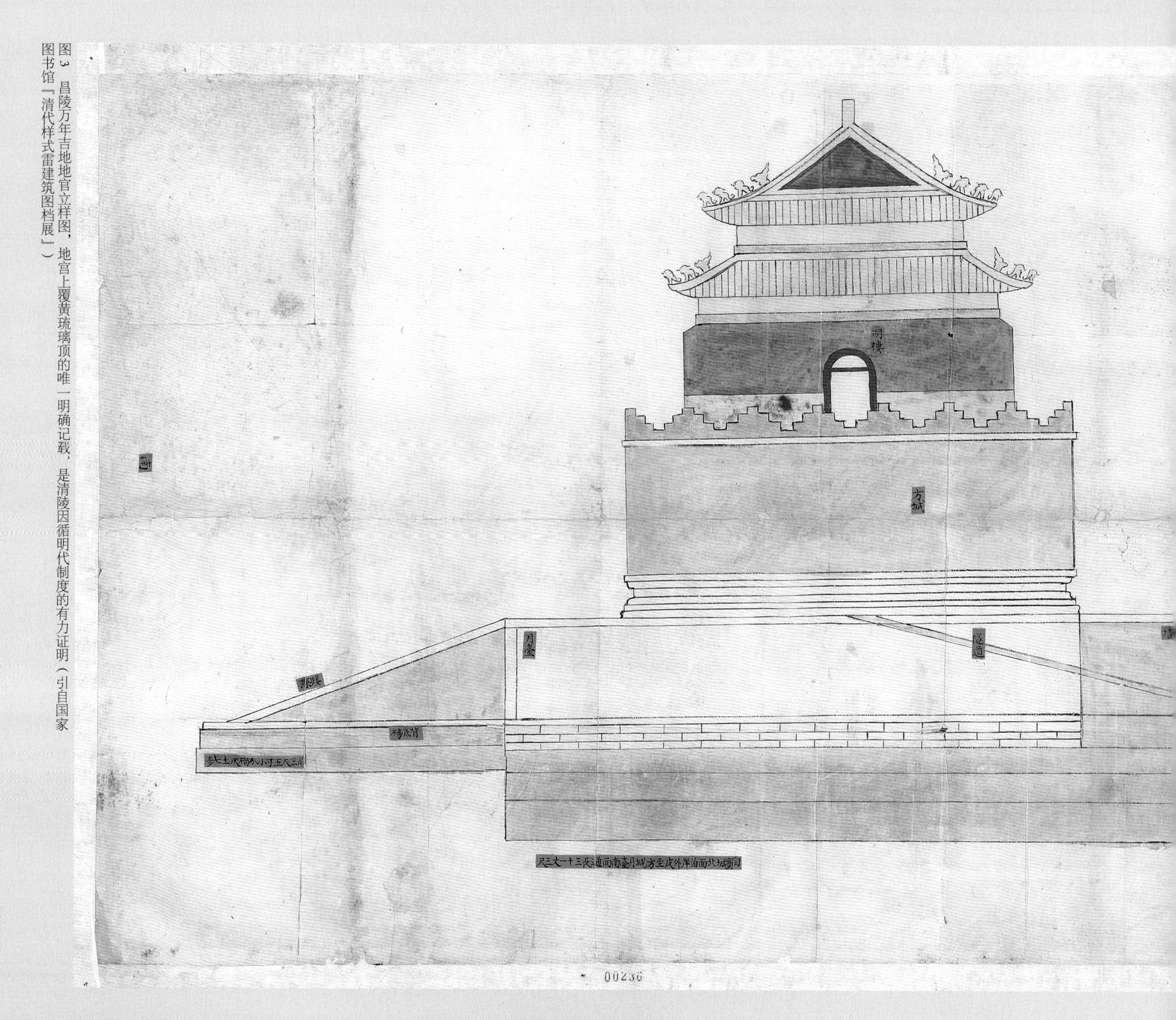

图3 昌陵万年吉地地官立样图，地宫上覆黄琉璃顶的唯一明确记载，是清陵因循明代制度的有力证明（引自国家图书馆『清代样式雷建筑图档展』）

Fig.3 Elevation of the Underground Palace in forever propitious site burial site of Tomb of Good Fortune; the yellow glazed tile roof atop the Underground Palace demonstrates the use of Ming-dynasty tomb as design inspiration. (National Library of China, "Exhibition of the Qing-dynasty Yangshi Lei Archives")

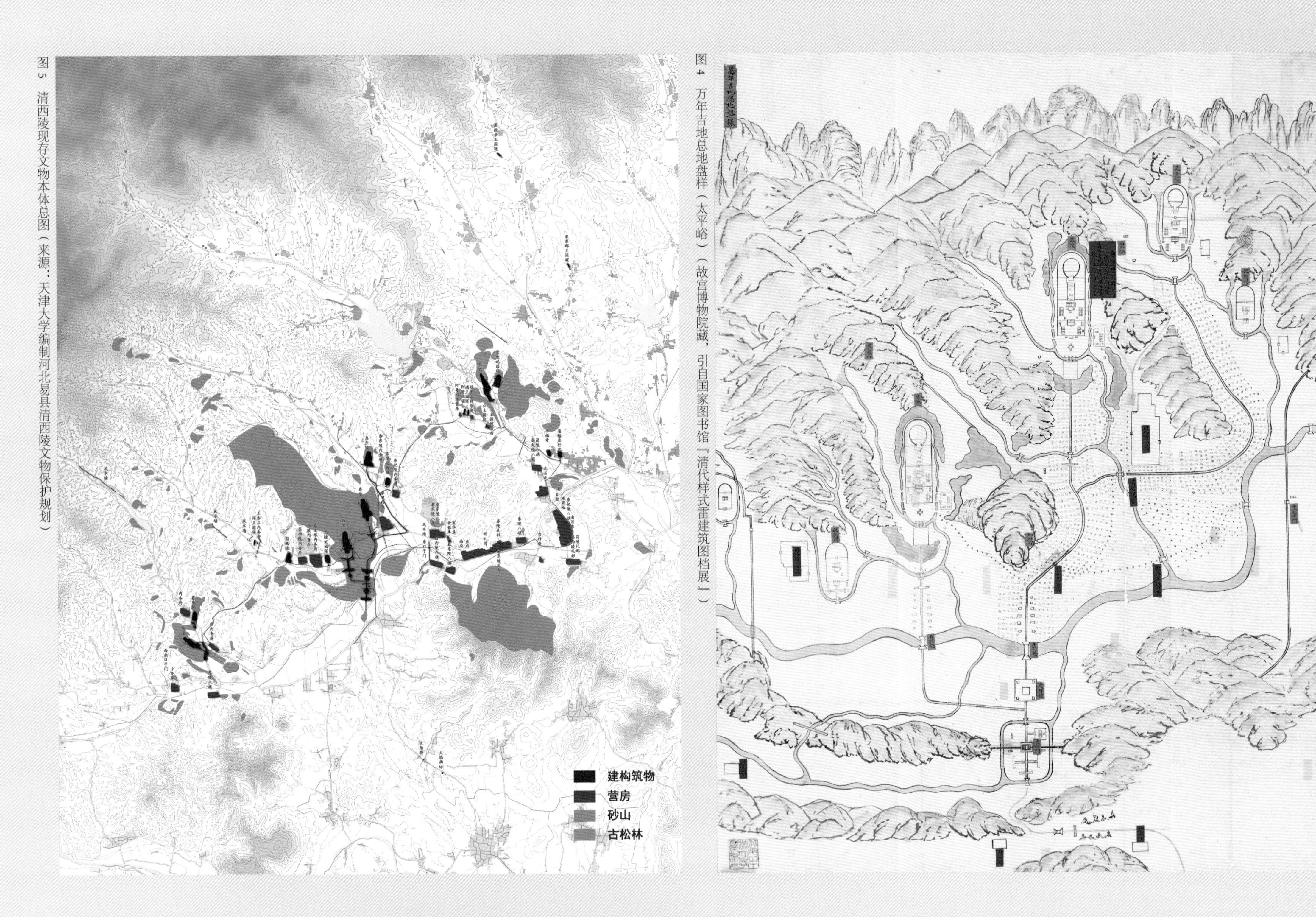

图4 万年吉地总地盘样（太平峪）（故宫博物院藏，引自国家图书馆『清代样式雷建筑图档展』）

图5 清西陵现存文物本体总图（来源：天津大学编制河北易县清西陵文物保护规划）

Fig.4 General layout of forever propitious site burial site (in Taiping Valley). (In the collection of in the Beijing Palace Museum; here from the National Library of China, "Exhibition of the Qing-dynasty Yangshi Lei Archives")

Fig.5 Location of extant cultural heritage. (Western Qing Tomb Cultural Relics Protection Plan in Yi County, Hebei Province, compiled by School of Architecture, Tianjin University)

the Western Qing Tombs in 1831, his eleventh reign year, and completed in 1835. The tomb of his concubines went through the same construction, demolition, and reconstruction process, and in 1855, the fifth reign year of emperor Xianfeng, it was named East Tomb of Veneration (the tomb of emperor Daoguang) because Daoguang's imperial noble consort Lady Borjigit was posthumously honored as empress.

In 1909, the first reign year of emperor Xuantong (born: Puyi)—after emperor Guangxu (temple name Dezong; born Zaitian) had passed away, Guangxu's Tomb (Tomb of Serendipity, the tomb of emperor Guangxu) was built in Jinlong Valley in the northeast of the Western Qing Tombs. After the founding the National Government, the Tomb construction was funded by the new government and completed in 1915. The same happened to the tomb of emperor Guangxu's concubines.

Furthermore, the tombs of the princes of the first rank (Heshuo *qinwang*) Duan and Huai and the tomb for a son of emperor Yongzheng were built from 1735, the thirteenth year of emperor Yongzheng's reign, to 1738, the second year of emperor Qianlong's reign. In 1803, the eighth year of emperor Jiaqing's reign, the tombs of the princess of the first rank (Gulun *gongzhu*) Hui Min and the princess of the second rank (Heshuo *gongzhu*) Hui An were built (Fig.5).

2. Individual Tomb Features

The Ming and Qing imperial tombs are in topographical settings carefully chosen according to principles of geomancy (Fengshui) and comprise numerous buildings of traditional architectural design and decoration. The tombs and buildings are laid out according to Chinese hierarchical rules and incorporate sacred ways lined with stone monuments and sculptures designed to accommodate ongoing royal ceremonies as well as the passage of the spirits of the dead. They illustrate the great importance attached by the Ming and Qing rulers over five centuries to the building of imposing mausolea, reflecting not only the general belief in an afterlife but also an affirmation of authority.

(Description by UNESCO World Heritage Committee)

The three imperial Tombs outside of Shanhai Pass, built during the early Qing dynasty, had set the standard for future burials, and had established a system of distinctive imperial Tomb construction. After their conquest of Mainland China, the Manchus brought to perfection the Tomb system, applying even stricter rules than the preceding Han-Chinese Ming dynasty (Fig.6, Fig.7).

宣统元年（1909年）清德宗崩逝后崇陵兴工于西陵东北部的金龙峪，1912年由国民政府继续出资修建，1915年完工。崇陵妃园寝于同期兴营。

雍正十三年（1735年）至乾隆三年（1738年）建和硕端亲王园寝、和硕怀亲王园寝、阿哥园寝。嘉庆八年（1803年）建慧愍固伦公主、慧安和硕公主园寝（图5）。

二、各陵特色

明清皇家陵寝依照风水理论，精心选址，由众多承载着传统建筑和装饰思想的建筑物组成。陵寝及附属建筑均依规制排布，前序设神道，两旁置有石碑、石像。如此序列以承载皇家祭奠仪式进行，并引导死者灵魂通过，展示了明清两朝统治者在长达五个多世纪的时间内对建造恢宏陵寝的推崇，这不仅是出于对死后生活的信仰，也是对皇权至上的肯定。

——联合国教科文组织世界遗产委员会

清初所建『关外三陵』形成独特的陵寝建筑规制与体系。入关后，清陵制度在明代陵寝制度基础上更臻完备，体系更严整（图6、图7）。

图6 明代帝陵比较图（来源：天津大学建筑学院测绘）

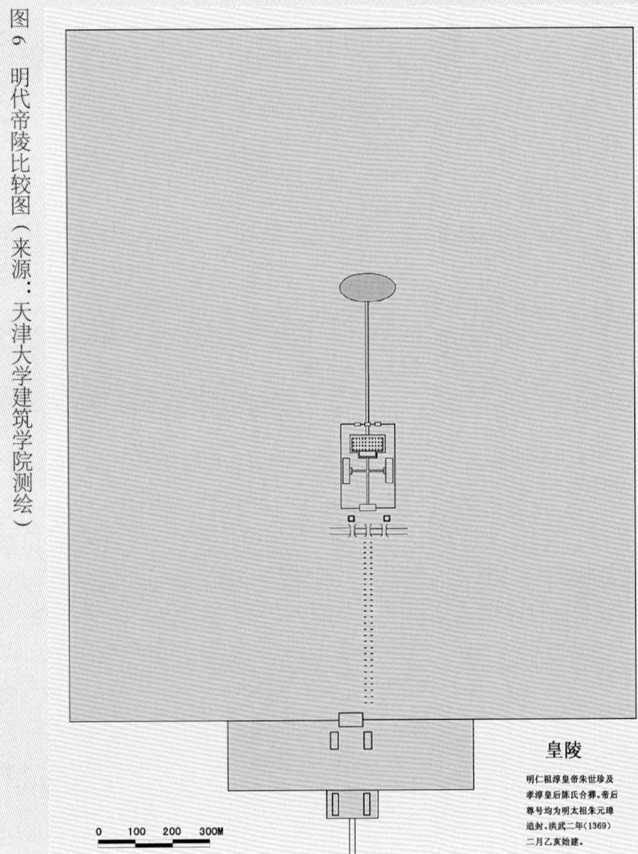

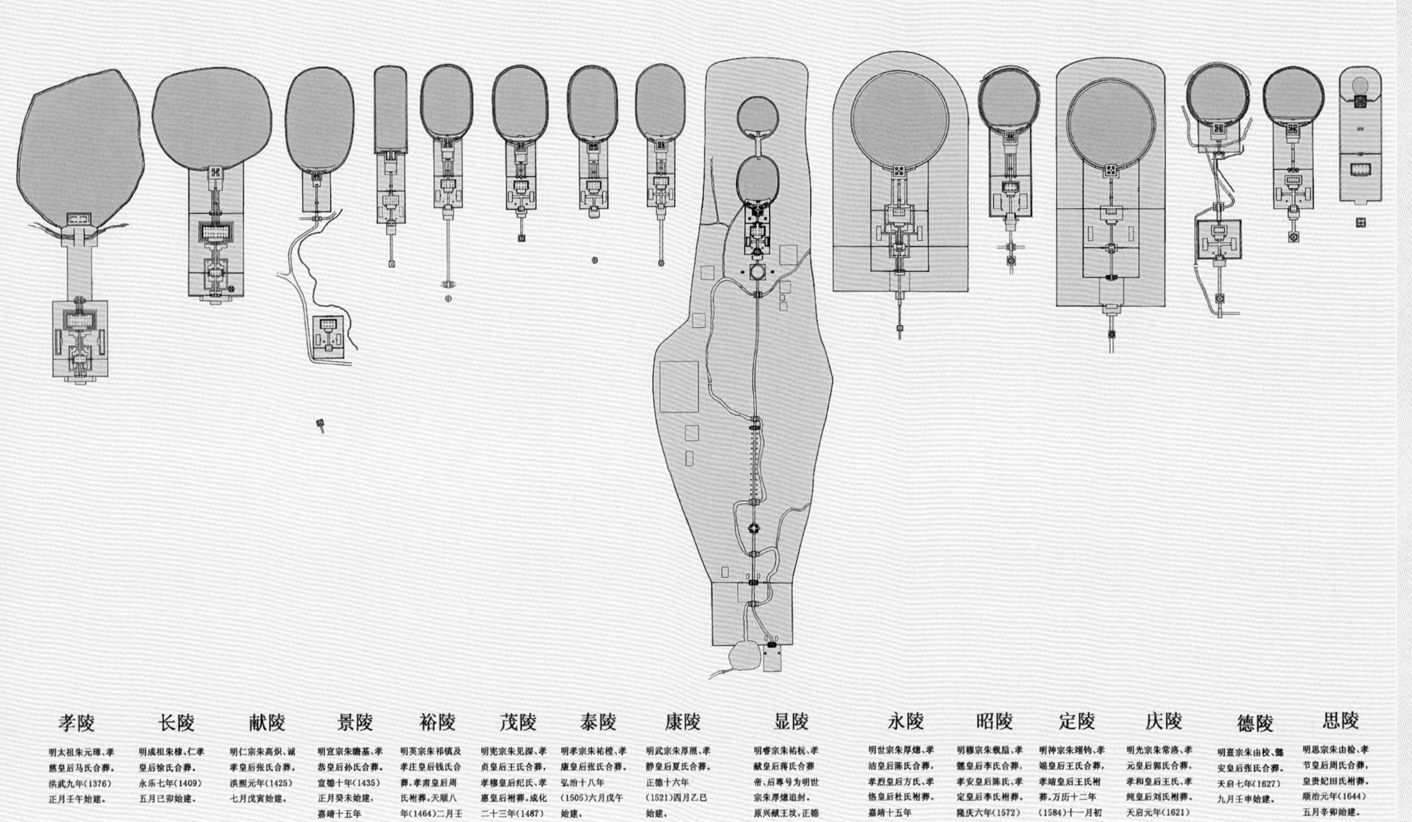

图7 清代帝陵比较图（来源：天津大学建筑学院测绘）

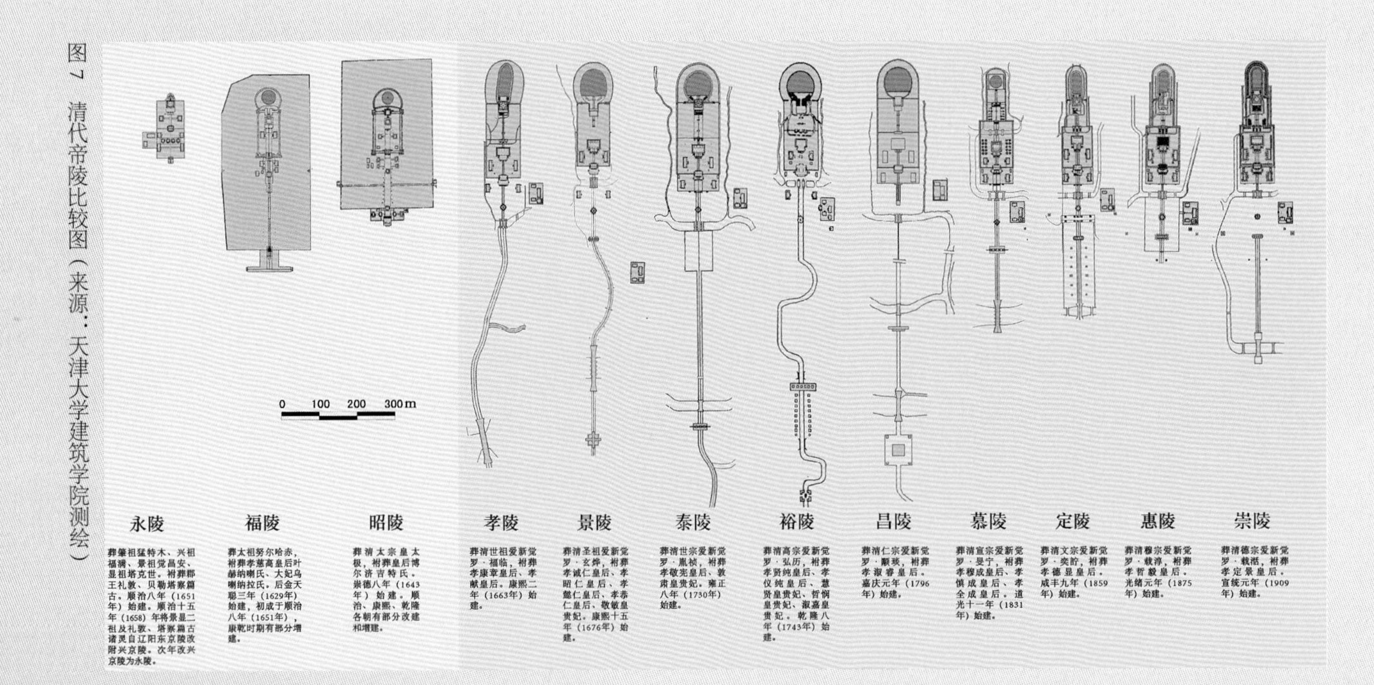

Fig.6 Comparison of Ming emperors' tombs (Mapping by School of Architecture, Tianjin University)

Fig.7 Comparison of Qing emperors' tombs (Mapping by School of Architecture, Tianjin University)

居于陵寝建筑体系至尊地位、被视为『关乎天运之发祥』的帝陵，自康熙初年鼎建清世祖福临的孝陵开始，承袭北京昌平的明陵规制，陵区南部神道随山水蜿蜒向北，参照明长陵列置石牌坊、下马牌、大红门、具服殿、神功圣德碑亭、华表、望柱、石像生和龙凤门等作为展谒的引导空间；陵区北部背倚昌瑞山建设陵宫作为祭祀空间；仿照明代后期定型的帝陵规制次第配置神道碑亭、神厨库、朝房、宫门、焚帛炉、配殿、享殿、陵寝门（琉璃花门）、二柱门、石台五供、方城明楼、哑巴院、月牙城、琉璃影壁和宝城、宝顶等，规模远比长陵节制；宫门、享殿也由明代的祾恩门、祾恩殿改称为『隆恩门』『隆恩殿』。大多由重檐庑殿、面阔九间缩减为重檐歇山、面阔五间，主陵方城明楼缩小，并增设高大升起台基。以景陵为范，各皇帝登基后要卜地预建陵寝，暂称『万年吉地』一并前缀地名。清世宗胤禛在易县永宁山下新辟兆域预建泰陵。这一局面，促发清高宗弘历『东西陵昭穆制度』构想，最终形成以孝陵为主陵的清东陵和以泰陵为主陵的清西陵。

清代陵制破除明代不设后陵的成宪，清西陵泰东陵在清东陵昭西陵和孝东陵基础上进一步完善，成为规制最标准的后陵。不但因循选址命名制度㊀，而且建置完备。陵宫前铺神道联系帝陵，设下马牌；隆恩门、隆恩殿、东西配殿、焚帛炉、陵寝门和石五供等规制类同帝陵，但以尊卑之谊，裁去二柱门，显著缩减并局部改易方城明楼和宝城、宝顶的规模。

㊀ 皇帝入葬后仍在世的皇后应在帝陵左或右另预建陵寝，暂称『万年吉地』并冠以地名，皇后薨逝才用帝陵名号缀以方位命名。

Starting with Tomb of Filial Piety, built for the third Qing emperor Shunzhi (temple name: Shizu; born: Fulin) in the first year of emperor Kangxi's reign, Qing imperial tomb came to represent the well-being, prosperity, and fortune of the state, and followed the pattern set up by the Ming tombs in Changping district of Beijing municipality. The layout comprised a Spirit Way (*shendao*) winding its way northwards, a Marble Memorial Gateway (*paifang*), Stela Marking the Place for Dismounting from one's Horse (*xiamapai*), Great Red Gate Building (*dahongmen*), Hall for Court Robes, Pavilion for the Stela of Sage Virtue and Divine Merit , Ceremonial Columns (*huabiao*), Ornamental Column (*wangzhu*), stone statues (*shixiangsheng*), and the Dragon and Phoenix gate (*longfengmen*), in reference to the oldest of Beijing's Ming tombs—emperor Yongle's Tomb named Tomb Chang (*chang*, translates as lasting). In the north, near Changrui Mountain, stood the "tomb palace" (*linggong*) where memorial ceremonies were hold. Then, in imitation of late-Ming design but on a much smaller scale, Qing Tombs featured a Pavilion for the Stela on the Spirit Way, an Culinary Courtyard for Sacrifices (*shenchuku*), Reception Halls for Court Officials (*chaofang*), a Gate of Tomb Palace (*gongmen*), Sacrificial Burner (*fenbolu*), Side Halls (*peidian*), the Hall of Ritual Sacrifice (*xiangdian*), the Gate with Glazed Roof Tiles (*liulihua men*) erected in the rear section of the Tomb, the Gate with Two Columns (*erzhumen*), a set of Stone Altar and Five Stone Ritual Vessels (*shitaiwugong*), the Square Walled Terrace (*fangcheng*), the Memorial Tower (*minglou*), a Mute Court (*yabayuan*), Crescent Wall (*yueyacheng*), a Screen Wall of Glazed Tiles Screen Wall of Glazed Tiles (*liuli yingbi*), Encircled Realm of Treasure (*baocheng*) and Tumulus (*baoding*). The Main Gate and the Hall of Ritual Sacrifice named The Gate of Bliss Grace (Ling'enmen) and the Hall of Bliss Grace (Ling'endian) were renamed as the Gate of Monumental Grace (Long'enmen) and the Hall of Monumental Grace (Long'endian). Most of the halls were downsized, from nine bays with double-eaves hip roof to five bays with double-eaves hip-gable roof. The Memorial Tower and the Square Walled Terrace understructure decreased in size, and a tall foundation platform was added.

Starting with Tomb of Admiration, the Qing emperors built their tombs right after they had ascended the throne, conducted procedures of siting and surveying to select the most auspicious site that was temporarily named forever propitious site preceded by the place name.

Emperor Yongzheng started choosing a piece of land at the foot of Yongning Mountain for the tomb (Tomb of Peace). Emperor Qianlong then established the Zhaomu System to unite the scattered sites into a unified whole, organizing the placement of tombs either in the eastern or western section, with the Tomb of Filial Piety as the center of the East Qing Tombs and the Tomb of Peace functioning as the center of the Western Qing Tombs.

承袭关外清陵附建妃嫔园寝的制度，摈弃明代妃嫔坟不入陵区、不予命名甚至一坟合葬的陋制。清陵妃园寝祔建帝陵左右，并以帝陵名号冠名，构成隶从分明的整体关联；参照明代妃嫔坟规制，宫门、享殿、琉璃花门等建筑覆盖绿琉璃瓦；按妃嫔册封位号，分建不同等级的地宫、宝顶及月台，尊卑有序地列布在园寝后部。

同明代比较，清代陵寝体系更臻完备，众多后陵和妃园寝统绪分明地分布各帝陵左右，成为清代陵寝的显著特点。再如乾隆朝《相度胜水峪万年吉地》题本概括，注重『遵照典礼之规制，配合山川之胜势』，遵循『事死如事生』的传统观念，相关建筑如宫门、享殿、配殿、陵寝门、明楼等都要观照宗法礼制的统绪关系和尊卑等级，采用宫殿建筑体制。因此，清西陵的各帝、后陵和妃园寝，在传承明代陵寝制度的同时，也鲜明展现了北京皇宫建筑典雅凝重的风范。

另外，清西陵前部序列反映出继承下的创新。设置独特的前置仪门火焰牌楼，作为『白差』进入陵区必经之地；主入口大红门采用三座石牌坊，每间呈竖长方形，雀替下设贴柱，每楼檐口起翘曲线

But the Qing also broke with Ming rules. After improvement of the design of the West Tomb of Brightness and the East Tomb of Filial Piety located at the East Qing Tombs, East Tomb of Peace, one of the Western Qing Tombs, became the most typical example of the empress's tomb; it followed the emperor's tomb system of site selection and naming[1] and because of its architectural completeness. Before the tomb palace where was paved a Spirit Way linked the tomb of the wife to that of her husband, the Stelas marking the place for dismounting from one's horse, the Gate of Monumental Grace, the Hall of Monumental Grace, the Eastern and Western Side Halls, the Sacrificial Burners, the Burial Area Gate, and a set of Stone Altar and Five Stone Ritual Vessels. However, to mark the lower status, there was no Gate with Two Columns, and the Square Walled Terrace, Memorial Tower, Encircled Realm of Treasure and Tumulus were downscaled and partially modified.

After the conquest of Mainland China, the Manchus continued their tradition of building separate tombs for consorts and concubines near the emperor's tomb as they had started at the Three Outer Tombs north of Shanhai Pass. That is to say, they abolished the undesirable Ming habits of not incorporating consorts' and concubines' tombs into the imperial Tomb complex, of not naming their tombs, and of even burying several of them in the same tomb. The tombs of imperial consorts and concubines were placed to the left or right side of the emperor's tomb and named after the tombs of the husband or master, so that these tombs together with the emperor's tombs formed an interrelated whole but with a clear distinction of importance. For example, the Main Gate, the Hall of Ritual Sacrifice, and Gate with Glazed Roof Tiles were covered with green (not yellow) glazed tiles; Underground Palace, Tumulus and front platform *(yuetai)* were built in line with the consorts' or concubines' ranks and arranged in sequential order at the rear part of the Tomb area.

Compared with the Ming, the Qing Tomb construction system was more complex, regulating even the placement of consorts' and concubines' tombs to the left or right side of the emperor's tomb and distinguished according to the women's importance for their husband during his life. Just as summarized in the official report about auspicious tomb sites in Shengshui Valley (*Xiangdu shengshuiyu wannian jidi [tiben]*) from the Qianlong period, imperial Tomb construction should harmonize with the landscape and adhere to the traditional concept of "honoring the dead as they were alive" (*shisi ru shisheng*).That is to say, imperial tomb should

① The tomb of an empress still alive should be built to the left or right side of her deceased husband's tomb and named as forever propitious site (burial site) preceded by the place name. After her death her tomb should be named after the tomb of her deceased husband succeeded by a cardinal direction.

be built with reference to the architectural system for palace construction of the living—the Main Gate, the Hall of Ritual Sacrifice, Side Halls, Burial Area Gate, and Memorial Tower should reflect the Confucian relationships of master to subject and the correct sequential order according to the ritual system. Hence, the burial sites of emperors, empresses, consorts, and concubines at the Western Qing Tombs show not only the continuation of the Ming funerary system but also the elegance and dignity of the Qing imperial palace in Beijing.

However, the design of the Western Qing Tombs is also innovative. The front ceremonial gate that provided access to the Tomb for servants engaged in funeral or memorial service (*baichai*) was Flame Memorial Gateway (*huoyan paifang*); The main entrance, the Great Red Gate Building, consists of three Memorial Gateways arranged in U-shape each bay was shaped like a vertically positioned rectangle framed by tall pillars with small columns attached to their surface below the decorated bracket (*queti*); and every bay and flanking column was crowned by an own roof with upturned eaves. The two large Pavilion for the Stela (*beilou*) were built for the first two emperors. Two sets of five paired Stone Statues (*shixiangsheng*)—the figure of elephants carrying a treasure vase meaning peace (*taiping youxiang*) was used first time. Every emperor's tomb space sequence is started by a Five-Arch Stone Bridge (*wukong shiqiao*). Three tombs had Dragon and Phoenix Gates, Ceremonial Columns protected by stone fences, and a Pavilion for the Stela on the Spirit Way that separates and links the front section to the rear tomb palace section. Although smaller than the Ming tombs and the East Qing Tombs, the Western Qing Tombs have a closer and more complex, clustered relationship among individual structures, pushing the architectural art of tomb to a new high.

1) Breaking Fresh Ground—Tomb of Peace (the tomb of emperor Yongzheng)

Tomb of Peace or Tailing, the principal Western Qing Tomb is in Tianping Valley at the southern foothill of Yongning Mountain. It is the resting place of emperor Yongzheng, empress Xiaojing (of the Ula Nara clan), and his consort Dun Su (of the Nian clan). In 1741, four years after completion (in 1737, the second reign year of emperor Qianlong), Pavilion for the Stela of Sage Virtue and Divine Merit surrounded by four Ceremonial Columns at the corners was added. The *handicraft regulations and precedents published by the Ministry of Works* (*Gongbu zeli*) in 1748, the thirteenth reign year of emperor Qianlong, record that five pairs of stone statues and a pair of Ornamental Columns similar in number, size, and pattern with those at Tomb of Admiration were erected along the Spirit Way north of the Pavilion for the Stela by order of the emperor who valued filial piety and ritual.

明显；两座大碑楼、两组五对石像生；首次塑造『太平有象』的石象造型；以五孔石拱桥作为帝陵序列初始；设三座龙凤门；华表增设石围栏；各帝陵均设神道碑亭承接前部序列与陵宫区，使整体联系更为紧密。清西陵尺度上虽逊于明陵和清东陵，但组群关系上更为紧密与完善，将陵寝建筑艺术推向了一个新的高峰。

（一）别开生面的泰陵

清西陵主陵泰陵选址于永宁山南麓天平峪，合葬清世宗胤禛及孝敬皇后乌喇纳喇氏，祔葬敦肃皇贵妃年氏；乾隆二年（1737年）完工后又于乾隆六年（1741年）添建圣德神功碑亭及四隅华表；据乾隆十三年（1748年）《工部则例》记载，拘泥孝道和礼制的弘历还在泰陵圣德神功碑亭北的神道旁，补建了规制和景陵相同的五对石像生和一对石望柱。

图9 泰陵外部空间序列分析图（出自《风水理论研究》）

图8 泰陵龙凤门及陵宫鸟瞰（出自《中国美术分类全集·中国建筑艺术全集·清代陵墓建筑》）

Fig.8 Bird's-eye view from Dragon and Phoenix Gate to Tomb palace area at Tomb of Peace. (*Complete Series of Chinese Architectural Art, the authoritative volume on Qing dynasty funerary architecture*)

Fig.9 Layout analysis of structures outside Tomb of Peace. (*Fengshui lilun yanjiu*)

With Yongning Mountain as the protecting mountain behind the Tomb in the north (*houlong*), Yuanbao Mountain as a far distant mountain in the front i.e. south *(chaoshan*), and Zhizhu Mountain as the protecting mountain in the central part (*anshan*), Tomb of Peace was built in compliance with the geomantic landscape. The tomb palace complex in the northern part, along with the Pavilion for the Stela of Sagely Virtue and Divine Merit, Ceremonial Columns, Ornamental Columns, and five paired stone statues of the southern part were similar to those at Tomb of Admiration, while the Dragon and Phoenix Gate to the north of Zhizhu Mountain, which divided the area into front and rear section, was modeled after the gate at Tomb of Filial Piety. But the structures at Tomb of Peace were more exquisite than their counterparts at Tomb of Filial Piety and Tomb of Admiration. For example, the Pavilion for the Stela had stone carved arches. The keystones of the arched stone bridge were decorated with dragon heads (*chishou*), and the podium of the front platform in front of Square Walled Terrace was made of greenish white marble (Fig.8).

In reference to Tomb of Filial Piety, the first East Qing Tombs, Tomb of Peace as the principal West Qing Tomb also had a Great Red Gate Building, surrounding geomantic wall, Hall for Court Robes, Stela Marking the Place for Dismounting from one's Horse, and stone Memorial Gateways that fitted well with the scenic landscape in the front (southern) part of the site. The northern Yi River surrounded the open terrain of the front section. Jiulong Mountain and Jiufeng Mountain flanked the east and southwest sides, while Yuanbao Mountain and the East and West Huagai mountains created a grand, echoing vista in the far south so that the horizon was not empty (a *fengshui* principle).

But unlike Tomb of Filial Piety and even Ming-period Tomb of Good Fortune, Tomb of Peace had a pair of stone legendary auspicious animal (*qilin*) standing on both sides of the Great Red Gate Building and three five-bay (six-column) stone gateways with eleven roofs (here referred to as *lou*), rising abruptly from the ground south of the Great Red Gate Building. Their size and decorative carving surpassed that of Tomb of Filial Piety and Tomb of Good Fortune. Well integrated into the fengshui topography and landform, this layout spatially enhanced the Tomb entrance.

Tomb of Peace is a classic example of traditional Chinese spatial design. By looking at the physical features of the site (i.e. form or *xing*) and exploring how they are related with each other (i.e. configuration or *shi*), as done by the Form School of fengshui, individual structures were laid out properly to create a place full of positive energy. The careful sequential arrangement of structures along the Spirit Way from the Dragon and Phoenix Gate to the

泰陵北倚永宁山为后龙，向南遥对元宝山为朝山，中部横卧蜘蛛山为案山，陵寝建筑顺应风水形势布局。其中，北部的陵宫建筑群，南部的圣德神功碑亭、华表、望柱和五对石像生等，规制均仿效景陵；而蜘蛛山北建置龙凤门为陵区前后分界，却仿自孝陵。各建筑的细部处理，如碑亭添设石雕券脸，石拱桥券洞的龙门券雕出龙头，方城前月台的台明全部采用青白石等，则均比孝陵、景陵考究（图8）。

作为主陵，泰陵南端山水交会的胜境里，还祖述孝陵配置大红门、风水墙、具服殿、下马牌、石牌坊等建筑。在这里，北易水分岔环绕着一片开阔地，东西各有九龙山、九凤山迤逦拱抱，前有元宝山和左右对峙的东华盖和西华盖山构成气势雄浑的天然对景。和孝陵以至明长陵不同的是，大红门外两侧添置一对石雕麒麟，南面更呈品字形拔起三座五间六柱十一楼石牌坊，形制划一，尺度规模都超过孝陵和明长陵的石牌坊，雕饰也格外繁复。这一格局同风水形势完美结合，有力强化了陵区入口的空间艺术效果。

清西陵泰陵是中国传统建筑外部空间设计的经典案例。泰陵建筑群积形成势，聚巧形而展势，其匠心独运的组群空间处理十分得宜地利用了地势与山川景物，充分展现了形以势得、驻远势以环形的藻思。

Square Walled Terrace with the Memorial Tower allowed the approaching visitor to enjoy different views of the whole complex and of individual structures from a short, moderate, or long distance, and to experience static scenes alternating with dynamic scenes because of cause and effect. This reveals the extraordinary skill of the builders to capture the relationship between the whole (architectural complex) and its parts (individual structures) (Fig.9).

In Chinese architectural history studies, Tomb of Peace occupies a significant place. As mentioned before, Tomb of Peace is a key physical reference to the study of building standards as outlined in *Engineering manual of the Ministry of Works* (*Gongbu gongcheng zuofa*). Additionally, the three stone Memorial Gateways were built in accordance with one chapter of the handbook, *Precedents/Model Calculations of Memorial Gates: Six-column Five-bay Eleven-roof Stone Memorial Gateways* (*Pailou suanli, wujian liuzhu shiyilou shipailou fenfa*). The Memorial Gateways are the earliest examples of a new, modified eaves design for draining rainwater, replacing “hook tiles sitting in the middle” with “drip tiles sitting in the middle”. Furthermore, the Hall for Court Robes is the only one of its kind still extant and well preserved at a Qing Tomb (Fig.10).

The circular Encircled Realm of Treasure and Tumulus atop the Underground Palace located north of Square Walled Terrace and Memorial Tower were modeled after those at Tomb of Admiration. According to *Examples of Collected Statutes of the Great Qing* (*Daqing huidian shili*), despite reduction in height, their perimeter increased remarkably. For example, the Encircled Realm of Treasure perimeter at Tomb of Peace is over sixteen *zhang* (3.3 m) longer than the one at Tomb of Admiration. The Mute Court nestled between Square Walled Terrace and Encircled Realm of Treasure—including the Crescent Wall, Screen Wall of Glazed Tiles, and winder stairs—also reverberated a pattern known from Tomb of Admiration but at a larger size. For example, the Crescent Wall’s length is over three *zhang* longer than the one at Tomb of Admiration, making this courtyard with an entrance to Tomb of Peace’s Underground Palace to the biggest one of its kind at any imperial Qing Tomb (Fig.11).

2) East Tomb of Peace

East Tomb of Peace is the Tomb of an empress. The structure corresponds most closely to the standard for imperial Qing tombs and is the largest of its kind at the Western Qing Tombs. It was built for Yongzheng’s consort (of the Niohuru clan) who was posthumously honored as empress Xiaosheng because she had given birth to the heir to the throne—Yongzheng’s fourth son and successor, the later Qianlong emperor. Exhibiting the essen-

泰陵神道龙凤门迤北直至方城明楼这段空间序列的组织处理，使建筑群及建筑个体在远、中、近不同距离上的视觉感受，在步移景易和步移景滞中，动静相成、形势相因及相互转换的种种妙趣都得以充分表现；这段空间处理，表现出了古代哲匠把握建筑个体与群体关系深思熟虑和裕如运用的非凡功力（图9）。

泰陵建筑在中国建筑史学史中拥有重要的地位。如前文所述，泰陵是研究《清工部工程做法》的重要参照实物，可谓鲜活标本。三座石牌坊做法契符著名清代匠籍《牌楼算例·五间六柱十一楼石牌楼分法》，而且是中国古代建筑檐口线上『勾头坐中』转化到『滴水坐中』的现已知的第一个建筑实例（图10）。另外，泰陵具服殿是现今唯一完好的清代陵寝具服殿遗存。

泰陵方城明楼迤北掩蔽地宫的宝顶和宝城，承袭景陵的圆式平面。按《大清会典事例》记载，在宝顶和宝城高度略为降低的同时，径围却显著增大，如宝城周长超过景陵十六丈余。同时，围合在方城和宝城之间的哑巴院，包括月牙城、琉璃影壁和转向蹬道等，形制虽参照景陵，平面尺度却相应扩大，月牙城就比景陵展长三丈多。从而，泰陵哑巴院成为清代帝陵规模最大的一座（图11）。

（二）泰东陵

泰东陵是清代规制最标准的皇后陵，也是清西陵中规模最大的皇后陵。它是清代首座因墓主人生嗣帝而被尊封升格皇后的后陵，完全遵照皇后陵的规制设计并营建，既无陋简之设，也无明显越制之处，

tial characters of such tombs in a neither too simple nor to excessive manner, her Tomb is an architectural gem of best construction quality among the imperial Qing tombs (Fig.12).

East Tomb of Peace is the only Tomb of an empress with a Spirit Way connected to the Spirit Way of her husband, showing the close relationship between them (Fig.13). It also stands out for the longest construction history among Qing tomb, taking more than three decades to complete. It exhibits many unique features that were applied here for the first time. For example, the Underground Palace was decorated with Buddhist scriptures and images and texts in foreign and minority languages, which is unique among Qing empresses' tombs. The vase, one of Five Stone Ritual Vessels, used purple sandstone first time. The ceiling of the Hall of Monumental Grace was decorated with a rare design of three lotus. A Buddha shrine (*xianlou*), the only single-story example of its kind, was set up inside the east warming room. Finally, a bronze deer and a bronze crane stood on the front platform of the Hall of Monumental Grace, which made East Tomb of Peace the only Tomb of an empress featuring these animals in a pair.

3) Imperial Consorts' Tombs affiliated with Tomb of Peace

Located in Yangshugou one kilometer northeast of Tomb of Peace, the tomb of emperor Yongzheng's concubines is the resting place of twenty-one people, including the imperial noble consort Chun Yi (family name GENG) and consort Qi (family name LI). In 1730, the eighth year of his reign and the year when he decided to build his Tomb, emperor Yongzheng also selected a nearby site to build the tomb for his concubines, which was the first concubine tomb built at the Western Qing Tombs and the second one built in the Qing dynasty. There were fewer graves here than at the burial site of imperial concubines affiliated with Tomb of Admiration built for emperor Kangxi's consorts; the two sites are also distributed differently. Additionally, the architecture was partially downsized: the east and west watchhouses in front of the Main Gate were only two bays wide, and the east and west wing rooms (*xiangfang*) only three bays wide. But all the other structures were based on the configuration, pattern, and scale of Tomb of Admiration's concubine tomb. The platform in front of the Gate with Glazed Roof Tiles and the division (tomb palace) wall with side doors was built due to situational circumstances—to balance the height difference between the ground in the front and the rear of the platform—but not to alter the standard configuration (Fig.14).

4) Continuing the Established Pattern—Tomb of Good Fortune

Tomb of Good Fortune, or Changling, is the last of the Western Qing Tombs that was built

建筑质量属清代皇陵中的精品（图12）。

泰东陵是清西陵中唯一有神道与隶从帝陵神道相连的后陵，反映帝后陵之间的统绪关系（图13）。该陵建设时长三十余年，称冠清代所有帝后陵寝；地宫内首加经文、佛像、番字雕刻，是清代皇后陵的孤例；石五供花瓶上首次使用紫砂石；隆恩殿天花使用独特的三朵莲花图案，东暖阁首设仙楼，且是唯一的单层仙楼；首设铜鹿、铜鹤，是唯一成对设置的后陵。

（三）泰陵妃园寝

雍正帝纯懿皇贵妃耿氏、齐妃李氏及其他妃嫔、贵人等共21人的墓园，位于泰陵东北一公里杨树沟，隶从泰陵。

雍正八年（1730年），雍正帝经营泰陵的同时，选定泰陵东北杨树沟兴建妃园寝，成为清代第二座妃园寝，也是清西陵的第一座妃园寝。妃嫔墓穴数目远少于景陵妃园寝且排列有异，宫门前的东、西班房和东、西厢房缩减成三间，其余建筑配置及形制规模均仿照康熙朝的景陵妃园寝规制。琉璃花门、两

图12 泰东陵鸟瞰（易县清西陵文物管理处藏）

图10 泰陵五间六柱十一楼石牌坊（出自《中国美术分类全集·中国建筑艺术全集·清代陵墓建筑》）

图13 泰东陵方城明楼（王其亨摄）

图11 泰陵哑巴院（王其亨摄）

Fig.10 Six-column Five-bay Eleven-roof Stone Memorial Gateways at Tomb of Peace. (*Complete Series of Chinese Architectural Art, the authoritative volume on Qing-dynasty funerary architecture*)

Fig.11 Mute Court at Tomb of Peace. (Photo by WANG Qiheng)

Fig.12 Bird's-eye view of East Tomb of Peace. (In the collection of the Administration of Cultural Relic of the Western Qing Tombs)

Fig.13 the Square Walled Terrace and the Memorial Tower at East Tomb of Peace. (Photo by WANG Qiheng)

for a Qing emperor during the first stage of Ming-Qing imperial tomb construction. To put into practice the Zhaomu System, emperor Qianlong selected a burial site at the Western Qing Tombs for his son, the later Jiaqing emperor, in addition to his own site at the East Qing Tombs. Thus, Tomb of Good Fortune was the only imperial Tomb pre-selected by the owner's father and the first emperor's Tomb to implement the Zhaomu System.

Located southwest of Tomb of Peace in Taiping Valley, Tomb of Good Fortune is the second Western Qing Tombs and the resting place of emperor Jiaqing and empress Xiaoshu (of the Hitala clan) (Fig.15). The Tomb was completed in 1803, the eighth year of emperor Jiaqing's reign. In 1821, his first reign year, Jiajing's son and successor to the throne—emperor Daoguang—erected a Pavilion for the Stela of Sage Virtue and Divine Merit and Ceremonial Columns at his father's tomb.

So as not to challenge Tomb of Peace for the position of principal tomb, Tomb of Good Fortune had no stone Memorial Gateway, Great Red Gate Building, nor a Hall for Court Robes, but the other structures including the stone statues and Dragon and Phoenix Gate continued the layout, design, and scale of Tomb of Peace. Additionally, the Buddha shrine (*folou*) inside the east warming room imitated the one at Tomb of Prosperity, the Tomb of the Qianlong emperor. The Underground Palace was also decorated with images and scriptures of Tibetan Buddhism in the manner of Tomb of Prosperity. With the architecture aboveground emulating Tomb of Peace and the architecture underground Tomb of Prosperity, Tomb of Good Fortune continued the imperial tomb building tradition. The only exception was the modification of the Memorial Gateway with columns extending into the sky (*chongtian paifang*)—as used at Tomb of Admiration and Tomb of Prosperity—into a Dragon and Phoenix Gate, and here Tomb of Good Fortune demonstrates its pioneering role as the first non-principal Western Qing Tomb with such a gate.

Furthermore, the floor at the Hall of Monumental Grace was paved with piebald stone (*huabanshi*), not brick like other imperial tomb, creating a different impression thanks to its yellow surface permeated by purple veins.

5) West Tomb of Good Fortune

West Tomb of Good Fortune, the fourth empress's tomb of the Qing dynasty and the second one built at the Western Qing Tombs, is located west of Tomb of Good Fortune on the southern foothill of Wangxian Mountain. It is the resting place of the second empress con-

翼面阔红墙及随墙角门前建置景陵妃园寝所没有的通长平台，是由于前后地势高差较大所致，并非增崇规制（图 14）。

（四）因循守成的昌陵

清西陵昌陵是明清陵寝制度第一阶段的最后一座帝陵，乾隆帝为贯彻『东西陵昭穆制度』，将自己的陵寝设在东陵同时，亲自为嘉庆帝在西陵选定万年吉地，使昌陵成为明清皇家陵寝中唯一一座由太上皇选定陵址的帝陵，也是第一次完成『东西陵昭穆制度』的帝陵。

昌陵位于泰陵西南的太平峪，合葬着清仁宗颙琰及孝淑皇后喜塔腊氏，是清西陵的第二座帝陵（图 15）。嘉庆八年（1803 年）完工后，道光元年（1821 年）嗣皇帝旻宁援例添建圣德神功碑亭及华表。

尊奉泰陵为主陵，昌陵不另建石牌坊、大红门、具服殿等。石像生、龙凤门与其他建筑形制、规模和序列，都因循守成，类同泰陵；隆恩殿东暖阁效仿裕陵建佛楼；地宫内装修也效尤裕陵雕饰有藏传佛教的各类图像和经文。昌陵地面建筑参照泰陵，地宫参照裕陵，体现了传承关系，但是却将景陵与裕陵均采用的冲天牌坊改为龙凤门，成为明清皇家陵寝中第一座使用龙凤门的非主陵。

此外，昌陵隆恩殿使用花斑石墁地，与其他帝陵隆恩殿采用金砖铺地相异。花斑石表面呈黄色，缀以天然的紫色花纹，呈现出不同的气质。

图 14　泰陵妃园寝（易县清西陵文物管理处藏）

图 15　昌陵前区鸟瞰（出自《中国美术分类全集·中国建筑艺术全集·清代陵墓建筑》）

Fig.14 Imperial Consorts' Tombs affiliated with Tomb of Peace. (In the collection of the Administration of Cultural Relic of the Western Qing Tombs)

Fig.15 Bird's–eye view of the front part of Tomb of Good Fortune. (*Complete Series of Chinese Architectural Art, the authoritative volume on Qing–dynasty funerary architecture*)

sort of emperor Jiajing—empress Xiaohe (of the Niohuru clan)—who passed away at the end of the twenty-ninth reign year of emperor Daoguang. Since no tomb had been built during her lifetime, and because emperor Daoguang passed away soon after her, it was not until emperor Xianfeng was enthroned that siting for her tomb—West Tomb of Good Fortune—was carried out in 1851, the first year of Xianfeng's reign. Tomb construction lasted for the shortest period in the history of Qing tomb and was completed the next year. Facing social unrest and fiscal challenges in addition to the simplified construction of emperor Daoguang's own tomb (Tomb of Veneration), West Tomb of Good Fortune was built on a more modest scale than the previous tombs of Qing empresses. In stark contrast to the auxiliary structures watchhouses, the Reception Hall for Court Officials, Culinary Courtyard for Sacrifices, comparable in size to those at East Tomb of Peace, the Gate of Monumental Grace and the main buildings inside the gate were downsized dramatically. For example, the previously five-bay the Gate of Monumental Grace with side halls was now only three bays wide. The five-bay Hall of Monumental Grace featured a smaller hip-gable roof covered only by a single set of eaves, and, stripped of the stone fence surrounding the front platform and the large marble carving on the steps stone (*danbishi*) positioned in the middle of the front steps, it was to the lowest-ranking hall of its kind. The Gate with Glazed Roof Tiles retained only the central doorway in its original form, while the side doorways were simplified into small wall openings.

Square Walled Terrace and Memorial Tower that usually stood behind the Five Stone Ritual Vessels were now substituted by a square platform. The circular Encircled Realm of Treasure that enclosed the earth Tumulus covering the Underground Palace was built with a Mount-Sumeru-shaped podium (*xumizuo*) made of greenish white marble but without battlement and covered with yellow glazed tiles; inside were no sidewalks. With a perimeter of only ten *zhang*, eight chi, and seven *cun*, the perimeter of this Square Walled Terrace was less than a third of those at previous tombs of Qing empresses. The wooden plaque (*doubian*) inscribed with the name of the tomb that used to suspend from the southern eaves of Memorial Tower was replaced with a stone (greenish white marble) board horizontally inlaid into the lintel above the central doorway of the Gate with Glazed Roof Tiles (Fig.16).

Furthermore, West Tomb of Good Fortune has other unique features. It is the only empress's tomb where he ceiling of the Hall of Monumental Grace was painted with a pattern of red phoenixes spreading their wings and where three flat stone bridges crossed the Jade Belt River in front of the Gate with Glazed Roof Tiles. It is also the only tomb with extant bronze waterspouts installed at Encircled Realm of Treasure.

（五）昌西陵

昌西陵位于昌陵西侧望仙山南麓，安葬着清仁宗颙琰的孝和睿皇后钮祜禄氏，为清代的第四座后陵，也是清西陵中的第二座后陵。孝和皇后生前没有预建陵寝，道光二十九年（1849年）末薨逝，清宣宗旻宁未遑筹措其葬事就随即崩御，以至清文宗奕詝继统以后，才在咸丰元年（1851年）卜地建造，并荐号为昌西陵，转年落成，是修建时间最短的清代后陵。由于时局动荡，国家财政困难，加之慕陵改制的深刻影响，昌西陵的建筑规制远比先前清代后陵简陋。除班房、朝房、神厨库等辅助性建筑略如泰东陵外，隆恩门及迤里各主体建筑，则显著缩减规制。较之先前的后陵，隆恩门和配殿都由面阔五间改小为三间；隆恩殿虽保留五开间，但改为单檐歇山顶，尺寸规模也较小，殿前月台撤去周围石栏及丹陛石，成为规制最低的隆恩殿；琉璃花门由旧制三座改为仅保留了正中一座，左右两座改成随墙角门；石五供后例行建置的方城明楼完全被裁去，仅建方形月台；居中建围合宝顶以掩护地宫的圆式宝城一座，下施青白石须弥座；城顶宇墙不起雉堞，以黄琉璃瓦结顶；宇墙内无马道；周环十丈八尺七寸，尚不及其前各后陵宝城尺度的三分之一；向例悬挂在明楼南檐并题有陵寝名号的斗匾也改用青白石刊刻，横嵌在中座琉璃花门上（图16）。

昌西陵是清代唯一隆恩殿天花使用丹凤展翅图案、陵寝门（琉璃门）前有玉带河配三座石平桥的后陵；也是唯一现存宝城铜沟嘴的陵寝。

图16 昌西陵鸟瞰（出自《中国美术分类全集·中国建筑艺术全集·清代陵墓建筑》）

图17 昌陵妃园寝鸟瞰（易县清西陵文物管理处藏）

Fig.16 Bird's–eye view of West Tomb of Good Fortune. (*Complete Series of Chinese Architectural Art, the authoritative volume on Qing–dynasty funerary architecture*)

Fig.17 Bird's–eye view of the Imperial Consorts' Tombs affiliated with Tomb of Good Fortune. (In the collection of the Administration of Cultural Relic of the Western Qing Tombs)

6) Imperial Consorts' Tombs affiliated with the Tomb of Good Fortune

Located 1 km west of Tomb of Good Fortune, the tomb of Jiaqing's concubines is the resting place of seventeen people, including the imperial noble consorts Gong Shun and He Yu as well as other women of different ranks. Construction started simultaneously with Tomb of Good Fortune. In his first reign year, after he had decided to build his own Tomb, emperor Jiaqing selected a resting place for his consorts and concubines located southwest of Shuangshui Valley, which became the second concubine tomb at the Western Qing Tombs and the fifth one of its kind built in the Qing dynasty. The Hall of Ritual Sacrifice and auxiliary structures in front of it all complied with the standards set by the concubine tomb affiliated with Tomb of Peace, but there were no the Eastern and Western Side Halls. The east and west watchhouses in front of the Main Gate had three bays each, and the east and west wing rooms had three bays each—all of them being more moderate than those of the Imperial Consorts' Tombs affiliated with the Tomb of Admiration.

At both sides of the Hall of Ritual Sacrifice, a division wall with side doors was erected in the manner of the concubine tomb affiliated with Tomb of Prosperity but on a smaller scale (Fig.17). A low wall with green glazed tile top is to the north of the division wall. It had an passage in the middle that connected with the ramp leading to the large platform inside where the concubines' Underground Palaces, earth mounds, and front platforms were arranged according to the women's rank. The low wall was unique among Qing concubines' tombs.

7) Striking out a New Path—Tomb of Veneration

Construction at Tomb of Veneration or Muling, the third emperor's Tomb at the Western Qing Tombs, started in Longquan Valley in the western part of the complex in 1831, the eleventh year of emperor Daoguang's reign. Emperor Daoguang, (posthumously titled) empress Xiaomu Cheng (of the Niohuru clan), his first empress consort Xiaoshen (of the Tunggiya clan), and his second empress consort Xiaoquan (of the Niohuru clan) were buried here.

In 1821, his first reign year, emperor Daoguang began to build his tomb in Raodou Valley at the East Qing Tombs in accordance with the Zhaomu System. But in the next year, he selected a different site in nearby Baohua Valley and expressed his wish for "simplicity rather than extravagance" in design. The decoration for the Underground Palace was thus simplified, and the two-column gate was removed. However, all the other structures were still built in the same manner as Tomb of Admiration. The Tomb was completed after six years, but because

（六）昌陵妃园寝

昌陵妃园寝是嘉庆帝的恭顺皇贵妃和裕皇贵妃及其他妃、嫔、贵人等共17人的墓园，位于昌陵西一公里，与昌陵同期兴建，隶从昌陵。

嘉庆初年，嘉庆帝经营昌陵的同时，择地于陵西南隅双水峪，建造清代第五座妃园寝，也是清西陵域内的第二座妃园寝。昌陵妃园寝建筑规制，享殿及其前所配置的各建筑，均遵照泰陵妃园寝制度，不设东西庑（配殿），宫门前东西班房各三间，东西厢房各三间，逊于景陵妃园寝（图17）。享殿两侧建面阔墙，随墙东西各开角门一座，类同裕陵妃园寝，但远比裕陵妃园寝的琉璃花门简朴。其面阔墙北，增设一道绿琉璃瓦扣脊的矮墙，中央开口，铺墁礓礤上达墙内大平台，其上则依尊卑位次列置诸妃嫔的地宫宝顶和月台。这一矮墙的建置，在清代各妃园寝中是仅见的特例。

（七）自出机杼的慕陵

慕陵位于清西陵西部的龙泉峪，合葬清宣宗旻宁及其孝穆皇后钮祜禄氏、孝慎皇后佟佳氏、孝全皇后钮祜禄氏，道光十一年（1831年）始建，是清西陵第三座帝陵。

最初，旻宁曾恪遵东西陵昭穆制度，道光元年（1821年）在清东陵的绕斗峪预建陵寝，翌年改号宝

of water seeping into the Underground Palace, it was abandoned and demolished. Daoguang's tomb was then rebuilt in Longquan Valley at the Western Qing Tombs, and completed in 1835, his fifteenth reign year. Because of extra costs resulting from the relocation and reconstruction, and due to his failure to implement the Zhaomu System as originally promised, emperor Daoguang reproached himself and decided to build his new tomb as economically as possible. He followed the tomb building tradition of his ancestors at Shengjing, the former Qing capital from 1625 to 1644, showing his willingness to abide to former standards and his filial piety.

At the new Tomb of Veneration, stone statues and Ornamental Columns were omitted. The architectural components of the Pavilion for the Stela on the Spirit Way were decorated with simple but elegant polychrome painting where swirling lines were drawn with carpenters' ink in blue, green, red, white, and black without golden (*yawumo*). The Dragon and Phoenix Gate was modeled after that at Tomb of Good Fortune. The central stone bridge leading to the Gate of Monumental Grace retained its three arches, while the flanking side bridges were flattened. The wall surrounding the Tomb was made of tightly joint bricks. Tomb of Veneration's the Hall of Monumental Grace—a three-bay structure with surrounding corridor and single eave hip-gable roof—was similar in form to the Hall of Monumental Grace at the Tomb of Blessing and the Tomb of Brightness but without colored pattern. Each gable wall had a partition boards door in the middle. The iron chains on the ornaments (*zhengwen*) at the end of the principal roof ridge (*jianguang shida*) are the only examples still extant at Qing tomb. No stone fence was installed around the hall's front platform, but at the left and right sides were installed a square stone column and a stone sundial. Tomb of Veneration's the Hall of Monumental Grace and its the Eastern and Western Side Halls were built with Phoebe (*jinsi nanmu*), polished with melted wax (and not painted), and decorated with clouds and dragons in high relief. The burial area gate (*lingqinmen*) was a three-bay stone Memorial Gateway inscribed on its southern face with the name of the tomb—Tomb of Veneration—in Manchu, Mongolian, and Chinese languages by Yizhu, the fourth son and successor of emperor Daoguang known as emperor Xianfeng after accession to the throne. The northern face bears an inscription that explains emperor Daoguang's choice of name. ("Looking into the northeast with reverence, I really worship my ancestors infinitely forever. Their morality is like the giant, misty and endless mountains. Do I admire them? Yes, I do.")

No two-column gate stood north of the stone Memorial Gateway where only the Five Stone Ritual Vessels were placed centrally. On the stone alter, the vases and the incense burner were decorated with original, floating clouds instead of beast faces. This was emulated at West Tomb of Good Fortune and East Tomb of Veneration. Behind the Five Stone Ritual Vessels

华峪万年吉地并下旨『黜华崇实』，但只是简化地宫装修、撤掉二柱门而已，其他建筑规制仍一如景陵。该吉地历时六年建成，随即却以地宫浸水废弃，又改建在清西陵的龙泉峪，道光十六年（1836年）告竣。因重建工程额外耗费，而遵奉『东西陵昭穆制度』的信誓也落为覆水，旻宁曾引咎自责，并决意『概从撙节』，效仿盛京祖陵规制以标榜『敬绍先型，谨遵前制』和『孝思不匮』。

改制的慕陵裁掉了石像生和望柱，神道碑亭绘典雅朴素的雅伍墨彩画，其前效仿昌陵配置龙凤门。隆恩门前保留中路三孔石拱桥，两边改做石平桥；陵墙全为磨砖对缝。隆恩殿类同福陵和昭陵，三开间周围廊单檐歇山顶式样，却不施彩画，两山居中开设槅扇门，正吻遗存清陵仅有的吻链即所谓『见广识大』遗迹；月台撤除雕栏，左右分别陈设石雕方幢、日晷；隆恩殿及东西配殿使用金丝楠木，装修刻高浮雕云龙，原木烫蜡不施油彩。陵寝门改成三间石牌坊，南面镌刻嗣皇帝奕詝用满、蒙、汉字题写的慕陵名号，北面为旻宁有关慕陵命名的遗谕：『敬瞻东北，永慕无穷，云山密迩。呜呼，其慕与？慕也。』牌坊北没有二柱门，居中安设石五供，石五供的花瓶、香炉上不刻兽面纹而刻新颖流云图案，成为昌西陵与慕东陵蓝本；往后隆起两层台地，各三出陛，均安设雕栏；陵宫尽端罗圈墙围合的方形大月台中央，

was built a two-tier platform. Each tier had three flights of steps. Stone fences were mounted on both sides of each tier. A Encircled Realm of Treasure with bronze waterspouts was built in the center of the large front platform positioned in the rear of the tomb palace and enclosed by a curved wall. Huge cuboid stones with a height equal to that of the front platform (*hundun jietiao*) were leaning horizontally against it. There neither was a Square Walled Terrace, Memorial Tower, Mute Court, Crescent Wall, nor a Screen Wall of Glazed Tiles (Fig.18).

Emperor Daoguang had experienced socio-political problems like national betrayal and humiliation resulting from the Chinese defeat in the First Opium War during his lifetime. At the end of his life in 1850, his thirtieth reign year, he forbade to set up Pavilion for the Stelas and Ceremonial Columns and, ashamed before his ancestors, he said:

> *I hope that after my death, the posthumous title and temple name should be inscribed in Manchu and Chinese on the stone tablet standing in the Memorial Tower and, on the back of the tablet, give the name of the Tomb.*

Although, he had not built Square Walled Terrace and Memorial Tower, his fourth son and successor Yizhu—the later Xianfeng emperor—ordered officials to analyze the Tomb construction of preceding emperors and based on this, put forward the plan to add these buildings at his father's burial site. But eventually, being under constraint in many respects, emperor Xianfeng could only "compose an inscription for Tomb of Veneration… and inscribe it on the back of the stone tablet outside the Gate of Monumental Grace". And no other building regulations were changed. The Tomb of Veneration became the most exceptional Tomb among the Eastern and Western Qing Tombs.

8) East Tomb of Veneration

Located 1 km northeast of Tomb of Veneration at Two Peak Hill (Shuangfeng *xiu*), East Tomb of Veneration, is the resting place of emperor Daoguang's consorts and concubines of different ranks. It represents a combination of the empress and concubine tomb concepts. Originally intended as a concubines' tomb, it was simultaneously built with Tomb of Veneration. But in 1855, the fifth year of emperor Xianfeng's reign, after Lady Borjigin (of the Borjigit clan) was promoted from an imperial noble consort (as in emperor Daoguang's times) to the status of empress mother and posthumously entitled empress Xiaojing, the tomb was renamed East Tomb of Veneration to indicate its higher rank and affiliation with Tomb of Veneration. It is thus the third empress's tomb at the Western

圆台形的宝城孑然独立，宝城原设带铜挑头沟嘴，月台使用罕见的混沌阶条做法，无方城明楼及两翼面阔墙、哑巴院、月牙城和琉璃影壁等（图18）。

道光三十年（1850年）旻宁临终前，以鸦片战争失败而丧权辱国，愧对祖宗，遗命严禁建立圣德神功碑亭和华表，却含混提出『万年后著于明楼碑上镌刻大清某某皇帝清汉之文，碑阴即可镌刻陵名』。然而，慕陵营建时已遵旨罢建置方城明楼，奕詝继统后，曾为此派员核查各陵规制，拟出改建宝城、添建方城明楼的方案，但在各种因素制约下，只得『撰成慕陵碑文一篇……镌于隆恩门外碑石』，即刻在隆恩门前神道碑的背面，其他既有建筑规制全未改动，慕陵由此成为清东陵与西陵中最特殊的帝陵。

（八）慕东陵

慕东陵由妃园寝改建而来，是介于后陵与妃园寝规制之间的清代后陵特例。其为道光帝的孝静成皇后博尔济吉特氏（1812—1855）及道光帝其他皇贵妃、贵妃、妃、嫔、贵人等人的陵寝，位于慕陵东北一公里的双峰岫。与慕陵同期告竣。初为妃园寝，至咸丰五年（1855年），以博尔济吉特氏由道光帝

Fig.18 Bird's–eye view of Tomb of Veneration. (*Complete Series of Chinese Architectural Art, the authoritative volume on Qing–dynasty funerary architecture*)

Fig.19 Distant view of East Tomb of Veneration. (*Complete Series of Chinese Architectural Art, the authoritative volume on Qing–dynasty funerary architecture*)

图18 慕陵鸟瞰（出自《中国美术分类全集·中国建筑艺术全集·清代陵墓建筑》）

图19 慕东陵远眺（出自《中国美术分类全集·中国建筑艺术全集·清代陵墓建筑》）

Qing Tombs and the fifth one of its kind among all the tombs of Qing empresses.

Lady Borjigin was originally known as imperial noble consort Jing of the Daoguang emperor, and when he built Tomb of Veneration, he chose a position for her burial mound in the middle of the second row of his consorts' tombs, but later planned to rebuild her Encircled Realm of Treasure in imitation of his own in Longquan Valley. Unfortunately, he passed away before he could execute this plan. Afterwards, emperor Xianfeng issued an imperial edict that provided a practical solution to the problem of showing the newly elevated status of empress Xiaojing in comparison to Daoguang's other consorts and concubines, by building a ritual demarcation line in the form of a wall behind her Encircled Realm of Treasure. The concubines' cemetery was then renamed East Tomb of Veneration and altered accordingly soon afterwards.

Facing domestic upheaval and foreign aggression, the architecture at East Tomb of Veneration was modified efficiently and economically to fit the new requirements for the tomb of an empress. Only a set of Stone Altar and Five Stone Ritual Vessels, side halls, Culinary Courtyard for Sacrifices, Pavilion for the Well, and the Stelas marking the place for dismounting from one's horse were built from scratch in imitation of West Tomb of Good Fortune. The central doorway of the Gate with Glazed Roof Tiles was built. A new wall shaped like a three-part folding screen (U-shape) was added behind the empress's earth mound, while the green glazed tiles covering the other buildings and walls were replaced with yellow glazed tiles. Neither a Square Walled Terrace nor a Memorial Tower were erected. The wooden plaque inscribed with the name of the tomb was replaced with one made of greenish white marble and horizontally inlaid into the lintel above the central entrance of the Gate with Glazed Roof Tiles just like at West Tomb of Good Fortune (Fig.19).

9) Witnessing the End of the Qing Dynasty—Tomb of Serendipity (the tomb of emperor Guangxu)

Located in Jinlong Valley 4 km northeast of Tomb of Peace, Tomb of Serendipity marks the end of imperial tomb construction as the last Tomb built for a Qing emperor. Emperor Guangxu (temple name Dezong; born Zaitian) and empress Xiaoding (of the Yehe Nara clan; also known as empress dowager Longyu) were buried here (Fig.20).

From his accession to the throne until his death thirty-four years later, emperor Guangxu, as younger male cousin of his predecessor the Tongzhi emperor (temple name: Muzong;

原封静皇贵妃而尊谥孝静成皇后，遂将妃园寝升级改建，并更名『慕东陵』，隶从慕陵。成为清代第五座、清西陵第三座后陵。

博尔济吉特氏原为道光帝册封的静皇贵妃；在与慕陵同期建成的双峰岫慕陵妃园寝中，道光帝曾为其钦定位次于前层中座宝顶，嗣后旻宁曾拟『著照龙泉峪宝城式样改修宝城』，然而未及实施即已驾崩。奕詝下谕『拟以慕陵妃园寝作为山陵。惟宝城之后，必须筑墙一道，以崇礼制』。慕陵妃园寝随即更名为慕东陵，改建工程接踵完成。

迫于时局外患内忧，由妃园寝改建的慕东陵规制格外俭约。其中仿照昌西陵等添建石五供、配殿、神厨库、井亭和下马牌等，改建中座琉璃花门，增建宝城后的围屏墙，其他建筑及墙垣则以妃园寝旧物改覆黄琉璃而已，也没有建置方城明楼，题为陵寝名号的斗匾也像昌西陵那样改用青白石刊刻，横嵌在中座琉璃花门上（图 19）。

（九）历史终结的崇陵

崇陵是清代皇家工程的终点。其位于泰陵东北四公里的金龙峪，合葬清德宗载湉即光绪皇帝和孝定景皇后叶赫那拉氏，为清代最末一座帝陵（图 20）。

作为清穆宗载淳的堂弟，载湉继统及在位三十四年，均被慈禧太后把持。光绪十三年（1887 年）

图20 崇陵远眺（陈书砚摄）

图21 梁格庄行宫大殿工作图（中国文化遗产研究院藏）

Fig.20 Distant view of Tomb of Serendipity. (Photo by CHEN Shuyan)
Fig.21 Old photograph with scenery of the main hall at Lianggezhuang traveling palace during construction. (With courtesy of the Chinese Academy of Cultural Heritage)

born: Zaichun), was controlled by his aunt—empress dowager Cixi. On March 20, 1887, two months after emperor Guangxu had factually begun to rule the country in his own right (although formally his thirteenth reign year), he and Cixi paid a formal visit to the Western Qing Tombs. During their visit, Cixi selected a site in Jiulong Valley, where Guangxu's Tomb should be built, and renamed the valley Jinlong Valley. But only after Cixi and Guangxu both had passed away in December 1908, Guangxu's thirty-fourth reign year, was his tomb named Tomb of Serendipity and built "in strict accordance with the standards set by Tomb of Benevolence for emperor Tongzhi". Although construction began in the next year, it came to a sudden stop during the Xinhai Revolution of 1911 (that brought an end to the Qing dynasty) but continued in 1913 upon request of then-former imperial family. It was finally completed by the national government established in March 1915.

The Tomb of Serendipity shared almost the same layout and design with Tomb of Benevolence except from the covered drainage channels (*longxugou*) of the Underground Palace and the waterspouts of Square Walled Terrace and Memorial Tower. Furthermore, neither stone statues nor a Pavilion for the Stela of Sage Virtue and Divine Merit were built at Tomb of Serendipity. However, Tomb of Serendipity modified the Memorial Gateway used at emperors' tombs in the eastern complex since the construction of Tomb of Admiration, blending the structure with the Dragon and Phoenix Gate that was traditionally built at the Western Qing Tombs (to which Tomb of Serendipity belongs). This then becomes the only five-bay-six-into-the-sky-extending-column Memorial Gateway (*wujianliuzhu chongtianshi pailoumen*) at the Western Qing Tombs.

Tomb of Serendipity utilized modern technologies in its building. For example, the engineering department (Gongchengchu) responsible for Tomb construction took photos of the work progress to document it including site selection, construction, completion, and temporary palace maintenance. Officials of all levels engaged in supervision came directly from Beijing to the construction site, taking the new Xin(cheng)-Yi(xian) line running between Gaobeidian in Xincheng (county seat) and Lianggezhuang in Yi County, both in Hebei province. Contractors had built an extended railway to transport building material including timbers and stones. The engineering departments in Beijing and Lianggezhuang, where the construction site was located, were furbished with electric lamps and telephones to facilitate communication (Fig.21). When the Underground Palace, Square Walled Terrace, and Memorial Tower were built, hard stones delayed the construction process because they were difficult to cut; the traditional waterproofing originally planned to use turned out to be impracticable; the new building material of cement was used to solve the problem. At the time,

三月十二日，慈禧太后由亲政仅两月的载湉侍奉瞻谒清西陵，曾躬亲择定九龙峪为载湉万年吉地，并改称『金龙峪』。然而情随事迁，到载湉和慈禧太后相继死去，光绪三十四年（1908年）十二月才确定崇陵名号，并『恭照惠陵规制敬谨兴修』。翌年开工，历经辛亥革命，经由逊清皇室依《优待条件》再三催促，推至1913年方才接续施工，直到1915年3月才由国民政府宣告完竣。

崇陵建筑规制以惠陵为蓝本，从局部到整体都几乎如出一辙，除却在地宫龙须沟与方城明楼排水沟嘴的设计略有不同外，石像生和圣德神功碑亭也均未建置，而牌楼门采用景陵以来清东陵各帝陵的通行式样，则改变了以往清西陵各帝陵均为龙凤门的格局，成为清西陵仅有的一座五间六柱冲天式牌楼门。

崇陵工程大量使用了现代技术，例如清室崇陵工程处使用摄影术记录陵寝选址、营建、行宫修缮与竣工的工程进度；督工的各级官员均由北京乘新易铁路直抵工次；承建厂商接搭延长铁轨运输木石等建筑材料；京畿与梁格庄行宫两工程处请报架设电灯、电话，以便相互联络（图21）；崇陵头段地宫与方城明楼的施工过程中，在大槽开挖后发现地下均为难以打穿下桩的硬石，无法按拟定古法进行，随即改用国产水泥来解决防水问题，并大量应用于其他附属建筑的基础部分。

以上情况均充分体现出，我国古代的工程管理与技术人员在时事变迁、技术更迭的广阔时代背景下，针对具体实践充分发挥创造力，灵活变通，积极探索新工艺、新方法的时代精神与应变能力，也为中国古代皇家建筑工程的终结，奏出精彩的绝响。

（十）崇陵妃园寝

清代最后一座妃园寝，规制与景陵、定陵和惠陵妃园寝一脉相承。崇陵妃园寝经营期间，清朝政权被辛亥革命推翻，国民政府曾遵照《关于大清皇帝辞位之后优待之条件》宣布『光绪帝陵寝如制妥修』，然而最终却因经费奇绌，东西平列在崇陵园寝后部的两座地宫，并未援例建造石券，改修成更简陋的砖券。随瑾妃被逊清皇室按旧制晋封为端康皇贵妃，为她预留的砖券被改修成石券，外部宝顶和月台的规模也就明显超出了西面珍妃的砖券宝顶和月台（图22）。

（十一）宗室、公主园寝

清西陵陵区现存有落成于乾隆二年（1737年）的和硕端亲王、和硕怀亲王和阿哥园寝及嘉庆八年（1803年）的慧愍固伦公主、慧安和硕公主园寝，另有果恭亲王和果毅郡王陪葬园寝遗址，是使清西陵成为最完整的皇家陵墓群的重要组成部分（图23、图24）。

cement was then also widely used for the foundations of auxiliary buildings.

In essence, given the rapid historical change in the execution of construction work and in building technology, the builders of imperial China were capable of immediately responding to changing circumstances, exploring new methods and technologies, and applying them in a flexible and creative way. The wisdom and effort of these builders is what ultimately made this last imperial Tomb construction project possible—a testimony to the end of the glorious history of China's feudal dynasties.

10) Imperial Consorts' Tombs affiliated with Tomb of Serendipity

The last tomb of concubines built during the Qing dynasty followed the same standard as those affiliated with Tomb of Admiration, Tomb of Stability, and Tomb of Benevolence. During the construction, the Qing government was overthrown in the Xinhai Revolution. The national government in observance with the resignation agreement with the imperial family entitled *Favorite Treatment of the Imperial Family of the Great Qing dynasty after Abdication of the Emperor (Guanyu daqing huangdi ciwei zhihou youdai zhi tiaojian)*, announced that "the tomb of emperor Guangxu would be built appropriately and in accordance with the original design". Nevertheless, because of the high costs, the Underground Palaces of the consorts Zhen and Jin positioned at the east and the west sides in the rear part of Tomb of Serendipity's tomb of concubines were built in brick instead of stone. Later, when the former imperial family promoted consort Jin to the status of imperial noble consort Duan Kang, the brick vaults of the Underground Palace were rebuilt in stone. Thus, her Tumulus and front platform exceeded their brick counterparts of consort Zhen in height, visually expressing her higher rank and importance (Fig.22).

11) Tombs of Princes, Princesses, and other Members of the Imperial Clan

The tombs for the princes of the first rank (Heshuo *qinwang*) Duan and Huai, one of emperor Yongzheng's son, were completed in 1737, the second reign year of emperor Qianlong. The tombs for the princess of the first rank (Gulun *gongzhu*) Hui Min and the princess of the second rank (Heshuo *gongzhu*) Hui An were completed in 1803, the eighth reign year of emperor Jiaqing. Together, with the subordinate tombs for the prince of the first rank (*qinwang*) Guo Yi and the prince of the second rank (*junwang*) Guo Gong, they formed an important part of the Western Qing Tombs, the most complete imperial Tomb cluster of any Chinese dynasty (Fig.23, Fig.24).

12) Temple of Perpetual Happiness

Temple of Perpetual Happiness is the only funerary temple built for an imperial Tomb that has survived from the Ming and Qing dynasties (Fig.25).

Associated with Lamaism, it highlights the importance Tibetan Buddhism played in the Qing dynasty. Construction began in 1787, the fifty-second reign year of emperor Qianlong and the tenth anniversary of the death of his mother Xiaosheng Xian. The inscription on the temple tablet composed by Qianlong tells us that his reason for building the temple:

> *...lies in its close proximity to Tomb of Peace, and whenever I will visit my ancestors' tombs, I will be able to rest here first and show my respect to my ancestors through sacrifices. The role of Temple of Perpetual Happiness shall be just like that of Temple of Monumental Happiness at Tomb of Admiration.*

The architecture of Temple of Perpetual Happiness confirms the emperor's wish to follow the building traditions created by his ancestors and "to express his filial piety". Qianlong was the first of the Qing emperors who frequently made official visits to his ancestral tombs, especially to the Western Qing Tombs. From this we can also understand the depths of Qianlong's sorrows at his mother's passing.

13) Lianggezhuang Palace

Built in 1748, the thirteenth reign year of emperor Qianlong, but damaged, rebuilt, and occupied on several occasions after by the temporary security department and the engineering department of Tomb of Serendipity at the end of the Qing dynasty, Lianggezhuang Palace is a unique case of a traveling palace—a temporary residence of Qing emperors when traveling outside the capital—a structure still used today that had been built in close proximity to the Tomb complex. The palace and garden largely retain the original design and layout of the Qing traveling palace construction and highlight the integrity and completeness of the cultural heritage site of the Western Qing Tombs as inscribed on the World Heritage List (Fig.26).

3.Modern Protection and Research of the Western Qing Tombs

During the 1910s of China, the members of Commission for the Preservation of Western

（十二）永福寺

永福寺是明清皇家陵寝中唯一一座现存御用陵寝寺庙，它是一座专门为陵寝祭祀而修建的皇家喇嘛庙，是清王朝尊重、信仰藏传佛教的实物例证（图25）。

永福寺建于乾隆五十二年（1787年），时逢孝圣宪皇太后逝世十周年，乾隆帝在寺中碑文记述『寺之所以建者，以其地近泰陵，每遇祇谒桥山，先憩于此，斋心摄神以夙严对越之敬，亦如隆福之于景陵也』。永福寺的建立，表明乾隆帝遵承祖制，『以展孝思』之心，『上资寿母，福溥无边』，为其母祈福。在清朝皇帝之中，重视谒陵首推乾隆，特别是拜谒西陵次数最多，可见其对母后怀念之深。

（十三）梁格庄行宫

梁格庄行宫始建于乾隆十三年（1748年），经历清末为崇陵工程而改扩建暂安处和工程处的系列工程后，一直沿用至今，为历代陵寝行宫的孤例。现存行宫建筑园林基本保持始建时的格局，是清代行宫制度的实物体现，是清西陵遗产完整性的重要组成（图26）。

三、现代保护和研究

民国初期由清西陵陵寝古迹保管委员会自逊清手中全面接管清西陵保护管理工作。20世纪上半叶，

图23 端亲王园寝鸟瞰（易县清西陵文物管理处藏）

图22 崇陵妃园寝鸟瞰（易县清西陵文物管理处藏）

图25 永福寺鸟瞰（易县清西陵文物管理处藏）

图24 怀亲王园寝鸟瞰（易县清西陵文物管理处藏）

Fig.22 Bird's–eye view of Tomb of Serendipity. (In the collection of the Administration of Cultural Relic of the Western Qing Tombs)
Fig.23 Bird's–eye view of the tomb of prince Duan. (In the collection of the Administration of Cultural Relic of the Western Qing Tombs)
Fig.24 Bird's–eye view of the tomb of prince Huai. (In the collection of the Administration of Cultural Relic of the Western Qing Tombs)
Fig.25 Bird's–eye view of the temple of perpetual happiness. (In the collection of the Administration of Cultural Relic of the Western Qing Tombs)

Qing Tombs' Antiquities (*qingxiling lingqin guji baoguan weiyuanhui*) took over tasks of imperial Tomb protection and management from the former Qing dynasty. In the first half of the twentieth century, many foreign scholars visited China to investigate and study Chinese traditional architecture including Ming-Qing imperial tomb. In 1892, G. Deveria, a French sinologist and Tangutologist, published *Sépultures Impériales de la Dynastie Ta Ts'ing*, containing a brief introduction to Tomb of Peace and East Tomb of Peace in Western Qing Tombs. The book laid the foundation for Westerners to know the Tombs of Qing Dynasty for the first time. From 1900 to 1927, some scholars and officers left precious visual data of the Western Qing Tombs, including French sinologist Victor Segalen, guard commander of the French Legation Firmin Laribe, French military officer E. Fonssagrives, German architect Ernst Boerschmann, German ambassador to China Alfons von Mumm, American photographer C. E. Lemunyon, American scholar Sidney David Gamble and Japanese scholar Sekino Tadashi.

E. Fonssagrives was the commander of the occupying army in the Western Qing Tombs, so he has sufficient time and manpower as well as rich experience in photography and surveying and mapping. In 1907, he published the scientific investigation report *Si-Ling*, which was the only book in the early 20th century and before devoted to the Western Qing Tombs. The report is 180 pages, 196 images including photos, sketches and survey drawings. In 1920, he published *Conférence donnée à la Société Polymathique du Morbihan*, which was a review paper of the Tombs of Qing Dynasty following G. Deveria.

The German architect Ernst Boerschmann (1873-1949) visited the Western Qing Tombs during his trip to China in August 1907 with the goal to systematically document Chinese architectural works. He took seventy pictures and forty pages of illustrated notes including detailed survey maps. Afterwards, his photos and drawings became significant supporting material used for the nomination and formal inscription of the Western Qing Tombs on the World Heritage List (Fig.27, Fig.28).

In 1918, the Japanese architectural historian Sekino Tadashi (1868-1935) came to Yi County, Hebei province, to survey the Western Qing Tombs, taking many pictures. His collection of one hundred and ninety-one photos are now housed in the Institute of Oriental Culture at the University of Tokyo. Taken in the order of his visit, they can be grouped into three categories—overall; sub-items; and detail. The collection speaks to the completeness of his documentation, his scientific use of photographic techniques, and his rigorous logic (Fig.29).

伴随着国门开放，一批外国学者对包括清西陵在内的中国古代建筑展开前赴后继的考察与研究。

1892年，法国汉语言学家德微理亚（G.Deveria）发表《大清朝帝王陵》，为西方初识清陵奠定了基础，内含清西陵泰陵与泰东陵简介。1900—1927年间，法国汉学家谢阁兰（Victor Segalen）、公使馆少校菲尔曼·拉里贝（Firmin Laribe）、军官欧仁·冯萨格里维斯（E.Fonssagrives），德国建筑师恩斯特·柏世曼（Ernst Boerschmann）、驻华公使穆默（Alfons von Mumm），美国摄影师雷尼诺恩（C.E.Lemunyon）、学者西德尼甘博（Sidney David Gamble），以及日本学者关野贞，都留下了清西陵珍贵的影像资料。

其中，欧仁·冯萨格里维斯是清西陵占领军指挥官，时间、人力充足，摄影、测绘经验丰富。1907年，发表科学调查报告《西陵》，成为20世纪初及以前西方唯一一本专涉清西陵的著作，报告180页，照片、速写、测绘图合计196幅；1920年再著《清朝皇帝陵》，是继德微理亚后，专以清朝陵寝为对象发表的综述。

1907年8月，恩斯特·柏世曼（Ernst Boerschmann）的中国建筑系统考察研究之旅途经清西陵，完成拍摄70张照片及40页图文笔记，其中包括部分详实的测绘图。这些图档和照片成为清西陵申报世界文化遗产地的重要支撑材料（图27、图28）。

1918年，日本建筑学者关野贞调查了河北易县清西陵，现存191幅照片藏于日本东京大学东洋文化研究所。照片按谒陵流线拍摄，分整体、分项和细部三类。其记录完整，方法科学，逻辑严谨（图29）。

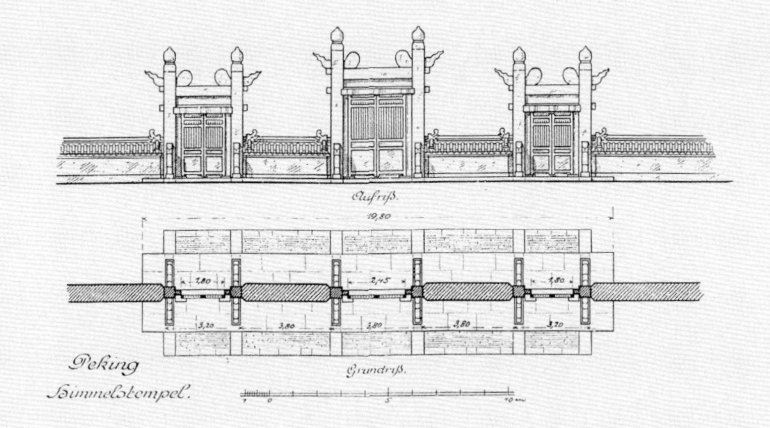
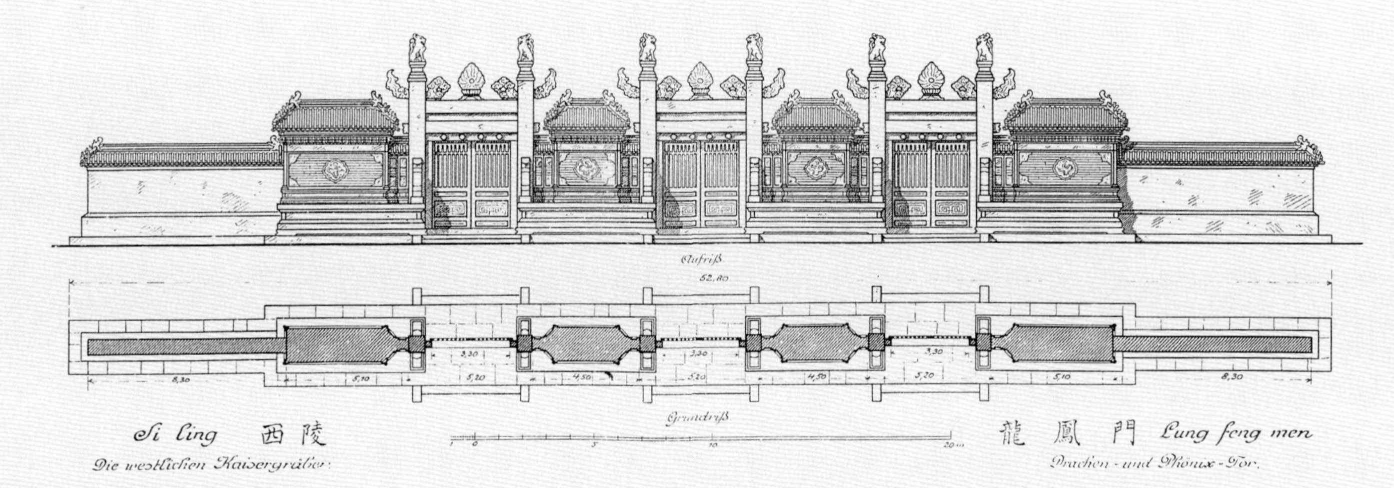

图28 昌陵龙凤门测绘图（柏世曼绘）

图26 梁格庄行宫宫门工作图（中国文化遗产研究院藏）

图27 昌陵龙凤门前神道（柏世曼摄）

图29 崇陵远眺（关野贞摄，日本东京大学东洋文化研究所藏）

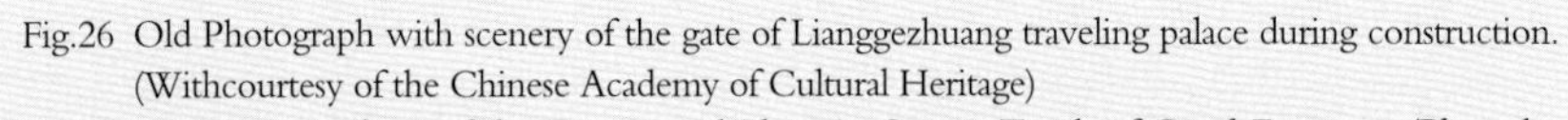

Fig.26 Old Photograph with scenery of the gate of Lianggezhuang traveling palace during construction. (Withcourtesy of the Chinese Academy of Cultural Heritage)

Fig.27 Spirit Way in front of the Dragon and Phoenix Gate at Tomb of Good Fortune. (Photo by Ernst Boerschmann)

Fig.28 Survey maps of the Dragon and Phoenix Gate at Tomb of Good Fortune. (Drawing by Ernst Boerschmann)

Fig.29 Distant view of Tomb of Serendipity. (Photo by Sekino Tadashi, stored at the Institute of Oriental Culture atthe University of Tokyo)

In 1934, during his investigation with MO Zongjiang and CHEN Mingda, LIU Dunzhen conducted preliminary surveying and mapping of the Western Qing Tombs and carried out site inspection of objects recorded on documents from the Yangshi Lei Archives to identify time and stage of design of the drawings. In his work *Western Qing Tombs in Yi County* (*Yixian Qingxiling*), LIU outlines the construction history, evolution of the layout of the emperors' and empresses' tombs, and the interior structure of Underground Palaces. Additionally, WANG Biwen discussed Tomb of Serendipity in his *Official-style construction methods for stone bridges of the Qing dynasty* (*Qing guanshi shiqiao zuofa*), providing original photos and survey maps while analyzing the engineering practices used in the light of today's knowledge.

Since the establishment of the People's Republic of China, particular attention has been given by the government at the national and local levels to the protection and management of the Western Qing Tombs. In 1954, the *Office for the Preservation of Western Qing Tombs' Cultural Relic* (*Qingxiling wenwu baoguanchu*) was founded. In 1987, the office was renamed to Administration of Cultural Relic (of the Western Qing Tombs) (*(qingxiling) wenwu guanlichu*) and became subject to local jurisdiction, namely Baoding, covering of the grave Prefectural Bureau of Cultural Affairs (*diqu xingshu wenhuaju*). Today, it is under the jurisdiction of the Yi county government. In 1961, the tombs were designated as National Priority Protected Sites (*quanguo zhongdian wenwu baohu danwei*). In 1996, they were inscribed on the World Heritage Tentative List, and in 2000, on the World Heritage List. In 2001, they ranked among the first batch of National 4A Tourist Attractions. In 2004, they passed the certification of ISO09001 and ISO14001 Quality and Environmental Management Systems. And most recently in 2006, they were honored as Advanced Group of World Heritage Site Protection by the Ministry of Culture of the People's Republic of China. In 2020, they upgraded to 5A Tourist Attraction.

Routine maintenance of the Western Qing Tombs is carried out in strict accordance with the principle of no change to the original appearance i.e., to preserve the authenticity of cultural relics to the most possible extent based on verified textual and archival sources. The tombs have thus maintained their historical authenticity about layout, design, building materials and methods and have become a best-practice example for organizing the repair and (tourist) visit of Qing tombs. Next to the preservation of individual monuments, attention was paid to the protection of nature and landscape including the more than 15,000 ancient pine trees.

Additionally, several generations of archaeological conservators have attached great

1934年，营造学社刘敦桢先生携莫宗江、陈明达考察清西陵，开展测绘记录工作；并以样式雷图档对照实物，来鉴定图档的设计阶段；作『易县清西陵』，对清西陵营建年代、平面变迁与地宫结构等进行了详细阐述。另外，王璧文《清官式石桥做法》以崇陵工程为重要案例，援引崇陵照片、测绘图和崇陵工程做法等一手资料完成一系列细致的研究。

中华人民共和国成立后，清西陵的保护和管理得到各级政府的特别重视。1954年西陵文物保管所成立；1987年文保所改为文物管理处，由保定地区行署文化局管辖，如今由易县县政府直接管辖。期间，文物保护工作持续稳步开展。清西陵于1961年列为全国重点文物保护单位，1996年列入世界文化遗产预备名单，2000年列入世界文化遗产名录，2001年评为国家旅游局第一批AAAA级旅游景区，2004年通过ISO 09001和ISO 14001质量与环境管理体系认证，2006年被国家授予世界文化遗产保护先进集体。2020年，升级为AAAAA级旅游景区。

在清西陵日常维修保护中，严格遵守『不改变原状』（即尽最大努力保存文物的真实性）的原则，以确凿文献和档案资料为依据，其设计、材料、工艺、布局等方面均保持了历史的真实性，成为修缮、参观清代陵寝的样本。在对清西陵建筑主体进行保护的同时，亦注重周围环境的保护。现存15000余株古松是清西陵环境风貌的真实写照。

清西陵几代文保人极尽可能努力保护清西陵文物遗存的同时，非常重视档案工作。纸质档案按照事由和时间归入相应排架，并编排可供电子检索的目录，成为现代清西陵保护研究的重要一手资料。其古松档案库包括每一棵古松的信息，使联合国教科文组织遗产评估专家都为之动容。

在此背景下，清西陵成为天津大学建筑学院最重要的教学和科研基地之一。1963年暑期，卢绳先生组织学生测绘泰陵、昌陵，并对周边陵寝进行系统考察，随后形成学术成果『清代陵寝建筑群造型的艺术分析』『易县古建筑考察记』。1983—1984年天津大学建筑系冯建逵、王其亨教授率领1981—1982级学生系统地测绘清西陵帝后陵寝，并基于测绘成果完成清代皇家陵寝研究的权威学术著作《中国建筑艺术全集·清代陵墓建筑》（图30）。

2008年，天津大学师生对妃园寝、宗室公主园寝及永福寺实施全面数字化测绘。前后共有179名师生投入，完成测绘图670幅。早期参加测绘的学生如今大多已成为建筑界的领军人物，甚至院士。作

importance to archival management. Historical documents were stored in corresponding shelves in terms of event and time and catalogued for convenient electronic retrieval. They constitute essential primary materials for protection and study of imperial Qing tombs. The Old Pine Archive has detailed records of every tree—a fact that impressed even the world heritage experts during their site evaluation.

Against this background, the Western Qing Tombs became a key teaching and research tool for the School of Architecture of Tianjin University. During the 1963 summer holidays, LU Sheng and his students undertook comprehensive surveying and mapping at Tomb of Peace and Tomb of Good Fortune and systematic investigation of nearby tombs, which provided the basis for their publications *Formal Artistic Analysis of the Architectural Complex of Qing-dynasty Tomb* (*Qingdai lingqin jianzhuqun zaoxing de yishu fenxi*) and *Notes on the Investigation of Traditional Buildings in Yixian* (*Yixian gujianzhu kaochaji*). Under the leadership of Professor FENG Jiankui, and graduate student WANG Qiheng, from the Department of Architecture at Tianjin University, students (Class of 1981/1982) undertook systematic surveying and mapping of emperor and empress tombs at the Western Qing Tombs from 1983 to 1984, *Complete Series of Chinese Architectural Art, the authoritative volume on Qing-dynasty funerary architecture (Zhongguo jianzhu yishu quanji, qingdai lingmu jianzhu)*.

In 2008, a team of one hundred and seven-nine teachers and students carried out digital surveying and mapping of the tombs for concubines, princes, princesses, and other male members of the imperial clan, as well as of Temple of Perpetual Happiness. As a result of this fieldwork, six hundred and seventy survey drawings were drawn. Today, many students who had participated have become leading figures in the field of architectural design and education. Over the past decades, as an important world heritage, the Western Qing Tombs have provided an excellent platform for the teaching and training of nation talent.

To better communicate and protect the Western Qing Tombs, and to demonstrate the beauty of the architectural heritage, the Administration of Cultural Heritage of the Western Qing Tombs and the School of Architecture of Tianjin University selected two hundred and sixty drawings from their past surveying and mapping activities and compiled this volume of the Traditional Chinese Architecture Surveying and Mapping Series. It contains in total two hundred and thirty eight pages of drawings in the following order: Tomb of Peace, East Tomb of Peace, The Imperial Consorts' Tombs affiliated with the Tomb of Peace; Tomb of Good Fortune, West Tomb of Good Fortune, The Imperial Consorts'

Tombs affiliated with the Tomb of Good Fortune; Tomb of Veneration, East Tomb of Veneration; Tomb of Serendipity, The Imperial Consorts' Tombs affiliated with the Tomb of Good Fortune, Tomb for Prince Duan, Tomb for Prince Huai, Tomb for a Son of Emperor Yongzheng, Tomb for Two Princesses, Temple of Perpetual Happiness, Lianggezhuang Palace, Flame Memorial Gateway. It is hoped that this book will allow the readers to better understand the cultural heritage of the Western Qing Tombs and be a boon to the protection, inheritance, and development of this outstanding site.

为重要的世界文化遗产地，清西陵不愧为培养我国文化遗产保护和建筑事业卓越人才的优秀教学平台。

为更好地宣传清西陵、保护清西陵，更形象地展示清西陵古建筑的华章，易县清西陵文物管理处和天津大学建筑学院密切合作，从相关测绘研究成果中精选了238幅图纸，按照泰陵、泰东陵、泰陵妃园寝、昌陵、昌西陵、昌陵妃园寝、慕陵、慕东陵、崇陵、崇陵妃园寝、端亲王园寝、怀亲王园寝、阿哥园寝、公主园寝、永福寺、梁格庄行宫、火焰牌坊的顺序，组织260页图版，编辑成书，奉献给公众，祈望能够裨益于这一优秀文化遗产的保护、继承、弘扬和借鉴。

图版

Figures

泰 陵
Tomb of Peace (Tailing)

泰陵组群剖面图
Site section of the building complex of the Tomb of Peace

1 五孔石券桥 Five-Arched Stone Bridge
2 石牌坊 Marble Memorial Gateway
3 石麒麟 Marble Qilin
4 下马牌 Stela Marking the Place for Dismounting from One's Horse
5 大红门 Great Red Gate Building
6 具服殿 Hall for Court Robes
7 三孔石平桥 Three-Arched Flat Stone Bridge
8 圣德神功碑亭 Pavilion for the Stela of Sage Virtue and Divine Merit
9 华表 Ceremonial Column
10 七孔石券桥 Seven-Arched Stone Bridge
11 望柱 Ornamental Column
12 石狮 Lion among the Stone Statues
13 石象 Elephant among the Stone Statues
14 石马 Horse among the Stone Statues
15 武官 Martial Official among the Stone Statues
16 文官 Literary Official among the Stone Statues
17 龙凤门 Dragon and Phoenix Gate
18 三孔石券桥 Three-Arched Stone Bridge
19 三路三孔桥 Three-Way Three-Arch Bridges
20 神道碑亭 Pavilion for the Stela on the Spirit Way
21 下马牌 Stela Marking the Place for Dismounting from One's Horse
22 石平桥 Flat Stone Bridge
23 神厨库门 Gate in the Culinary Courtyard for Sacrifices
24 神厨 Kitchen in the Culinary Courtyard for Sacrifices
25 神库 Repository in the Culinary Courtyard for Sacrifices
26 宰牲亭 Ritual Abattoir in the Culinary Courtyard for Sacrifices
27 井亭 Pavilion for the Well
28 朝房 Reception Hall for Court Officials
29 值房 Guard House
30 隆恩门 Gate of Monumental Grace
31 焚帛炉 Sacrificial Burner
32 配殿 Side Hall
33 隆恩殿 Hall of Monumental Grace
34 琉璃花门 Gate with Glazed Roof Tiles
35 二柱门 Gate with Two Columns
36 石台五供 Five Stone Ritual Vessels
37 方城明楼 Square Walled Terrace and Memorial Tower
38 宝城宝顶 Encircled Realm of Treasure and Tumulus

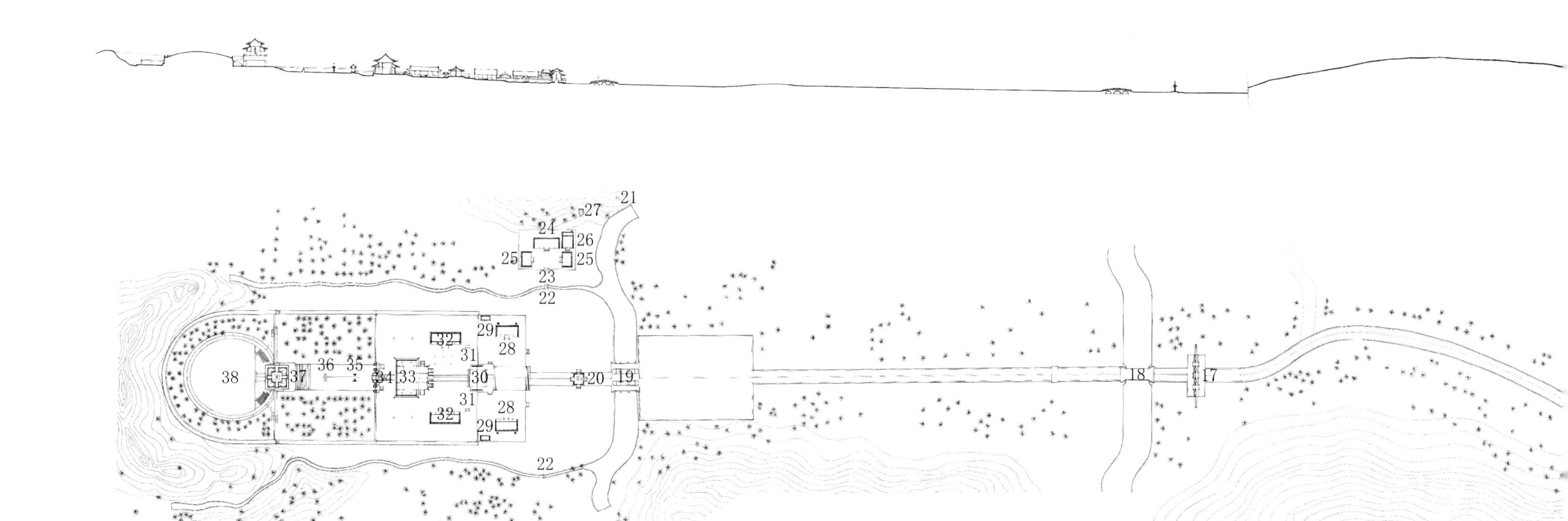

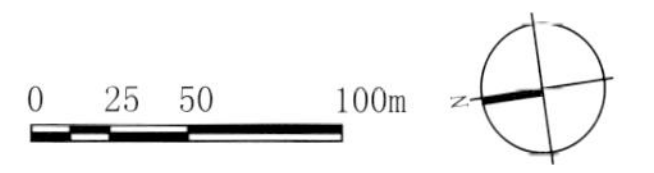

泰陵组群平面图
Site plan of the building complex of the Tomb of Peace

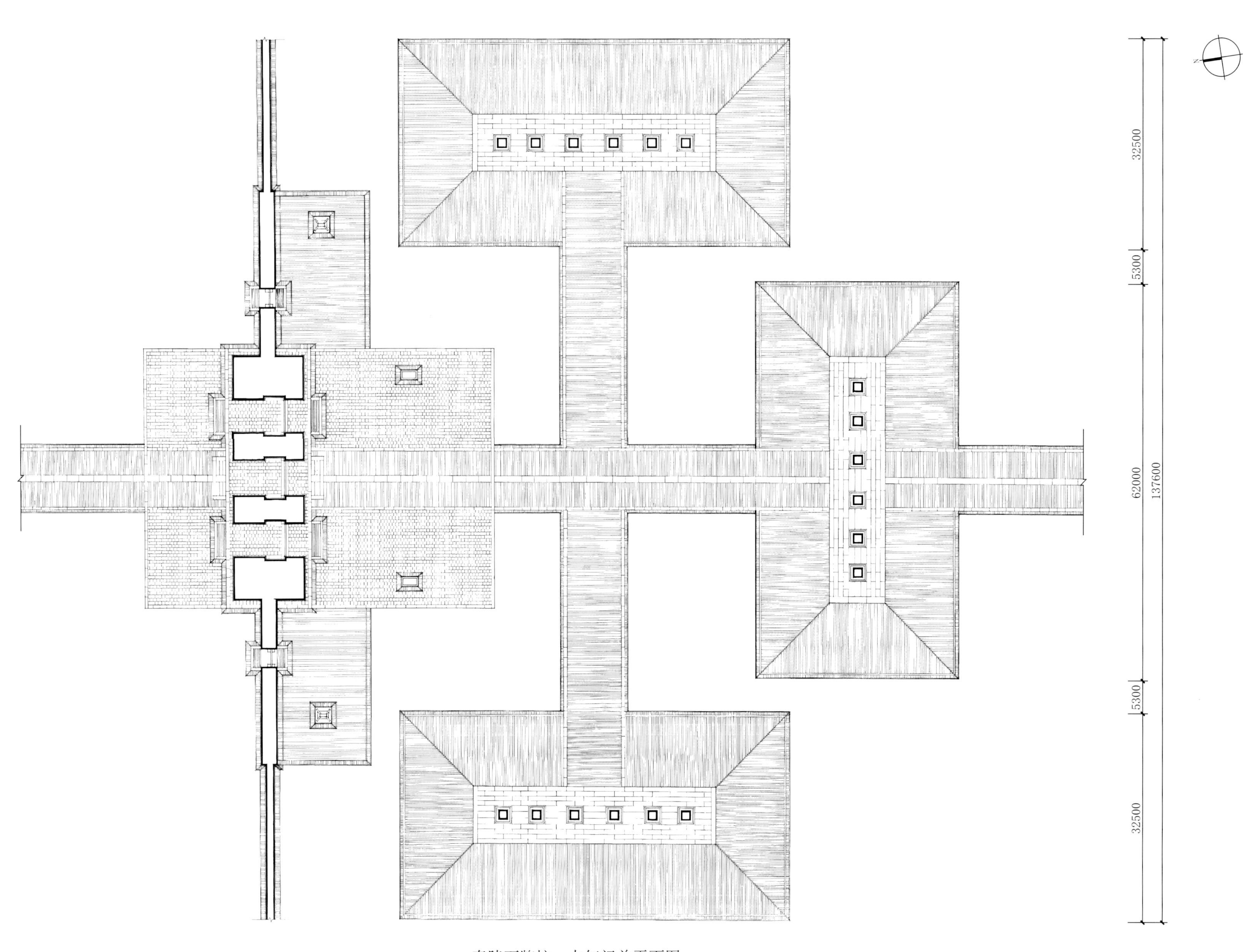

泰陵石牌坊、大红门总平面图

Site plan of the Marble Memorial Gateway and the Great Red Gate Building, Tomb of Peace

泰陵石牌坊正立面图
Front elevation of the Marble Memorial Gateway, Tomb of Peace

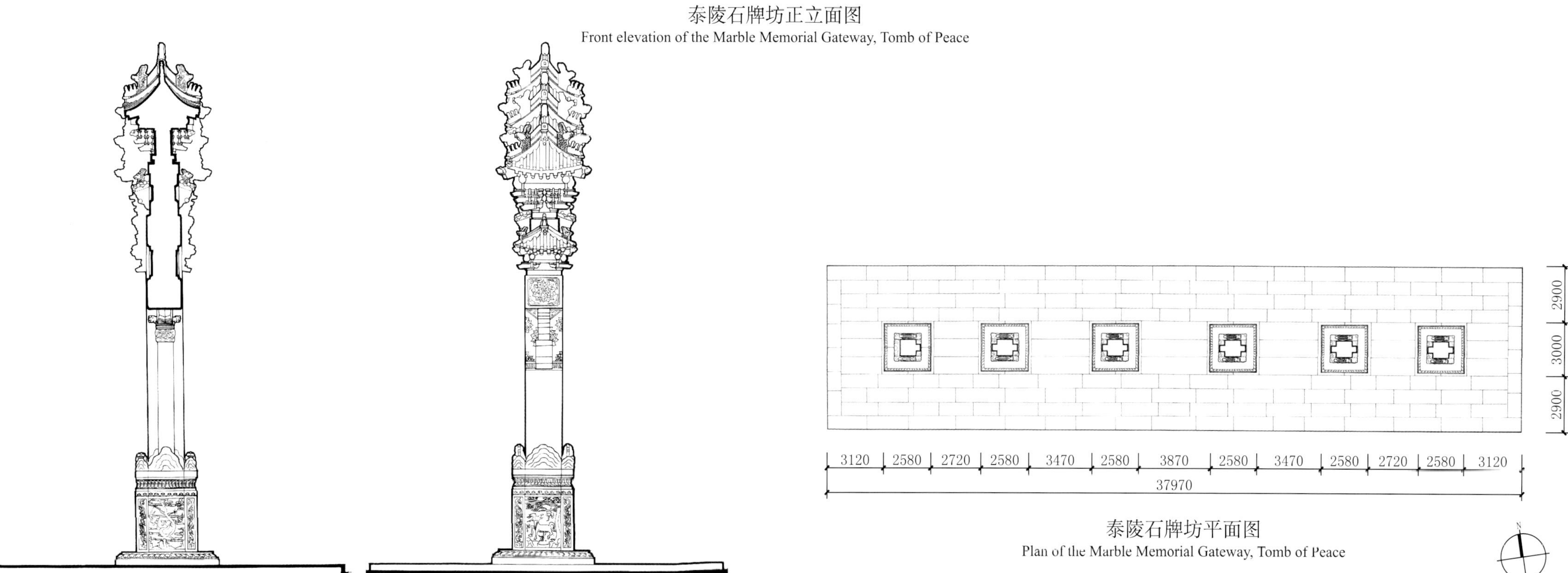

泰陵石牌坊平面图
Plan of the Marble Memorial Gateway, Tomb of Peace

泰陵石牌坊横剖面图
Cross section of the Marble Memorial Gateway, Tomb of Peace

泰陵石牌坊侧立面图
Side elevation of the Marble Memorial Gateway, Tomb of Peace

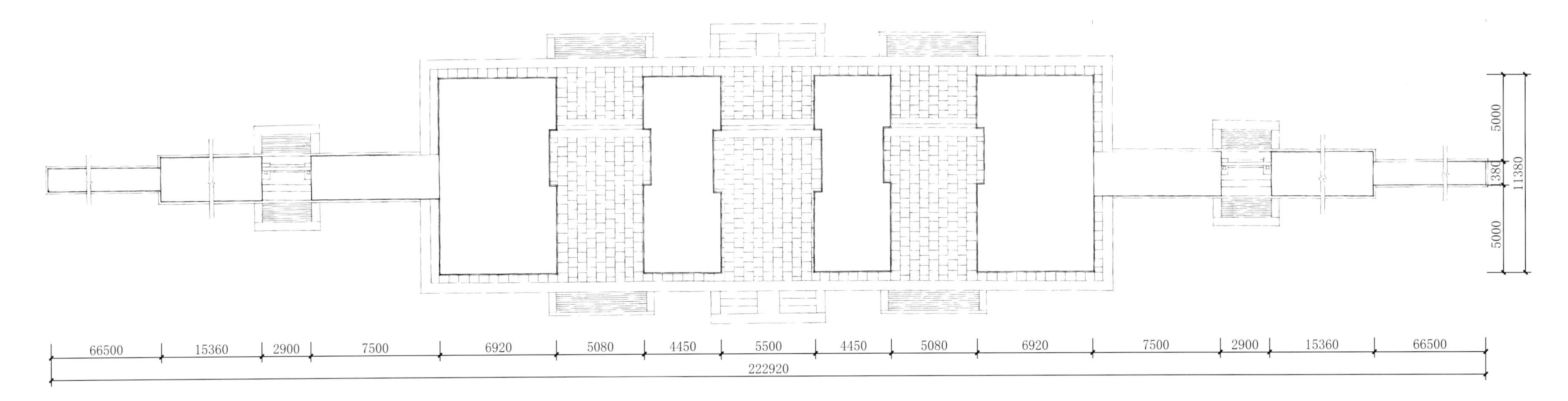

泰陵大红门平面图
Plan of the Great Red Gate Building, Tomb of Peace

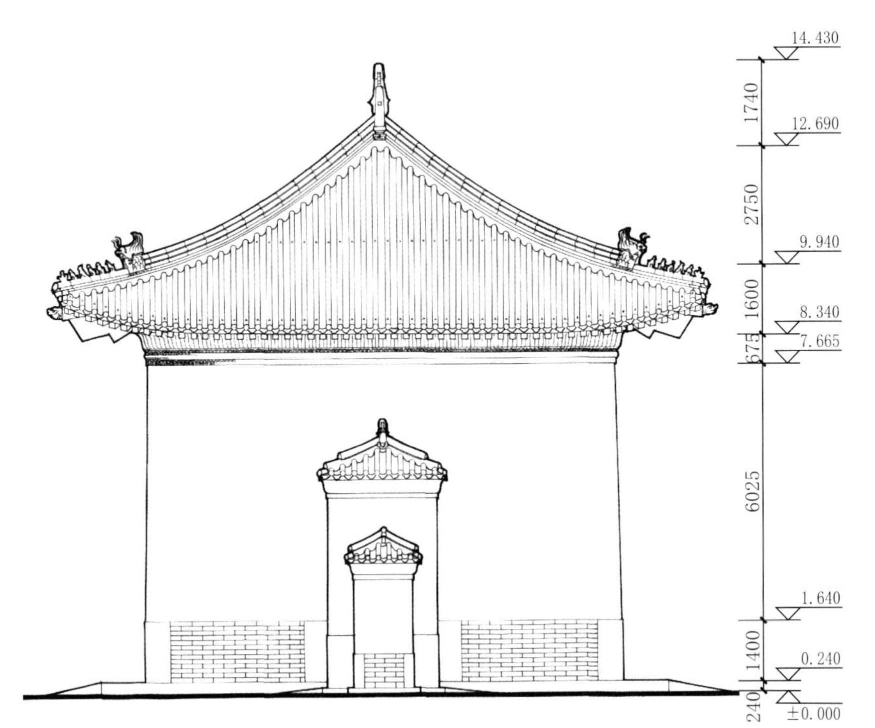

泰陵大红门侧立面图
Side elevation of the Great Red Gate Building, Tomb of Peace

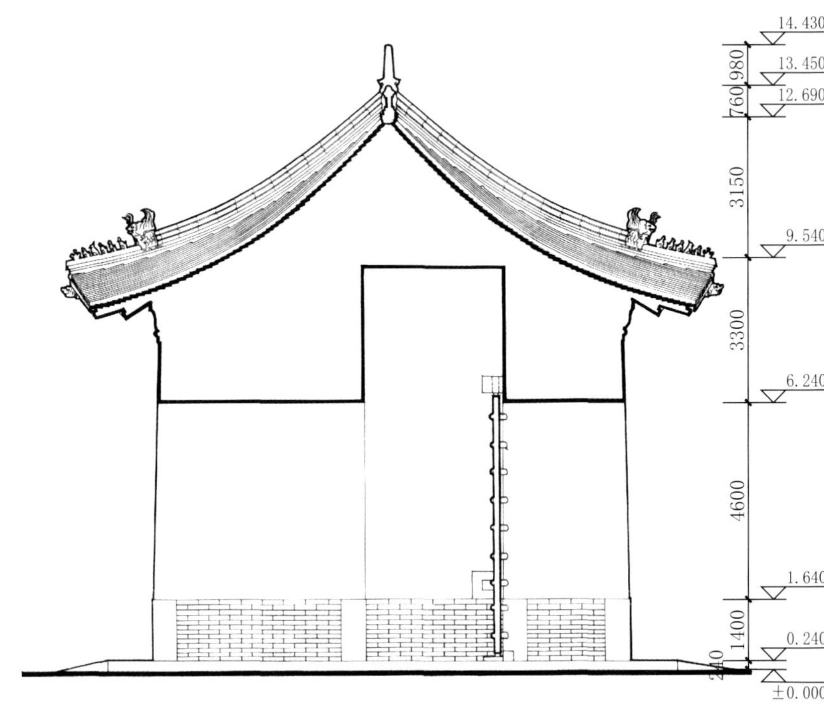

泰陵大红门明间横剖面图
Cross section of the Central Chamber within the Great Red Gate Building, Tomb of Peace

泰陵大红门麒麟正立面图
Front elevation of the Qilin of the Great Red Gate Building, Tomb of Peace

0 0.25 0.5 1m

泰陵大红门麒麟侧立面图
Side elevation of the Qilin of the Great Red Gate Building, Tomb of Peace

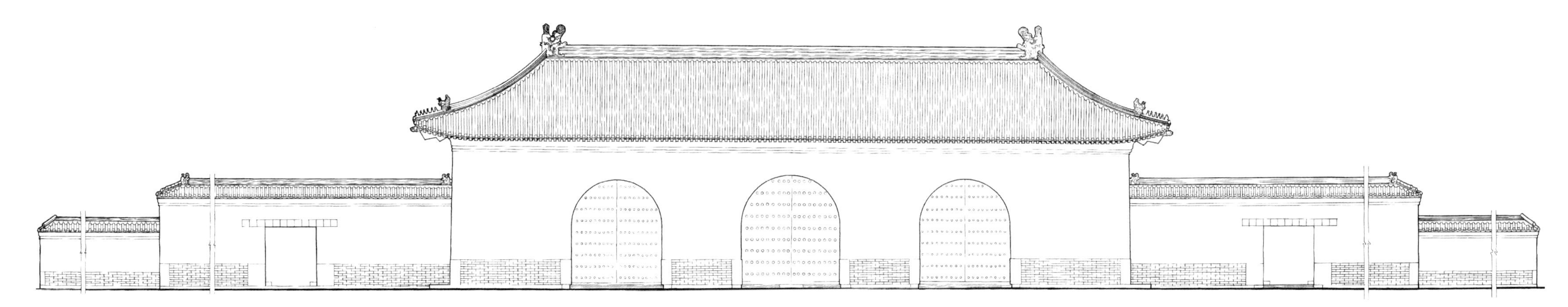

泰陵大红门正立面图
Front elevation of the Great Red Gate Building, Tomb of Peace

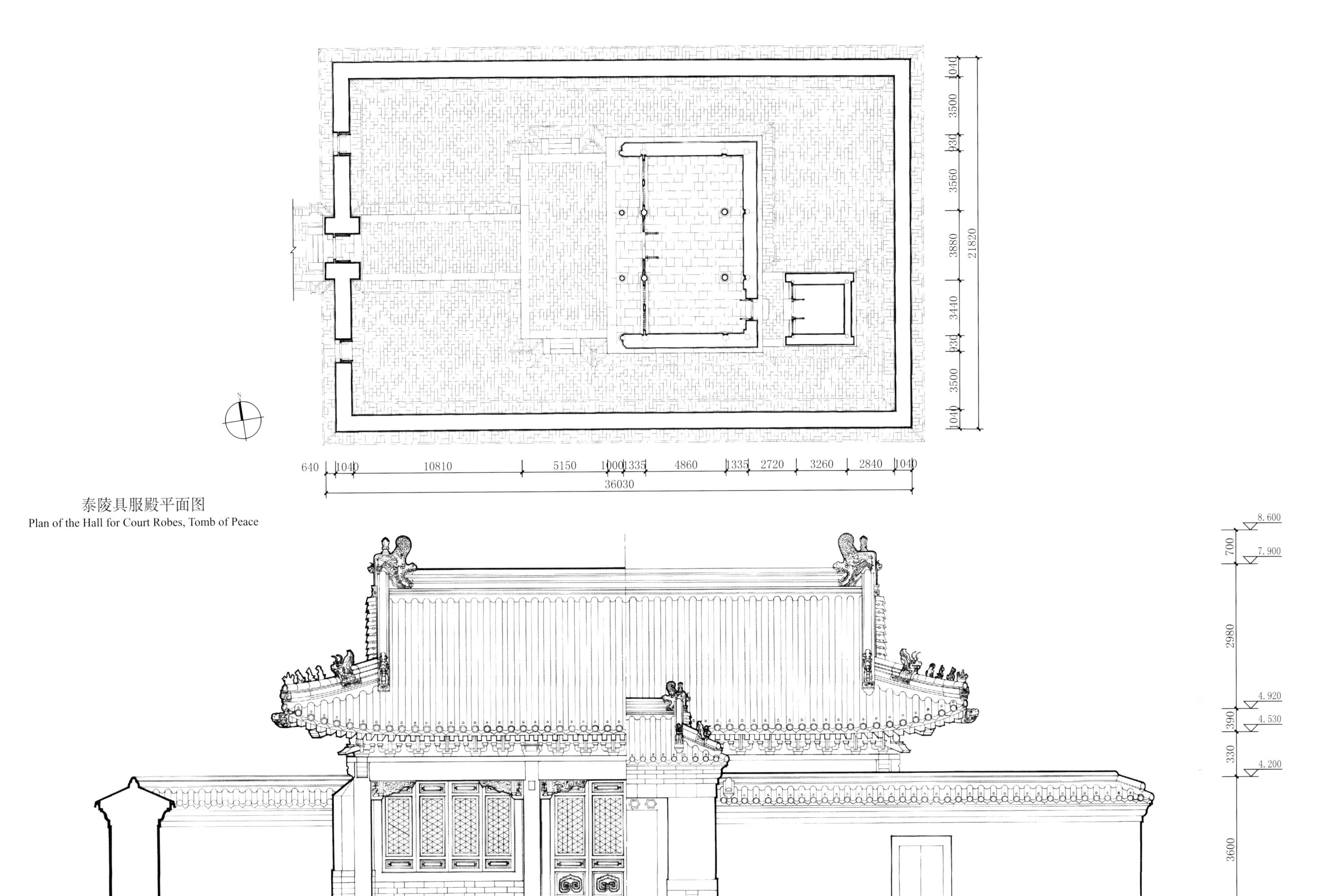

泰陵具服殿平面图
Plan of the Hall for Court Robes, Tomb of Peace

泰陵具服殿正背立面图
Front and back elevation of the Hall for Court Robes, Tomb of Peace

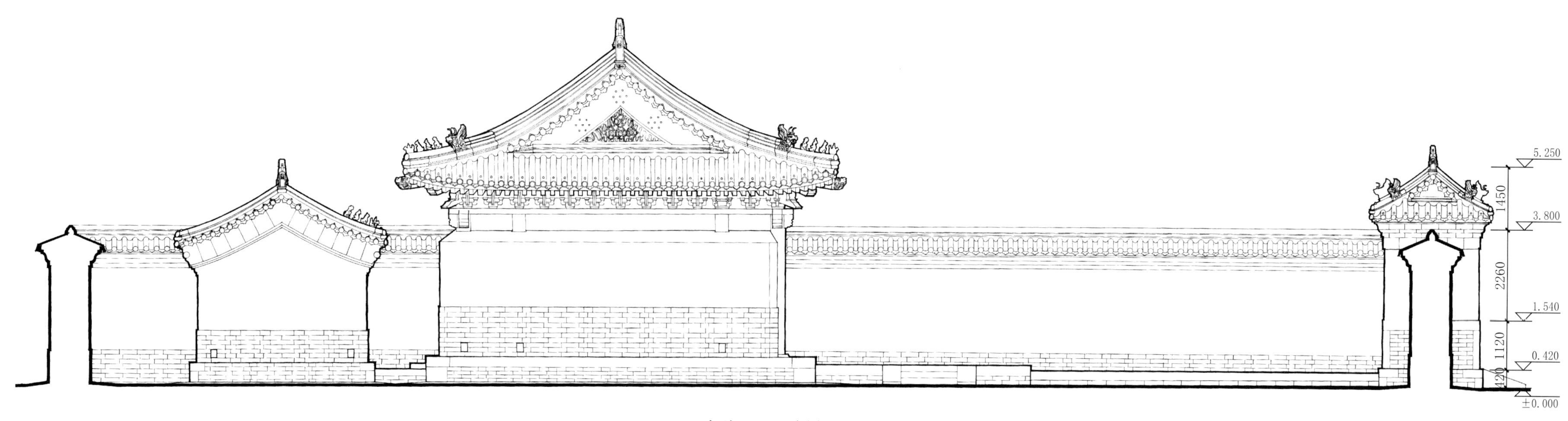

泰陵具服殿侧立面图

Side elevation of the Hall for Court Robes, Tomb of Peace

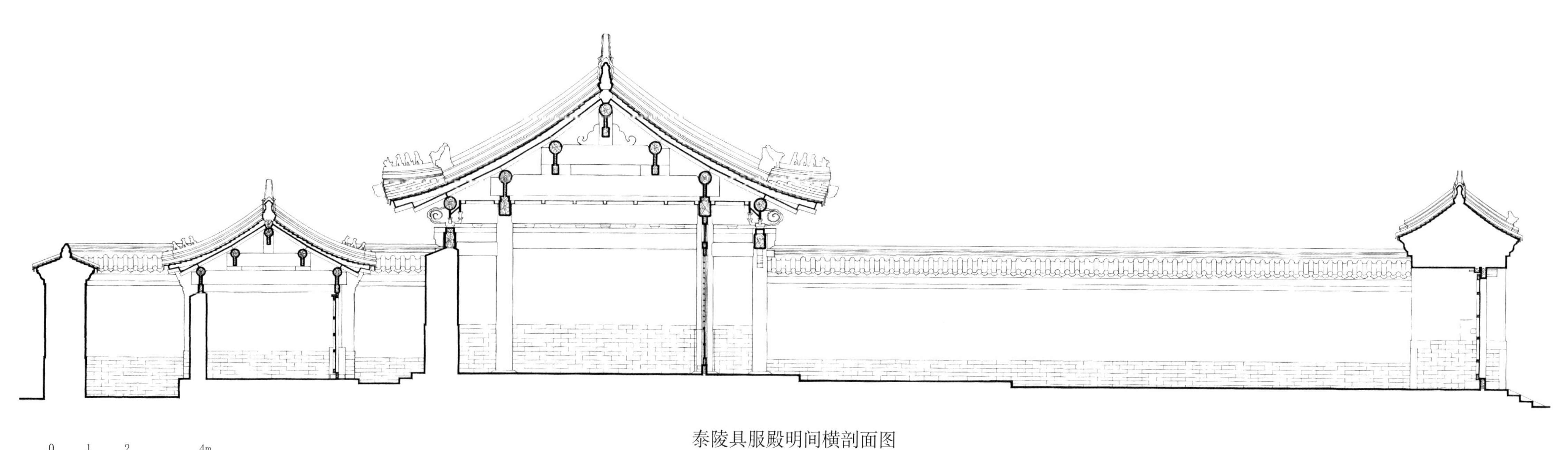

泰陵具服殿明间横剖面图

Cross section of the Central Chamber within the Hall for Court Robes, Tomb of Peace

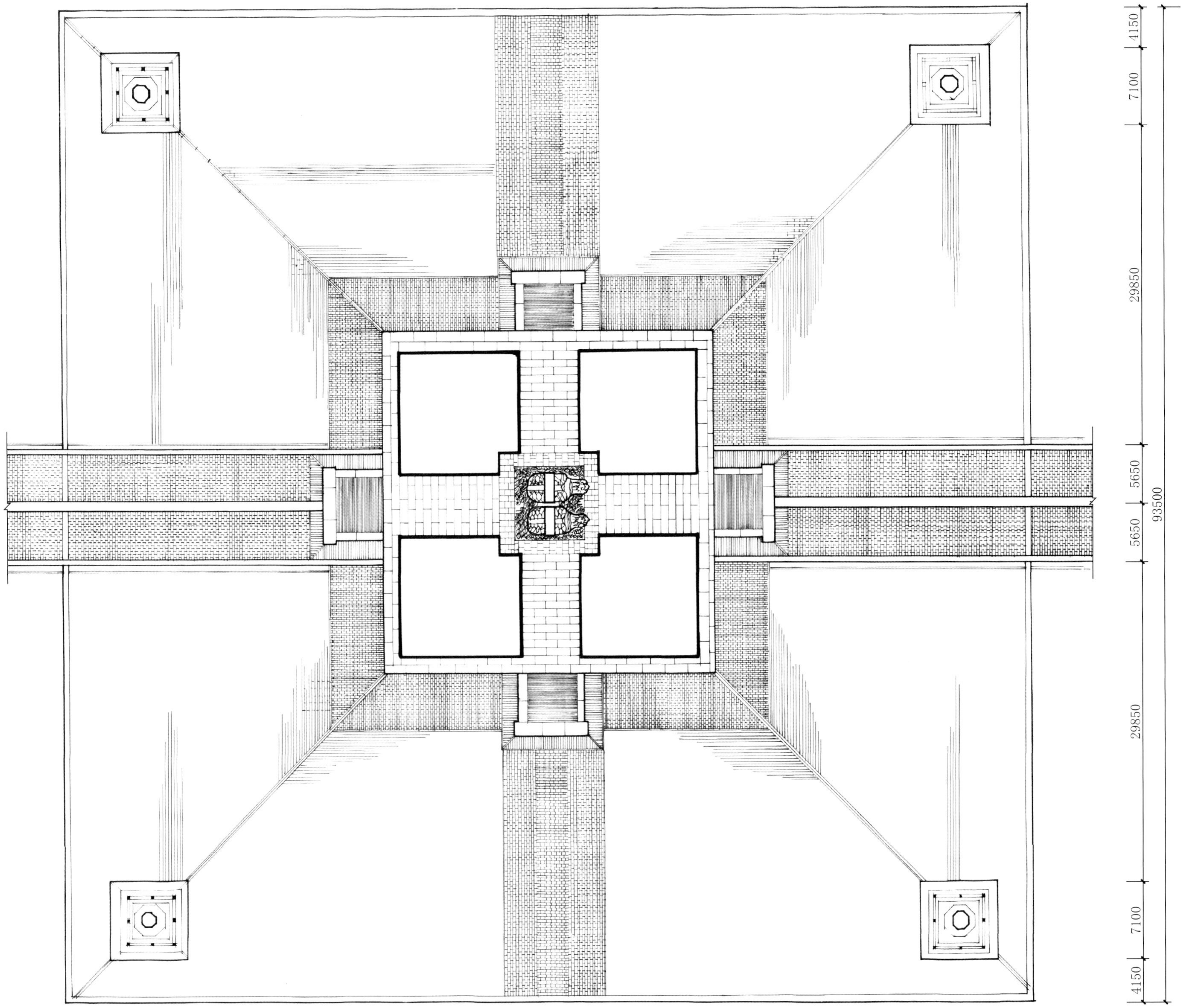

泰陵圣德神功碑亭总平面图
Site plan of the Pavilion for the Stela of Divine Merit and Sagely Virtue, Tomb of Peace

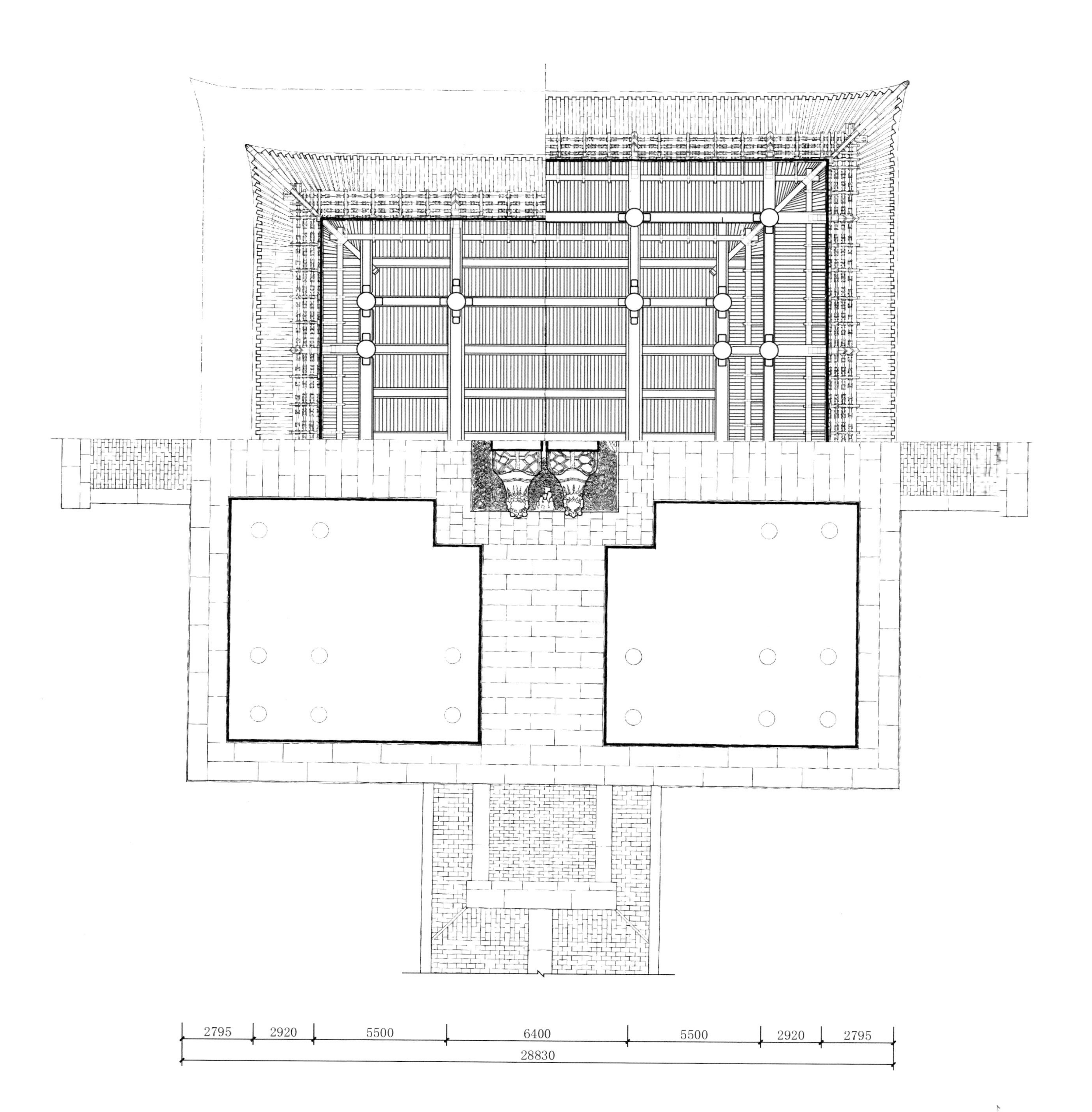

泰陵圣德神功碑亭平面图

Plan of the Pavilion for the Stela of Divine Merit and Sagely Virtue, Tomb of Peace

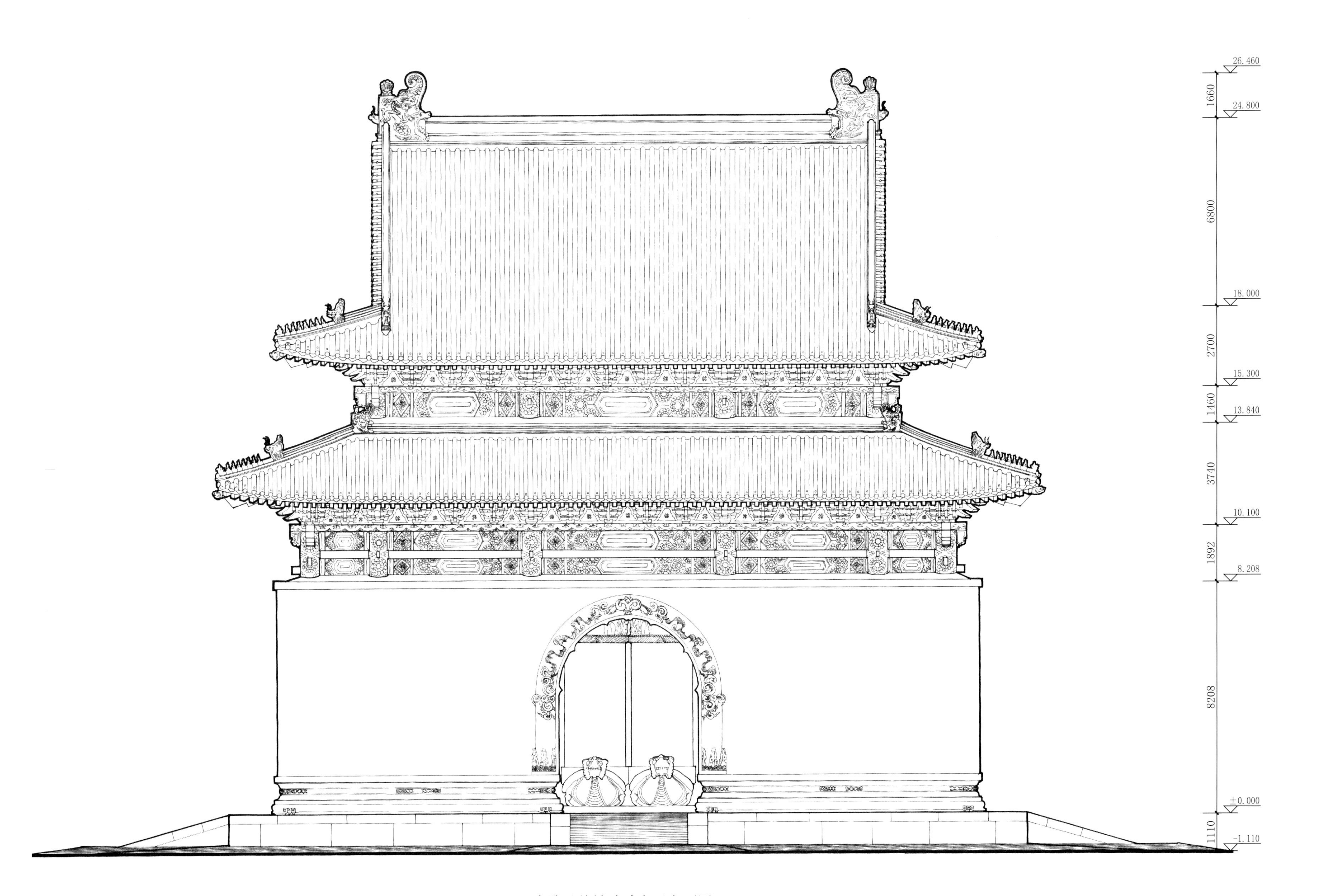

泰陵圣德神功碑亭正立面图

Front elevation of the Pavilion for the Stela of Divine Merit and Sagely Virtue, Tomb of Peace

泰陵圣德神功碑亭侧立面图

Side elevation of the Pavilion for the Stela of Divine Merit and Sagely Virtue, Tomb of Peace

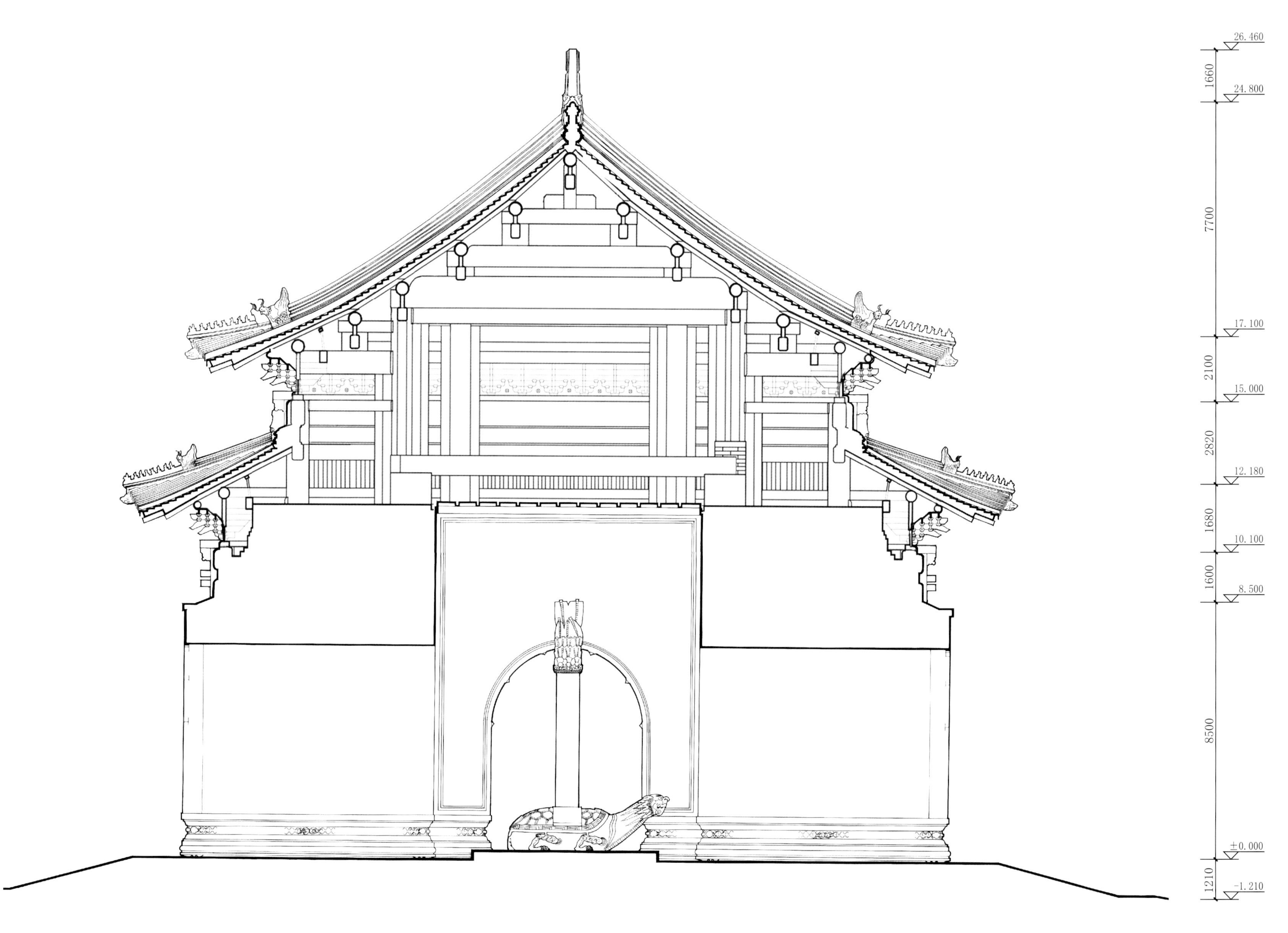

泰陵圣德神功碑亭明间横剖面图

Cross section of the Central Chamber within the Pavilion for the Stela of Divine Merit and Sagely Virtue, Tomb of Peace

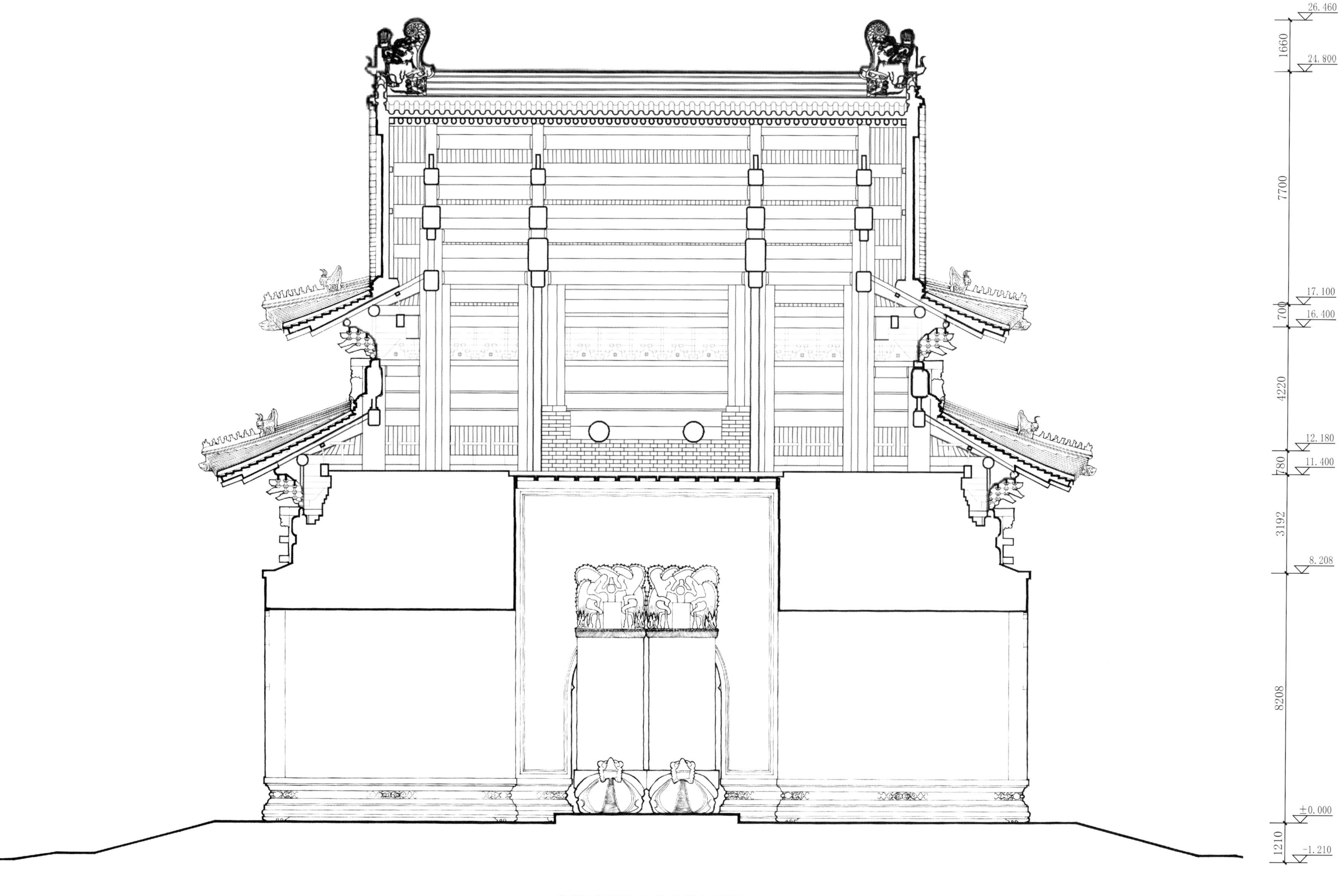

泰陵圣德神功碑亭纵剖面图

Longitudinal section of the Pavilion for the Stela of Divine Merit and Sagely Virtue, Tomb of Peace

泰陵华表正立面图
Front elevation of the Ceremonial Column, Tomb of Peace

泰陵华表侧立面图
Side elevation of the Ceremonial Column, Tomb of Peace

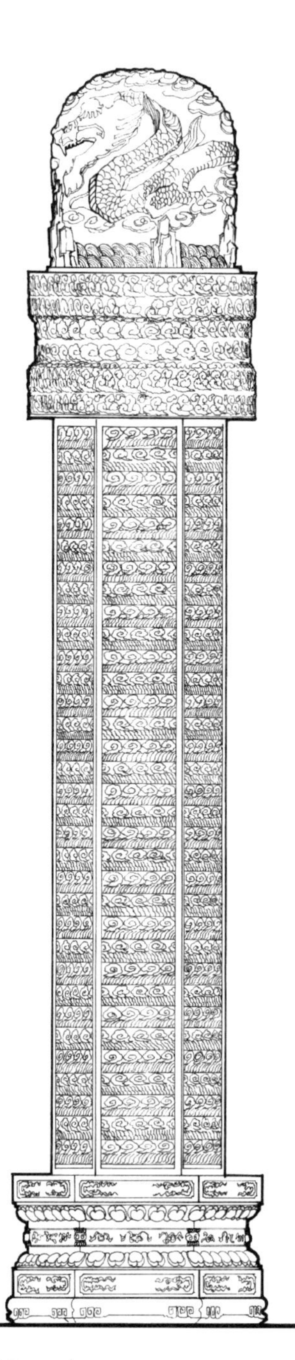

泰陵望柱正立面图
Front elevation of the Ornamental Column of the Spirit Way, Tomb of Peace

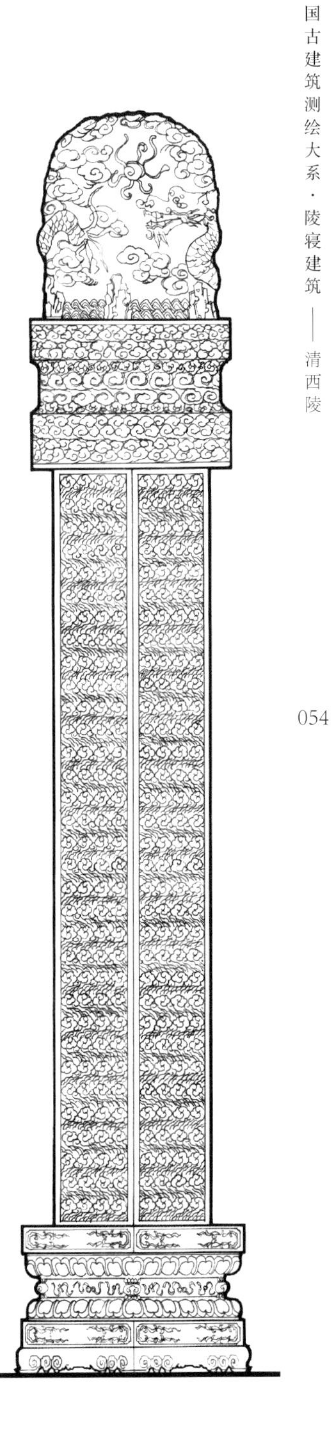

泰陵望柱侧立面图
Side elevation of the Ornamental Column of the Spirit Way, Tomb of Peace

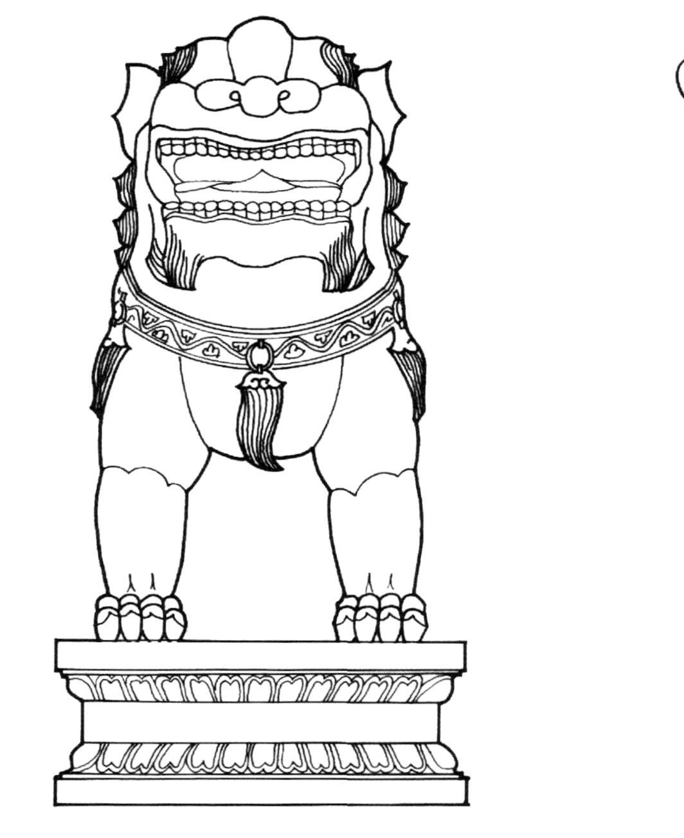

泰陵石狮大样图
Detailed drawing of the Stone Lion, Tomb of Peace

泰陵石马大样图
Detailed drawing of the Stone Horse, Tomb of Peace

泰陵文臣大样图
Detailed drawing of the Literary Official, Tomb of Peace

泰陵武将大样图
Detailed drawing of the Martial Official, Tomb of Peace

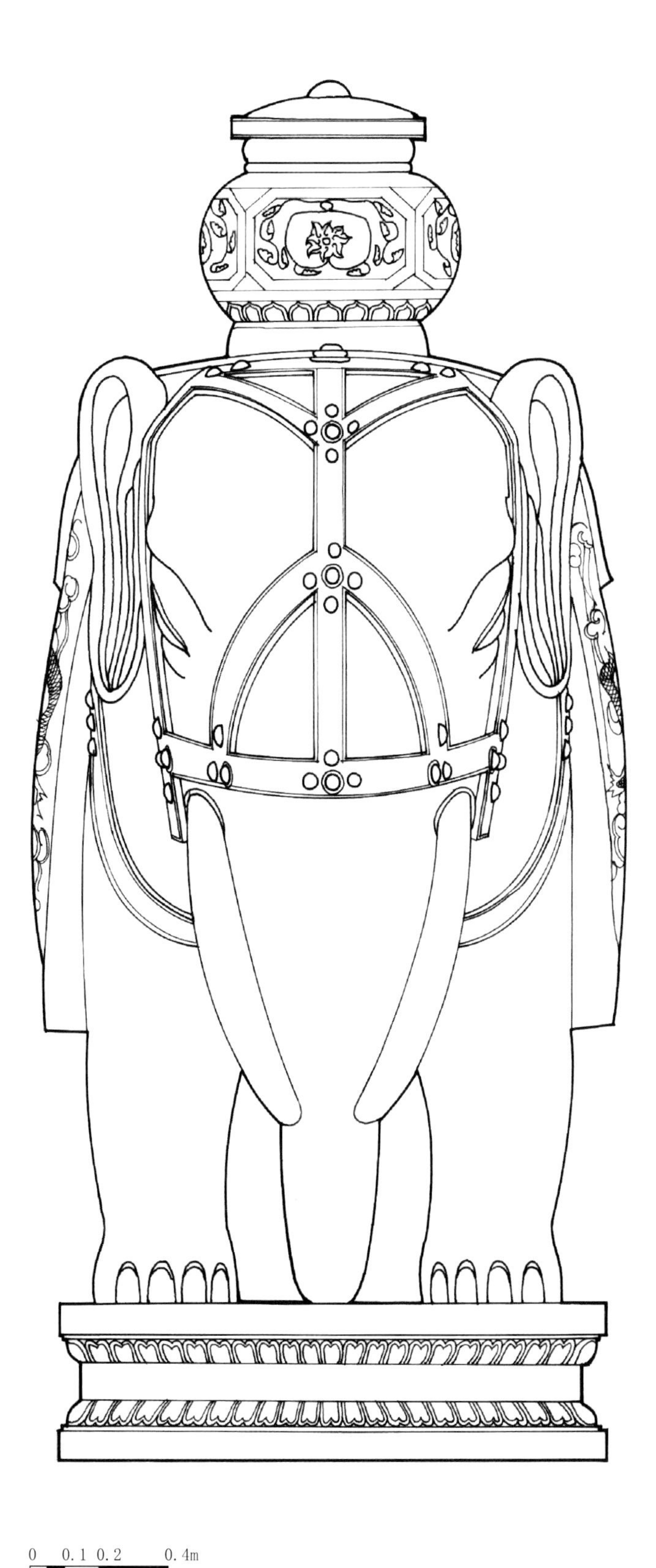

0　0.1 0.2　　0.4m

泰陵石象大样图

Detailed drawing of the Stone Elephant, Tomb of Peace

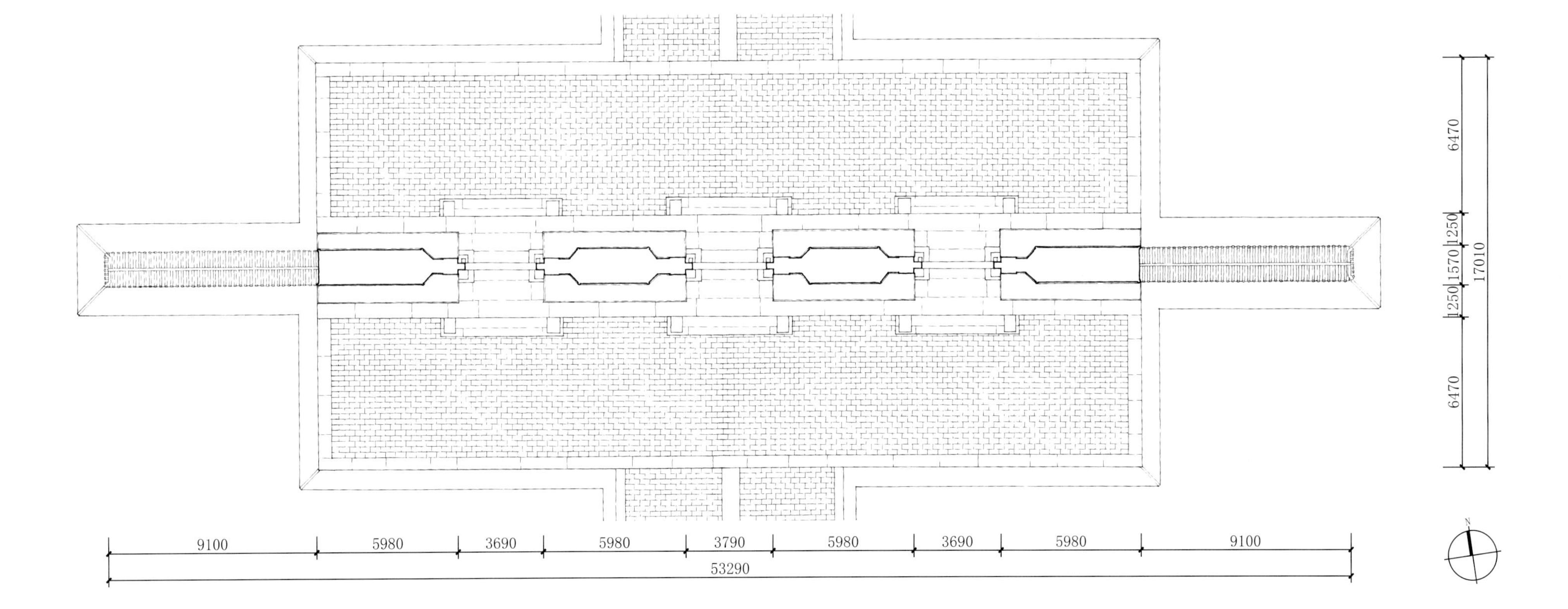

泰陵龙凤门平面图

Plan of the Dragon and Phoenix Gate, Tomb of Peace

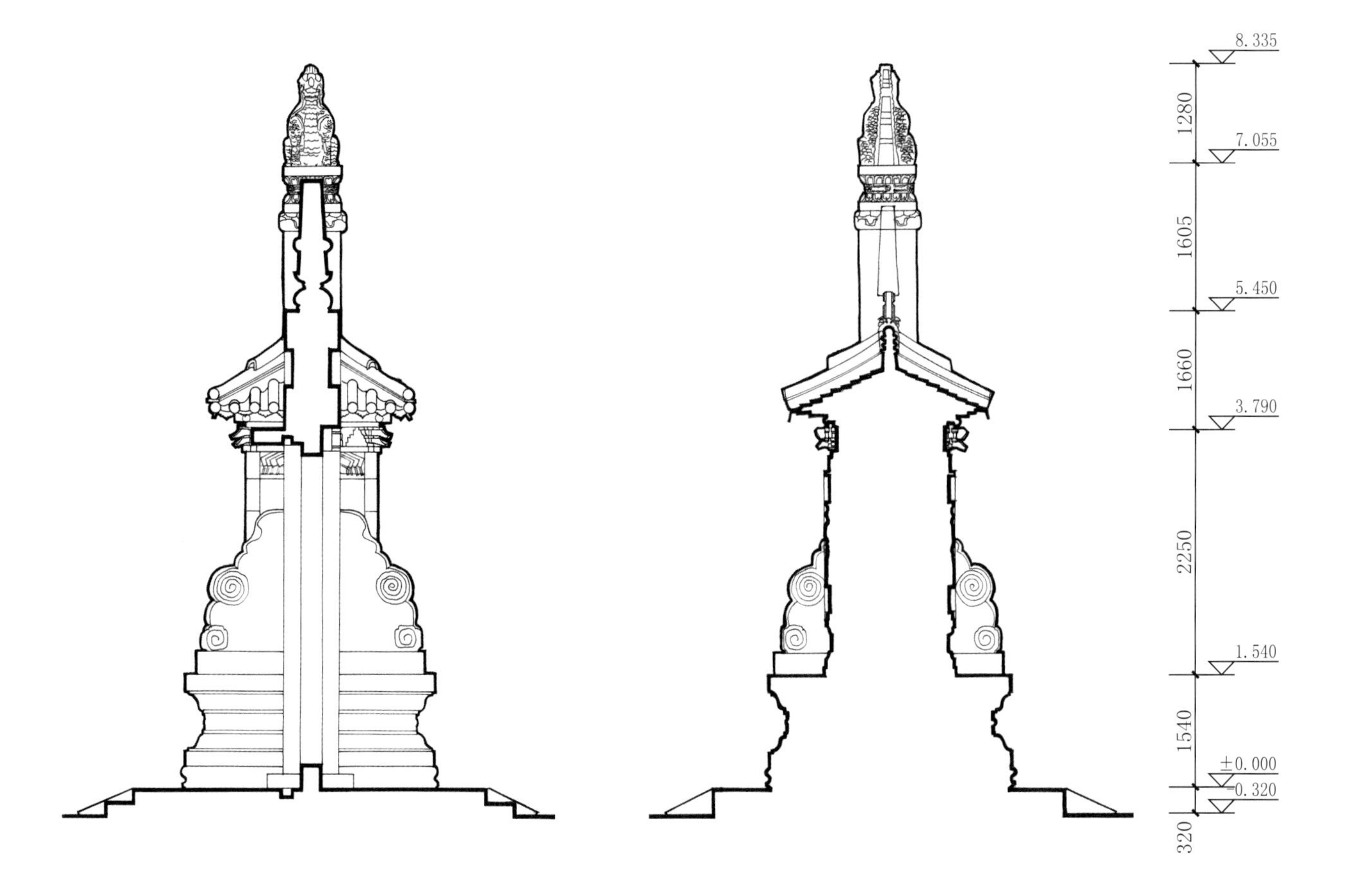

泰陵龙凤门横剖面图

Cross section of the Dragon and Phoenix Gate, Tomb of Peace

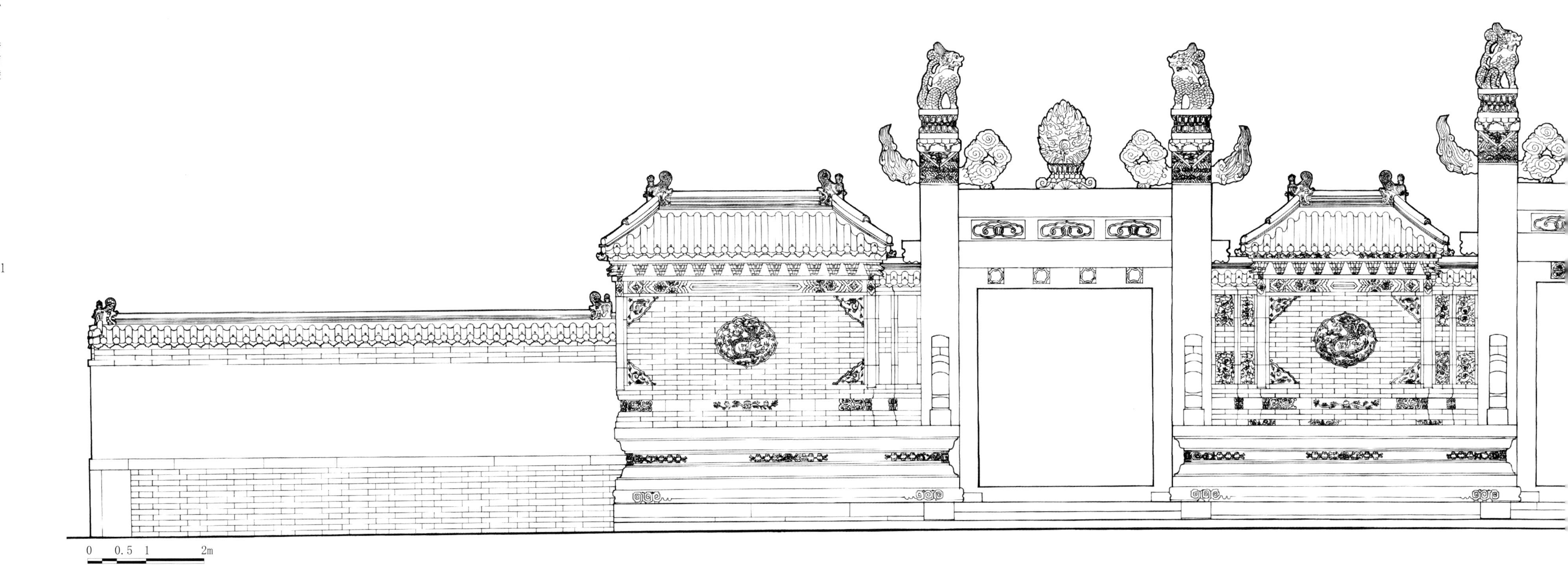

泰陵龙凤门正立面图
Front elevation of the Dragon and Phoenix Gate, Tomb of Peace

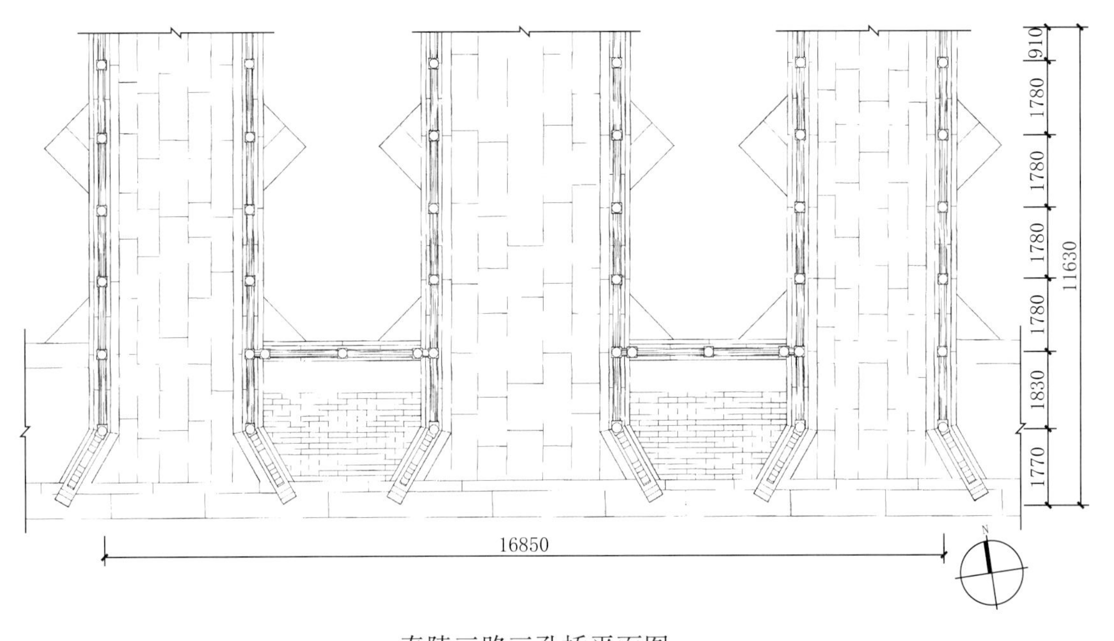

泰陵三路三孔桥平面图
Plan of the Three-Way Three-Arch Stone Bridge, Tomb of Peace

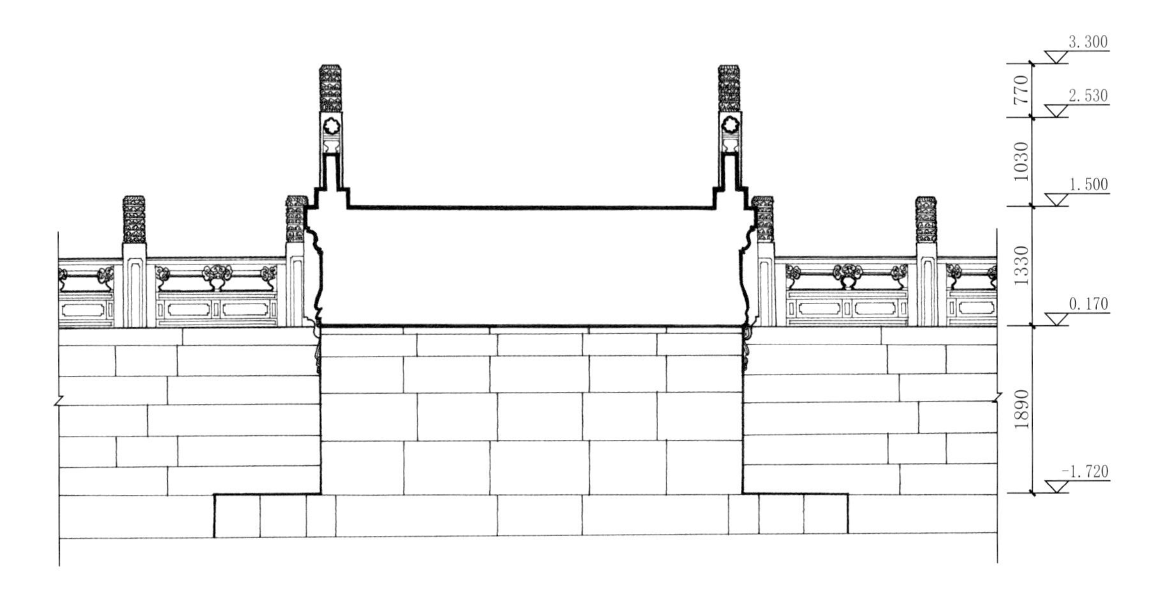

泰陵三路三孔桥横剖面图
Cross section of the Three-Way Three-Arch Stone Bridge, Tomb of Peace

0 0.5 1 2m

泰陵三路三孔桥侧立面图

Side elevation of the Three-Way Three-Arch Stone Bridge, Tomb of Peace

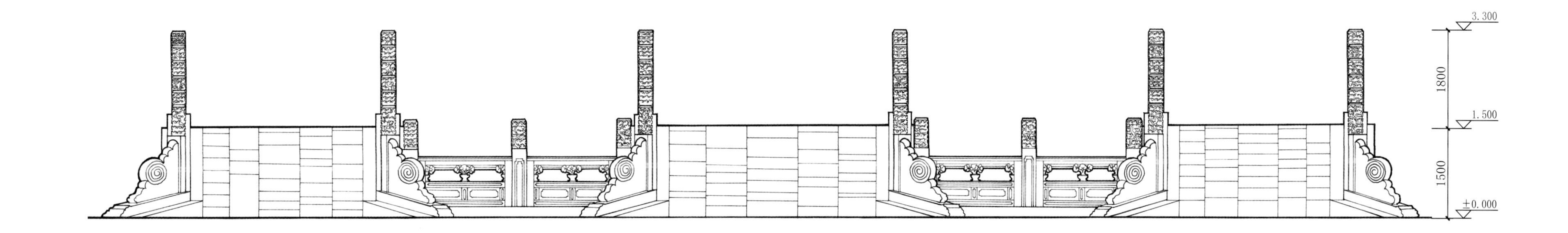

泰陵三路三孔桥正立面图

Front elevation of the Three-Way Three-Arch Stone Bridge, Tomb of Peace

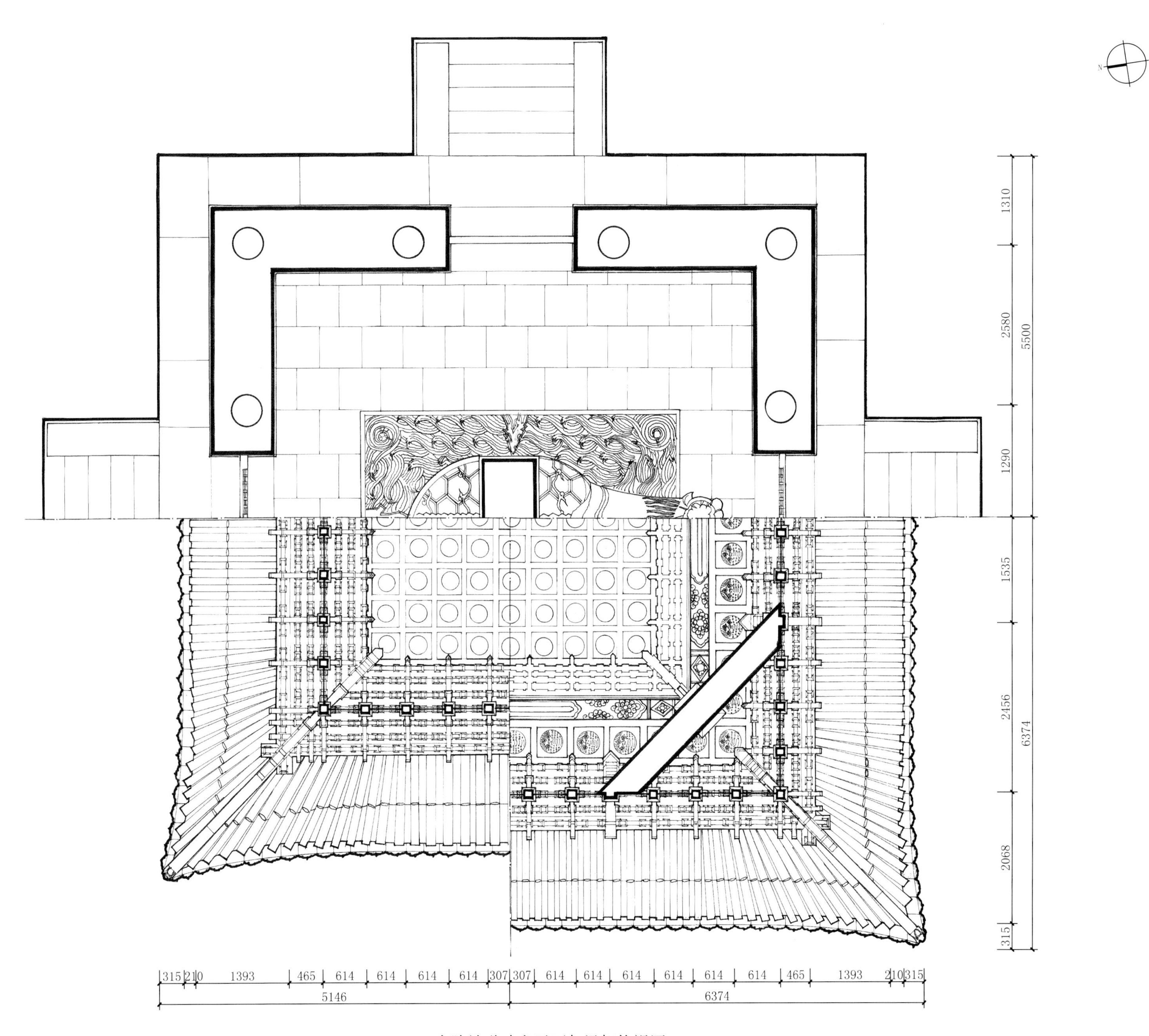

泰陵神道碑亭平面与梁架仰视图

Plan and looking up at the Truss and Ceiling of the Pavilion for the Stela on the Spirit Way,Tomb of Peace

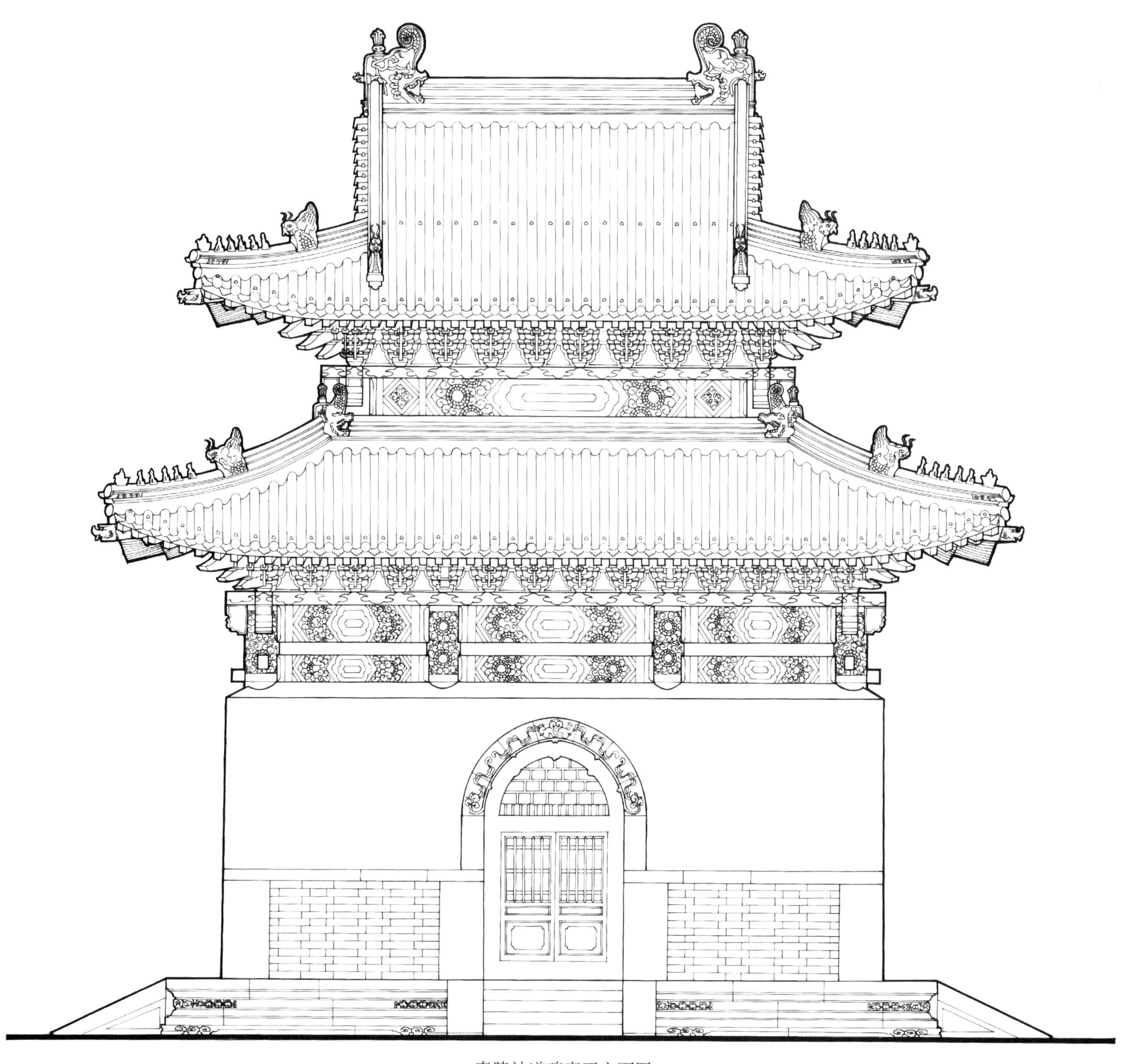

泰陵神道碑亭正立面图

Front elevation of the Pavilion for the Stela on the Spirit Way,Tomb of Peace

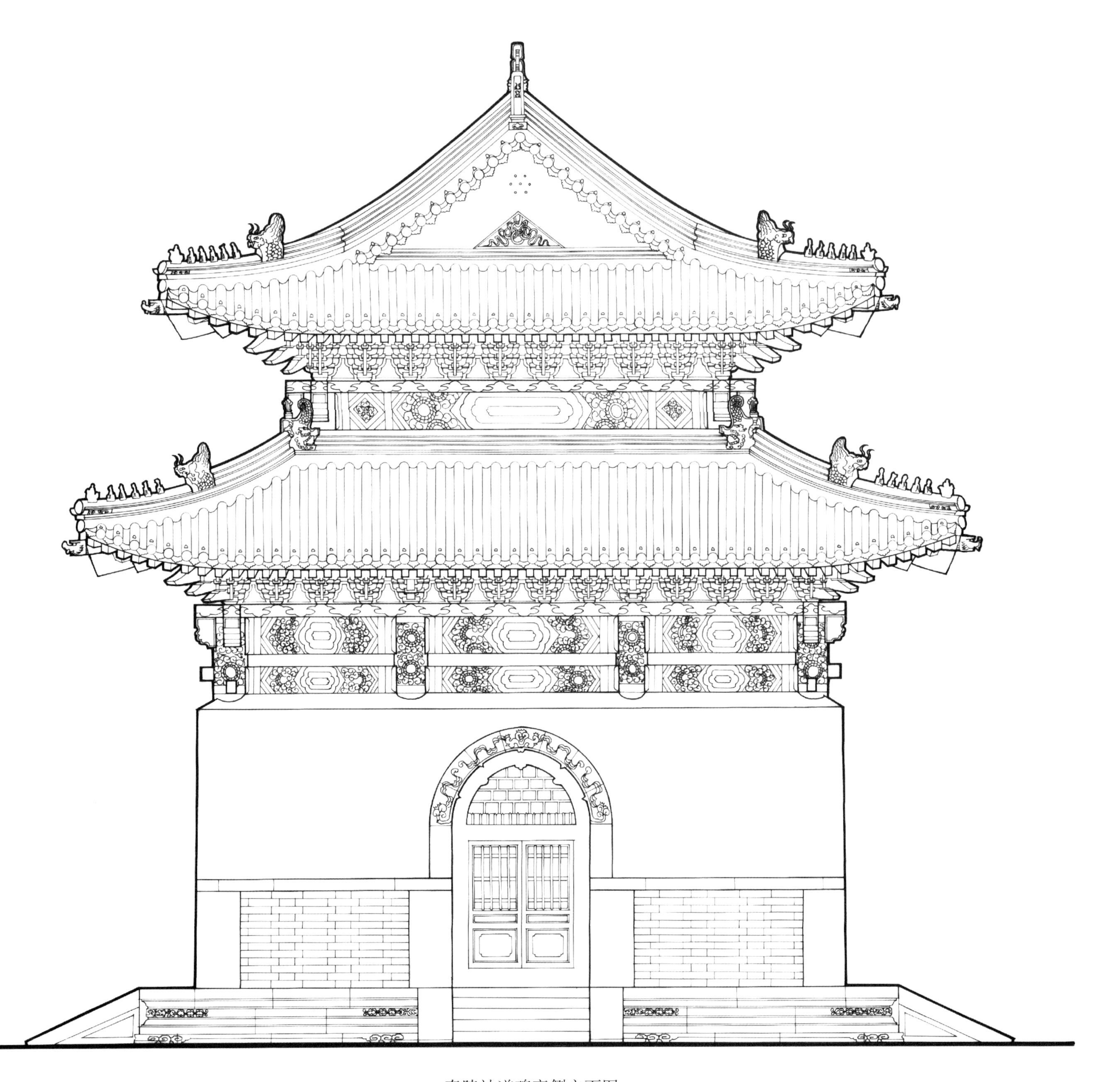

泰陵神道碑亭侧立面图
Side elevation of the Pavilion for the Stela on the Spirit Way,Tomb of Peace

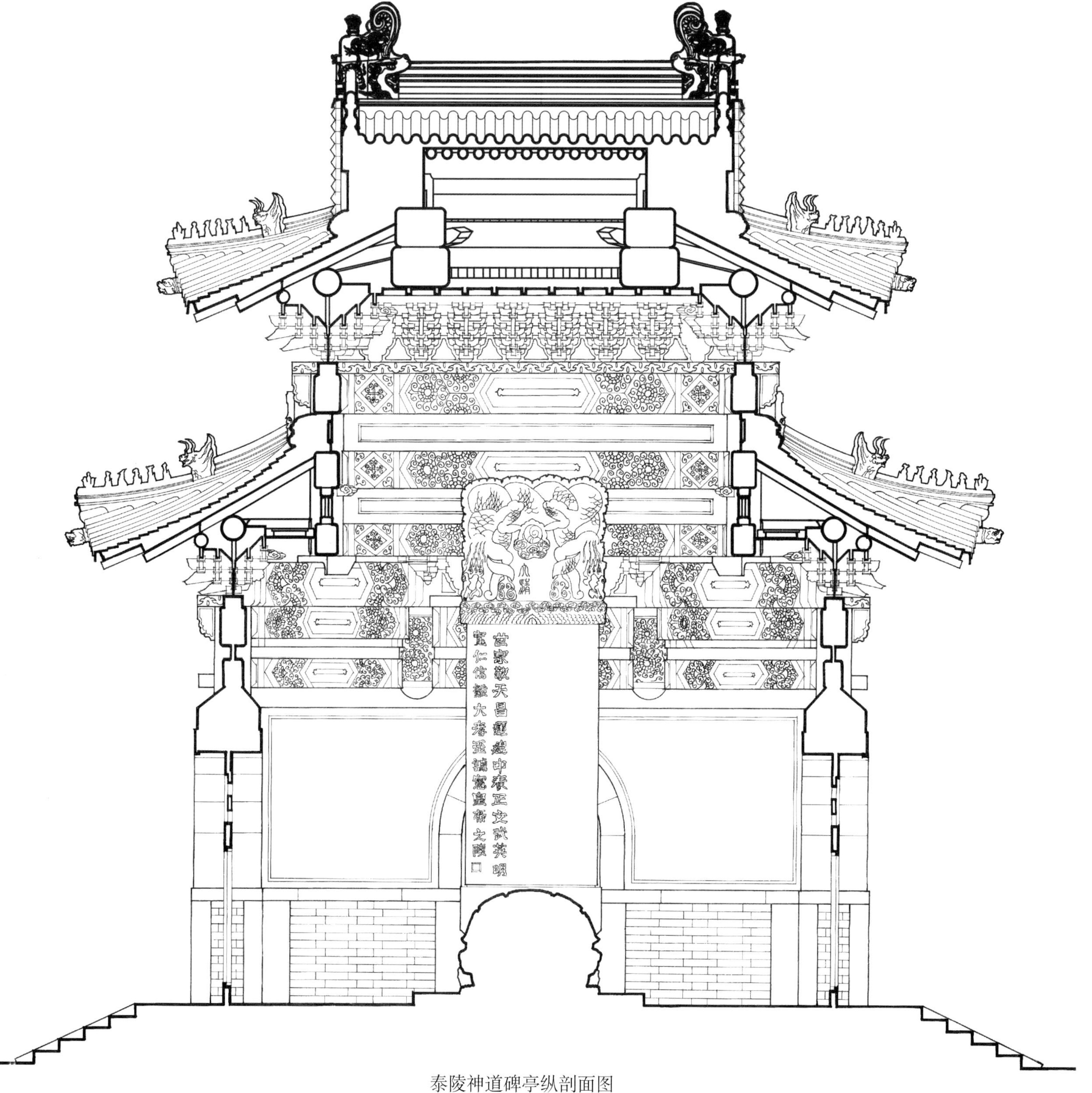

泰陵神道碑亭纵剖面图

Longitudinal section of the Pavilion for the Stela on the Spirit Way,Tomb of Peace

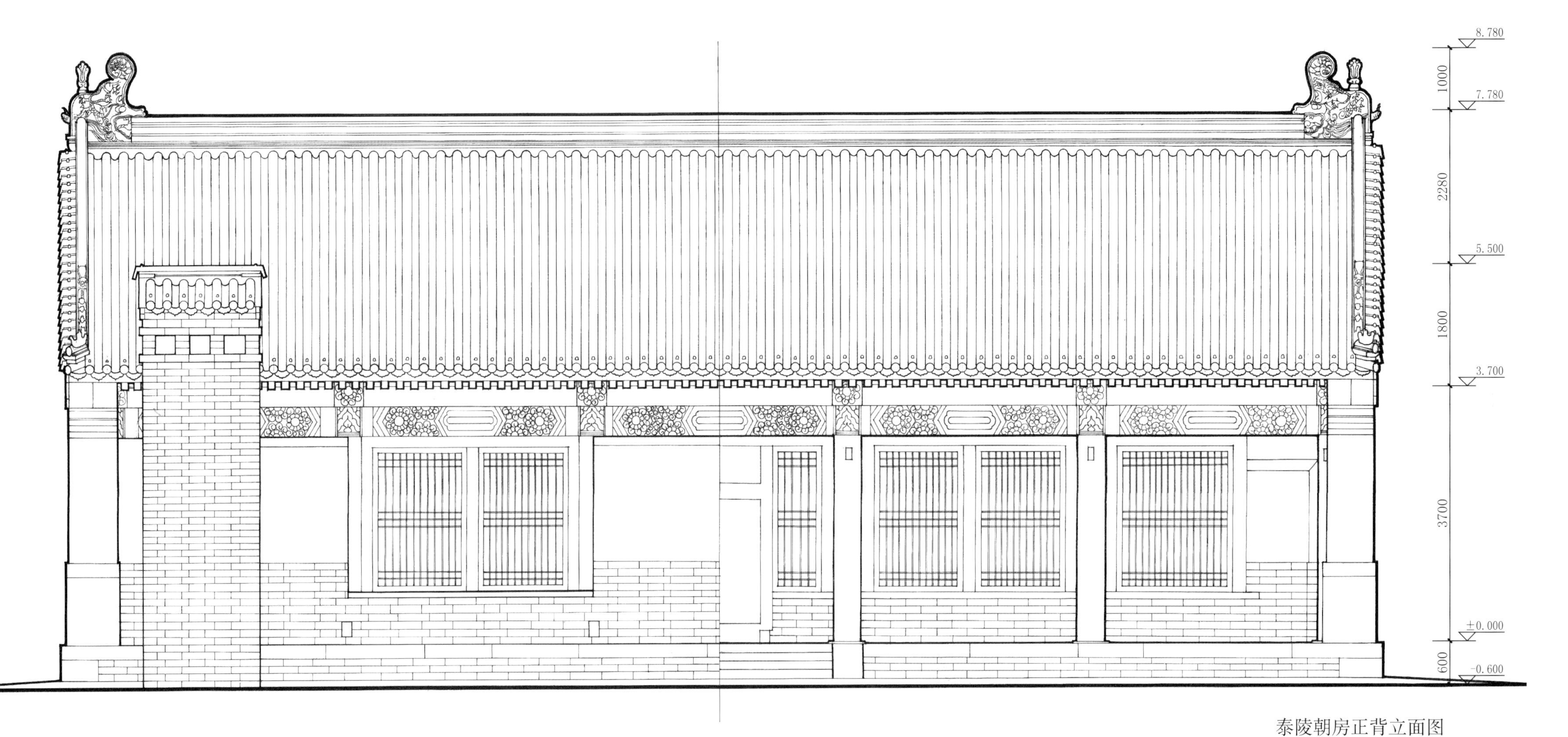

泰陵朝房正背立面图

Front and back elevation of the Reception Hall for Court Officials,Tomb of Peace

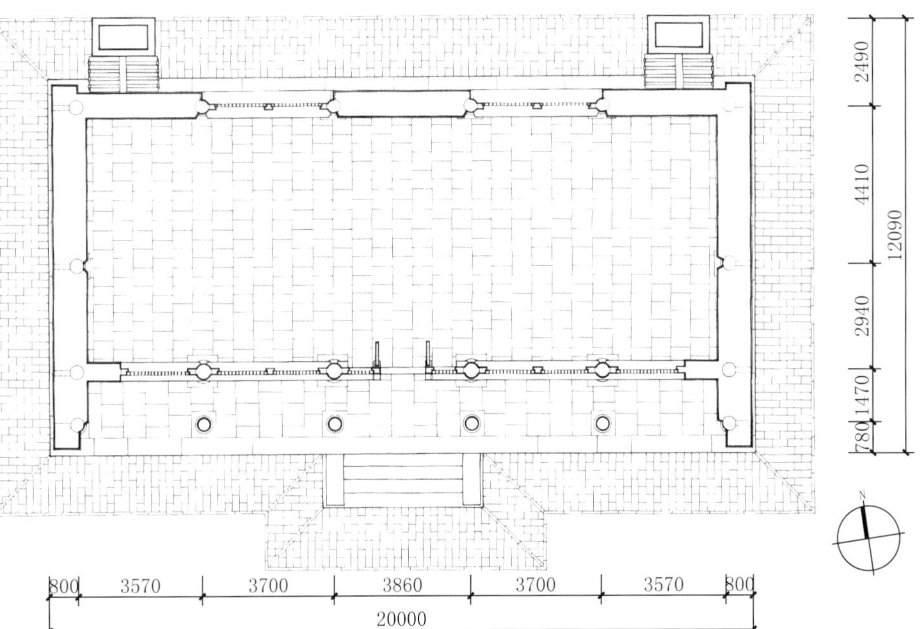

泰陵朝房平面图

Plan of the Reception Hall for Court Officials,Tomb of Peace

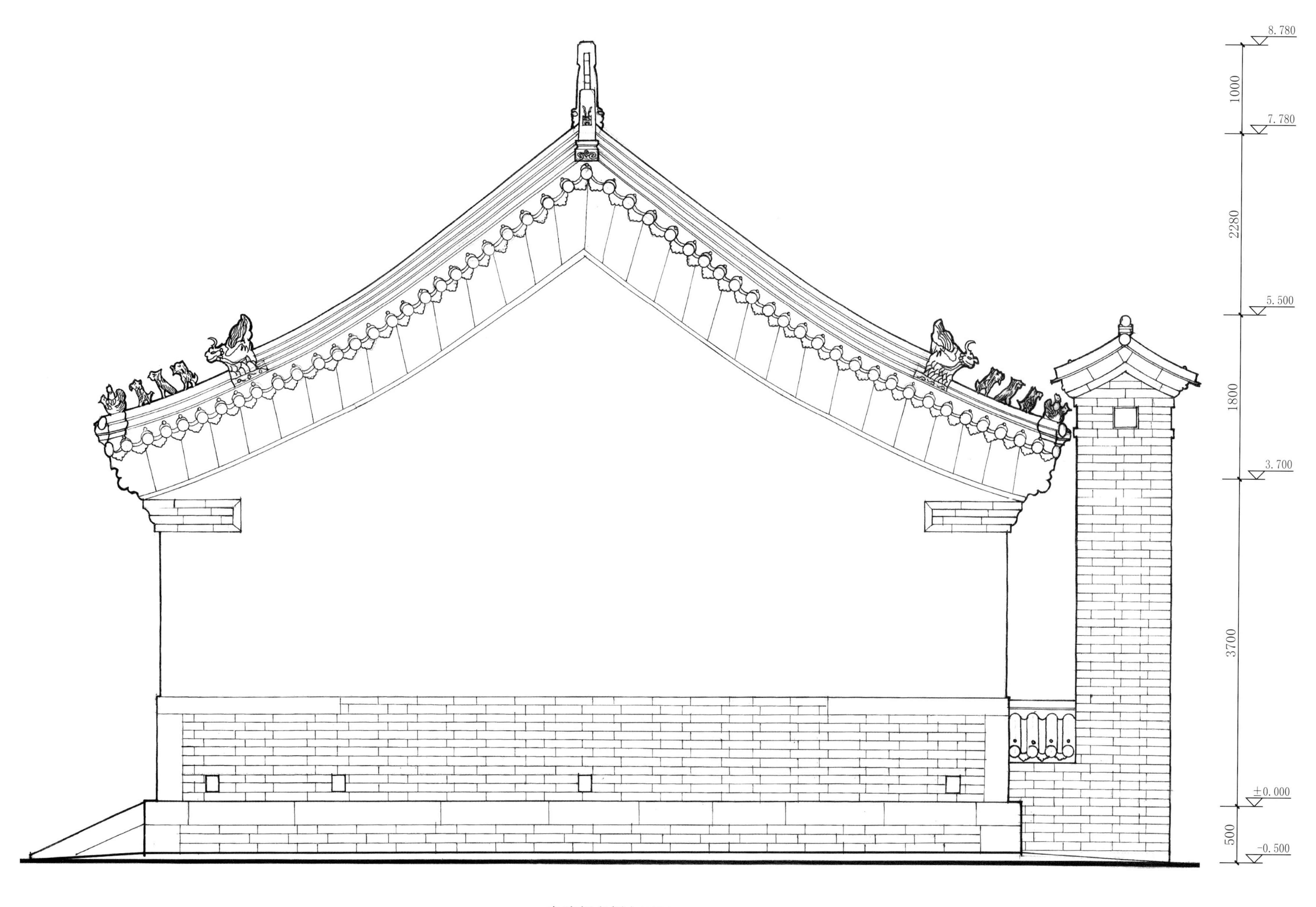

泰陵朝房侧立面图

Side elevation of the Reception Hall for Court Officials,Tomb of Peace

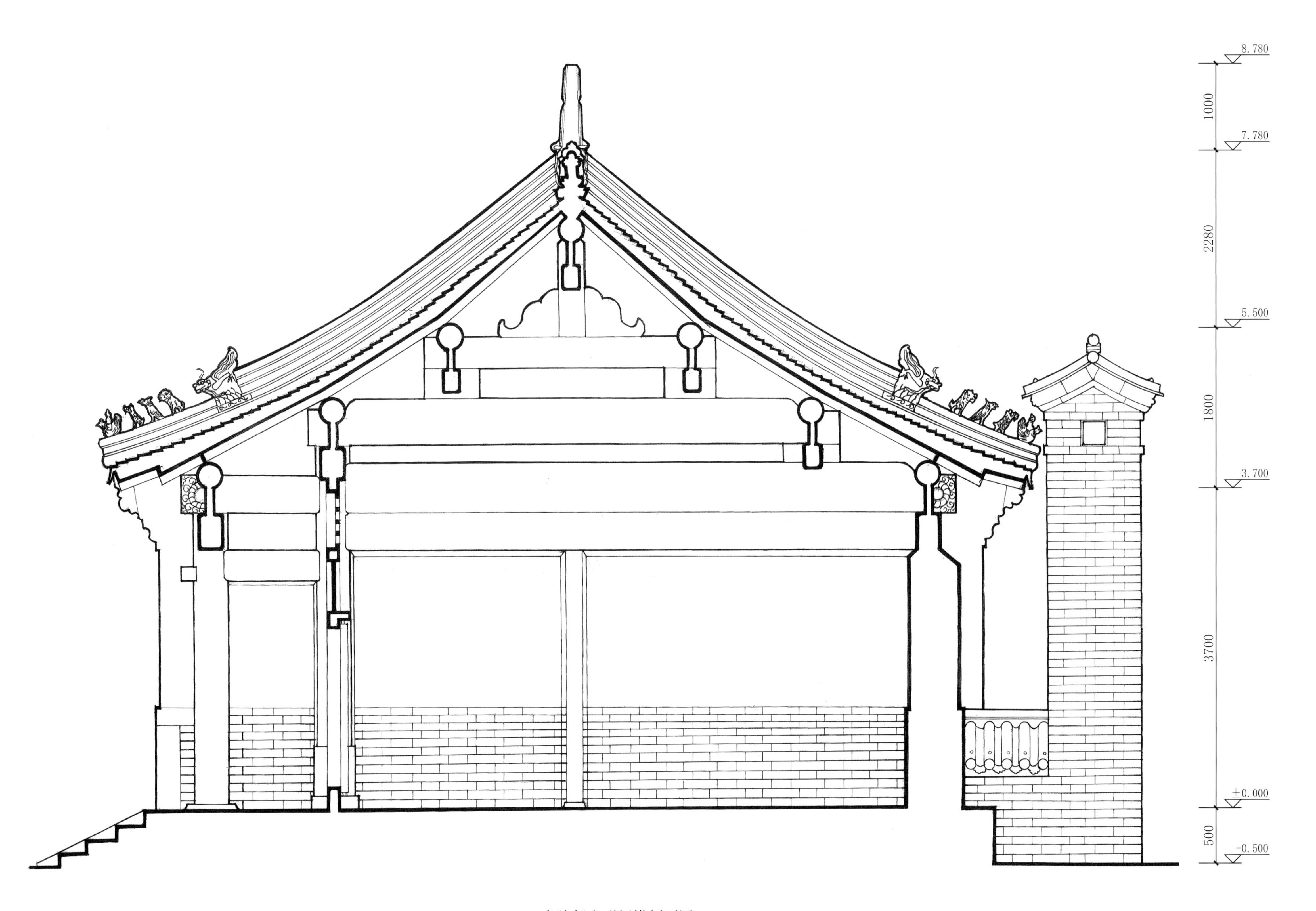

泰陵朝房明间横剖面图

Cross section of the Central Chamber within the Reception Hall for Court Officials,Tomb of Peace

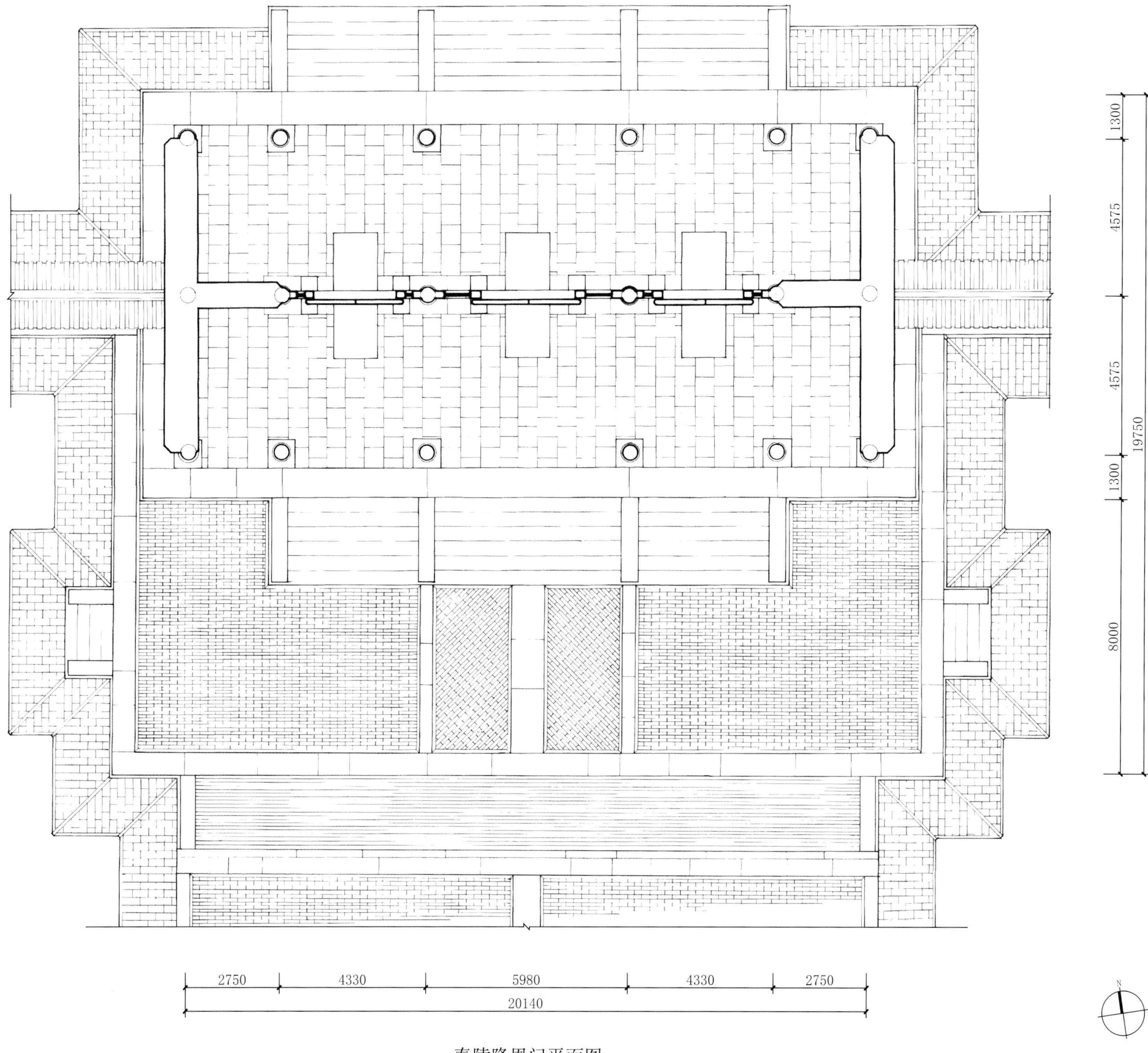

泰陵隆恩门平面图

Plan of the Gate of Monumental Grace, Tomb of Peace

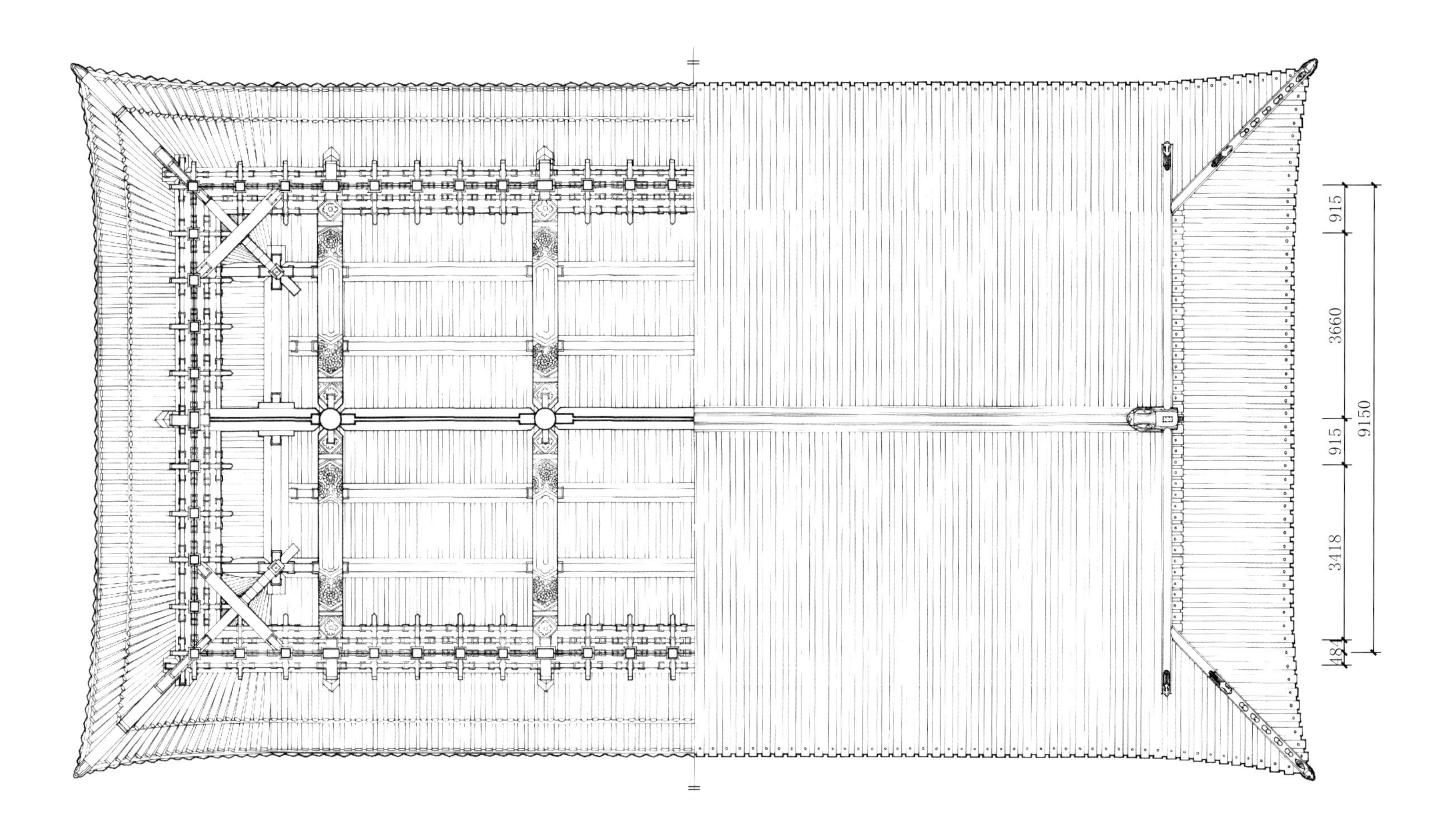

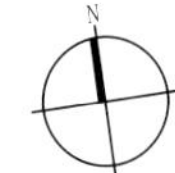

泰陵隆恩门梁架仰视与屋顶平面图

Roof plan and looking up at the Truss and Ceiling of the Gate of Monumental Grace,Tomb of Peace

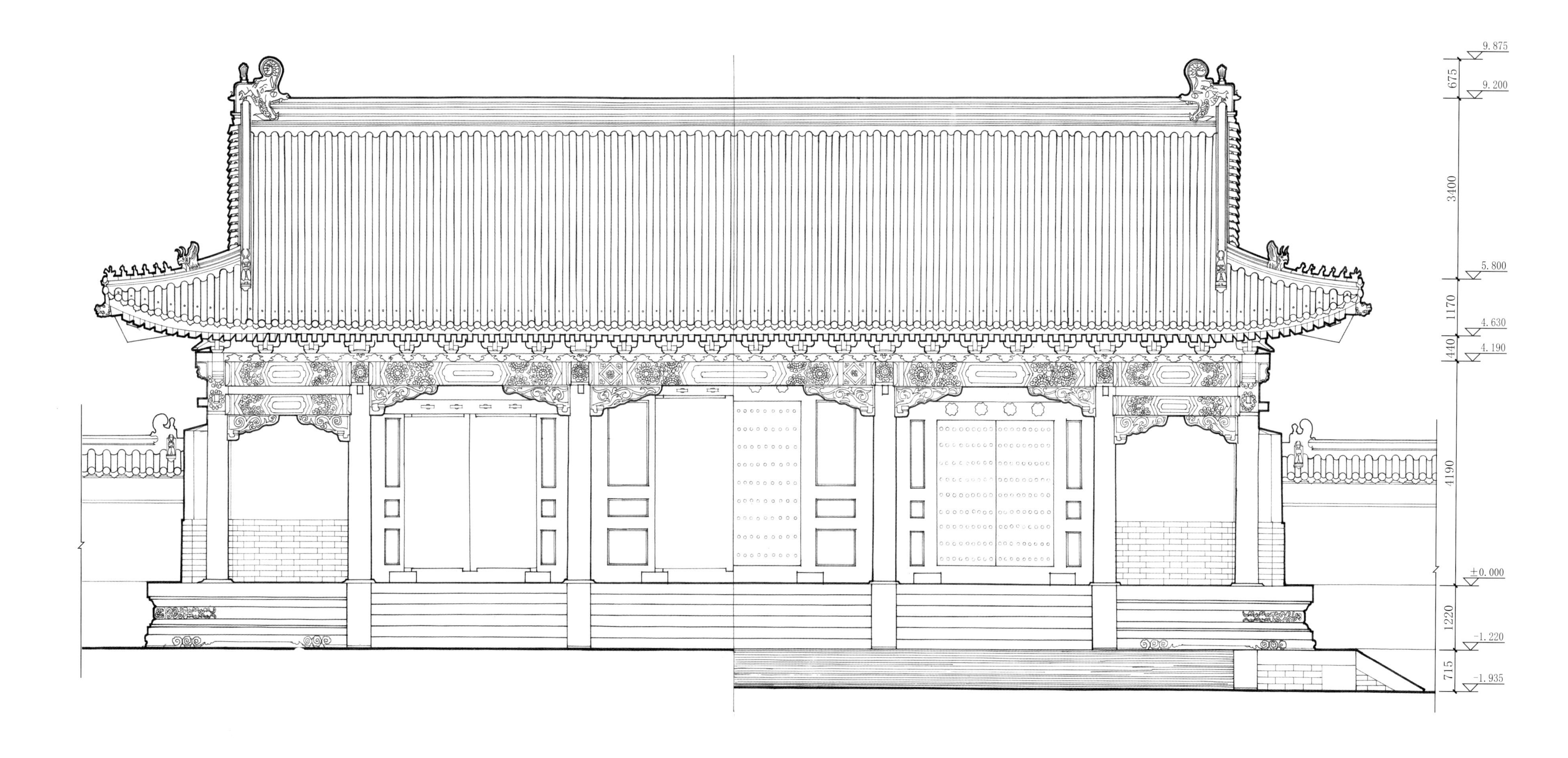

泰陵隆恩门正背立面图
Front and rear elevations of the Gate of Monumental Grace,Tomb of Peace

泰陵隆恩门侧立面图
Side elevation of the Gate of Monumental Grace,Tomb of Peace

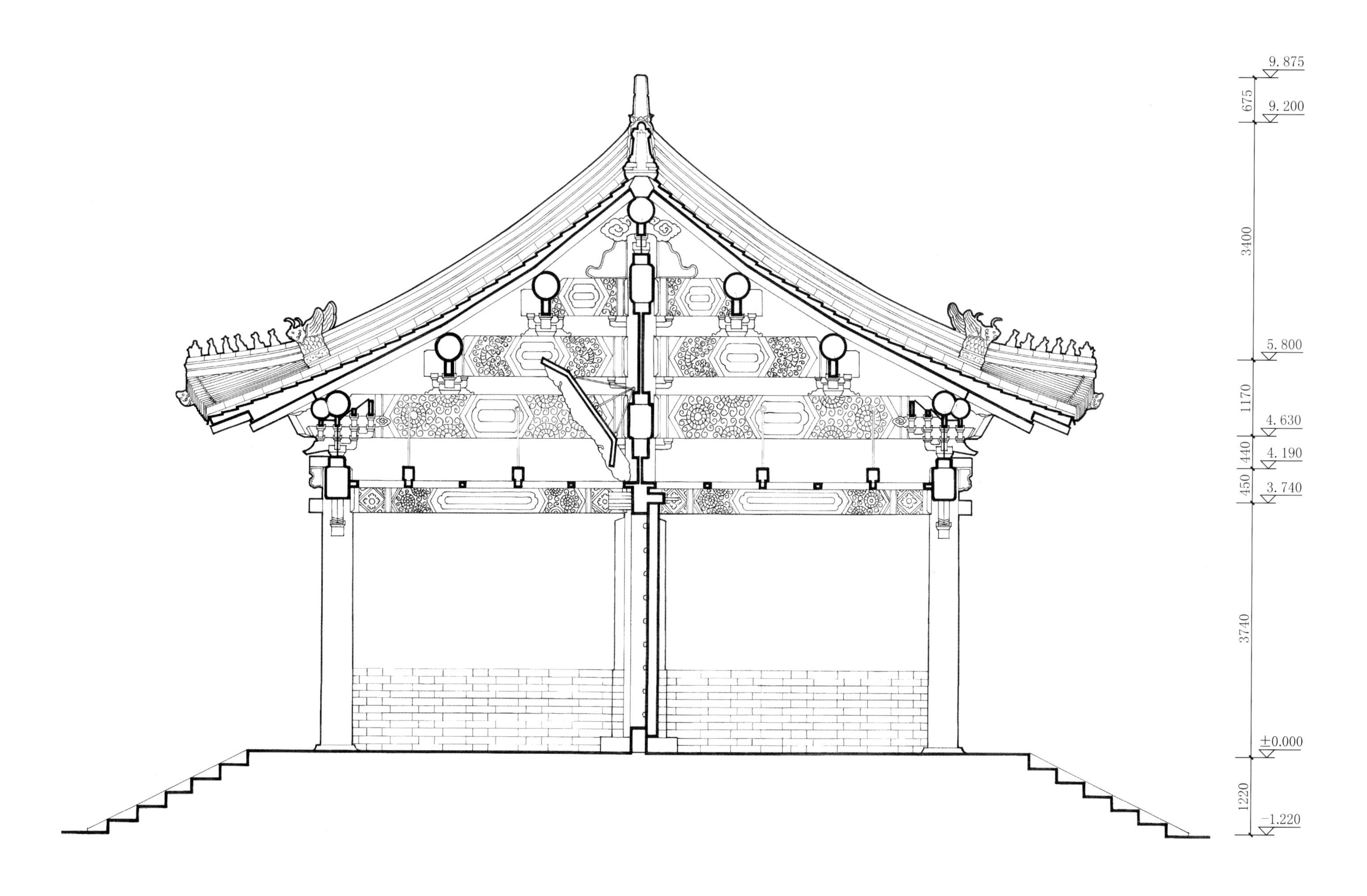

泰陵隆恩门明间横剖面图

Cross section of the Central Chamber within the Gate of Monumental Grace,Tomb of Peace

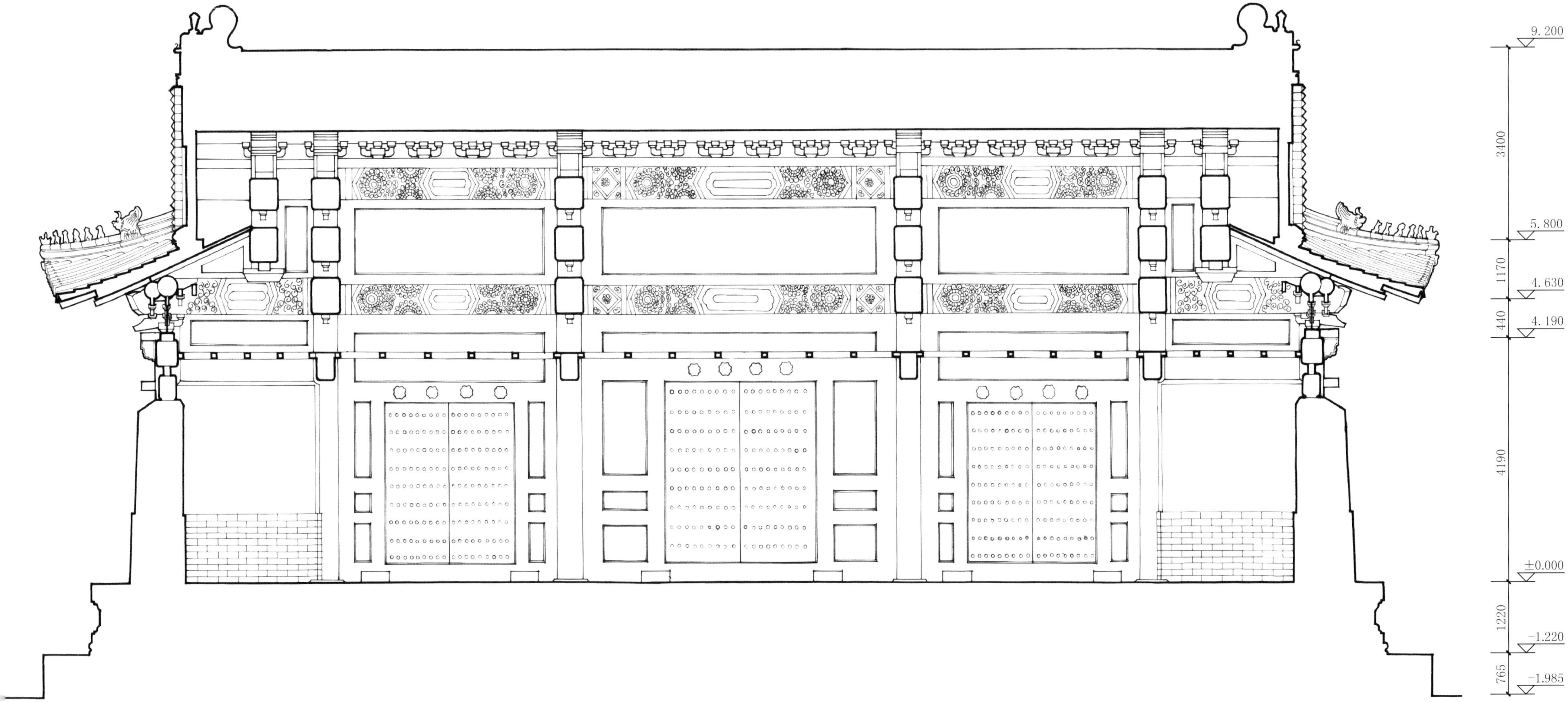

9.200
3400
5.800
1170
4.630
440
4.190
4190
±0.000
1220
−1.220
765
−1.985

泰陵隆恩门纵剖面图
Longitudinal section of the Gate of Monumental Grace,Tomb of Peace

泰陵焚帛炉正立面图
Front elevation of the Sacrificial Burner,Tomb of Peace

泰陵焚帛炉侧立面图
Side elevation of the Sacrificial Burner,Tomb of Peace

泰陵焚帛炉横剖面图
Cross section of the Sacrificial Burner,Tomb of Peace

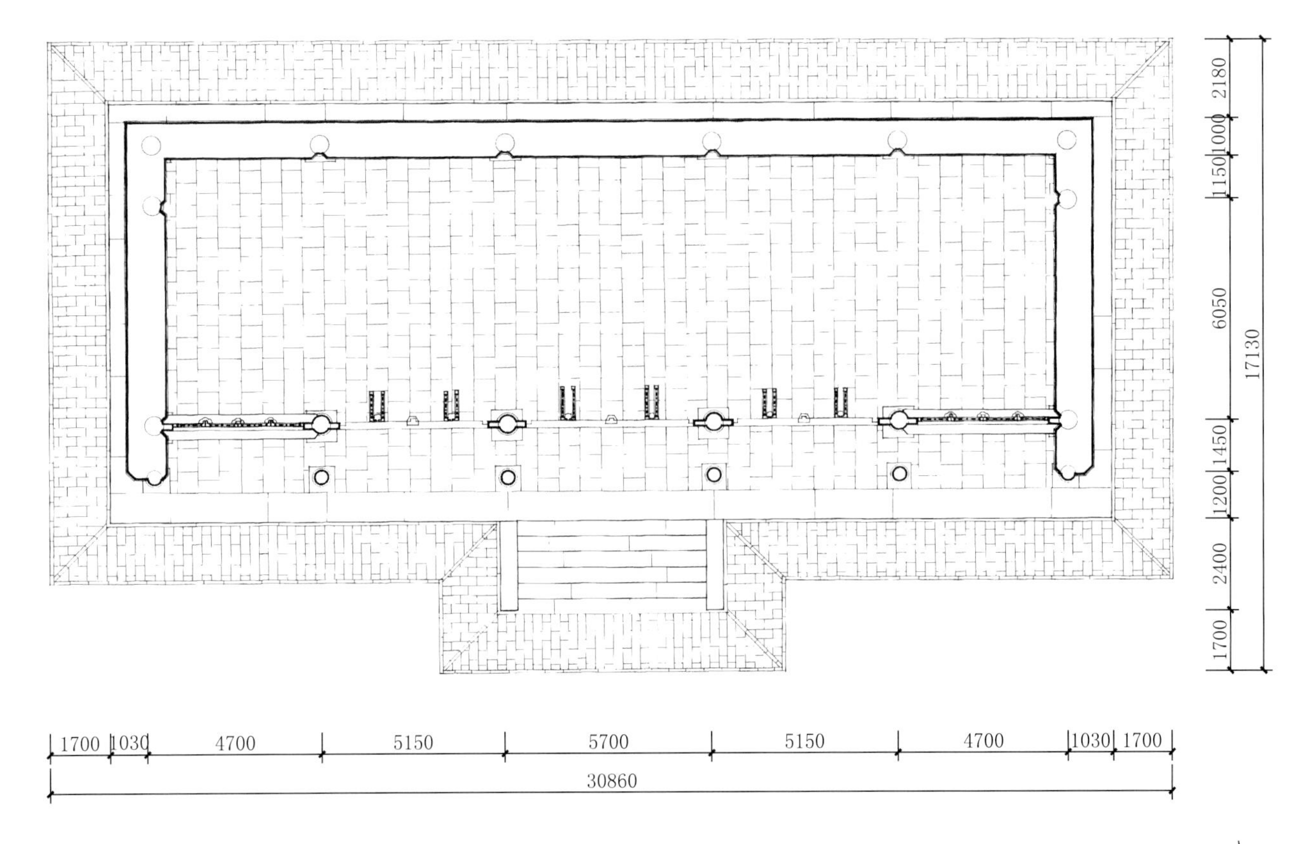

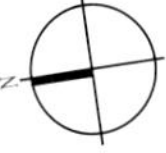

泰陵配殿平面图
Plan of the Side Hall,Tomb of Peace

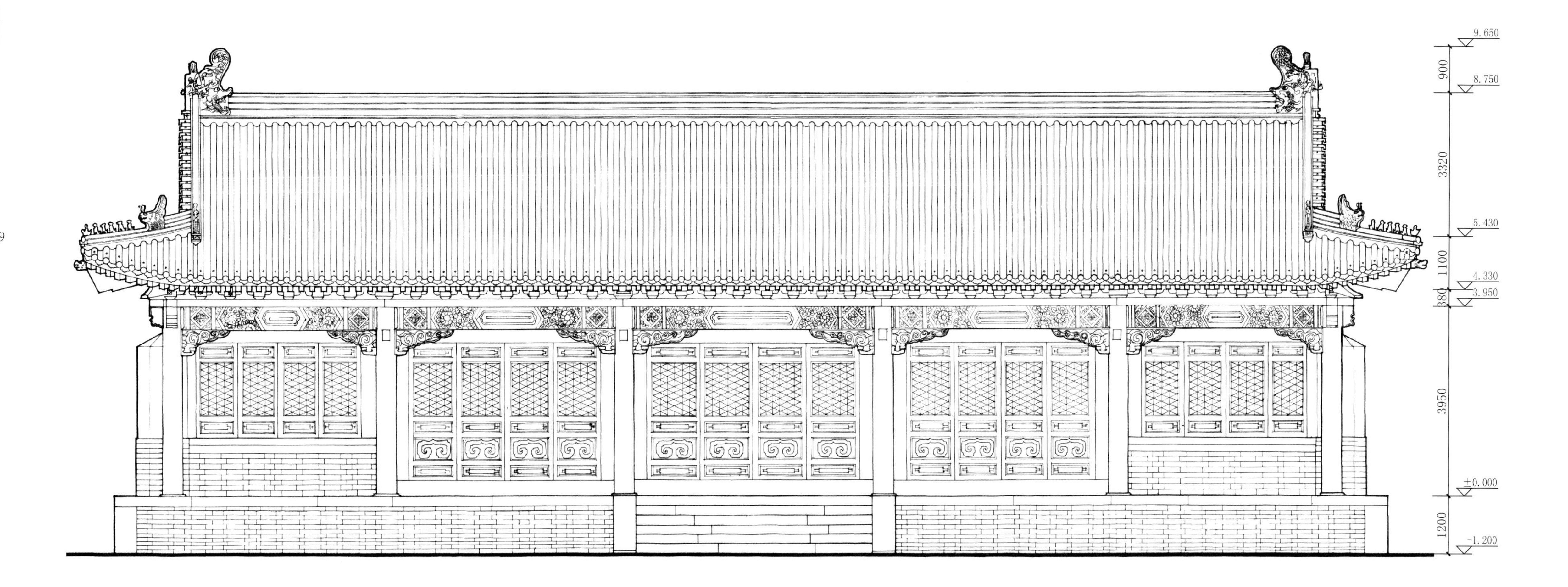

泰陵配殿正立面图
Front elevation of the Side Hall,Tomb of Peace

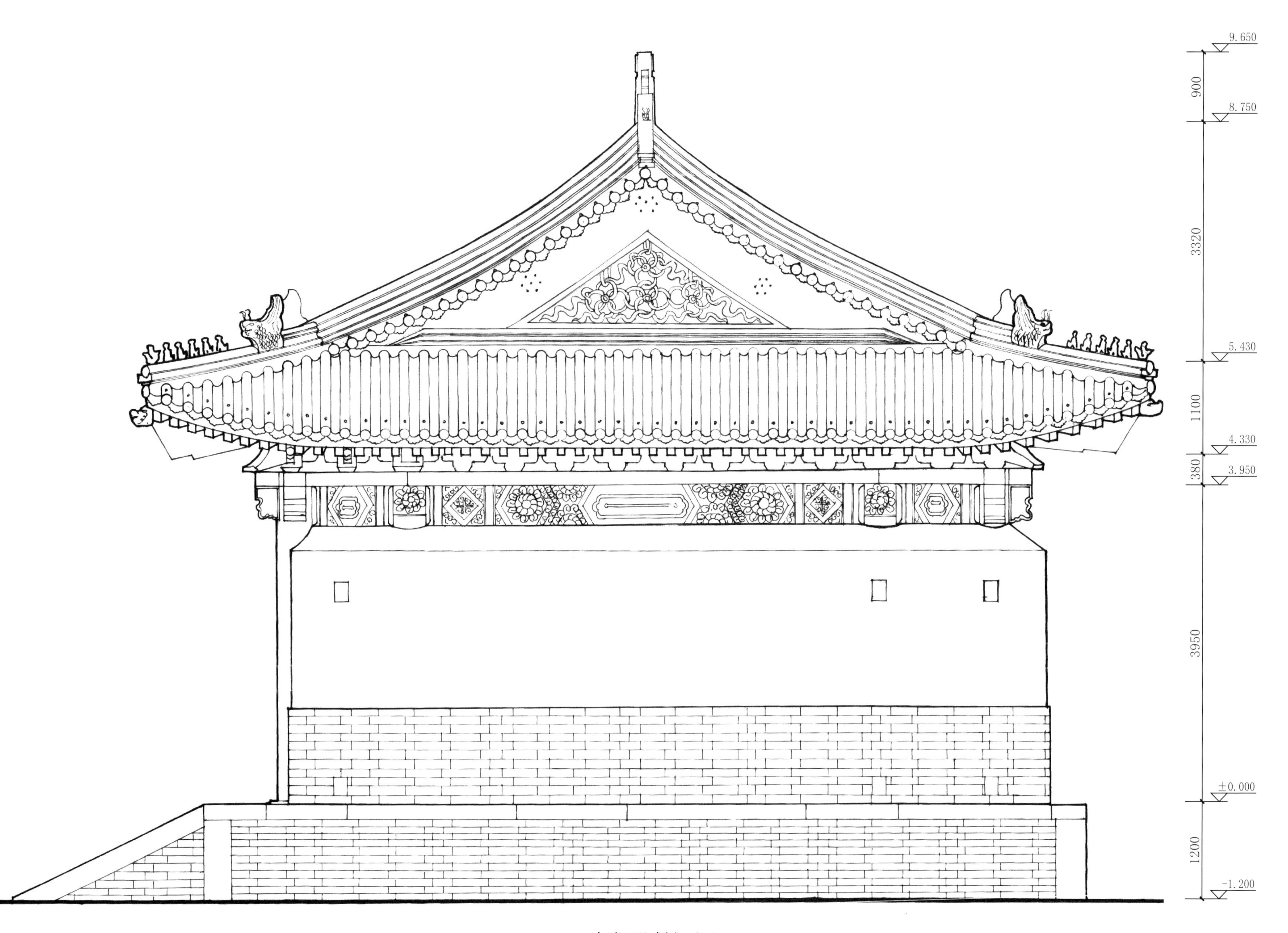

泰陵配殿侧立面图

Side elevation of the Side Hall,Tomb of Peace

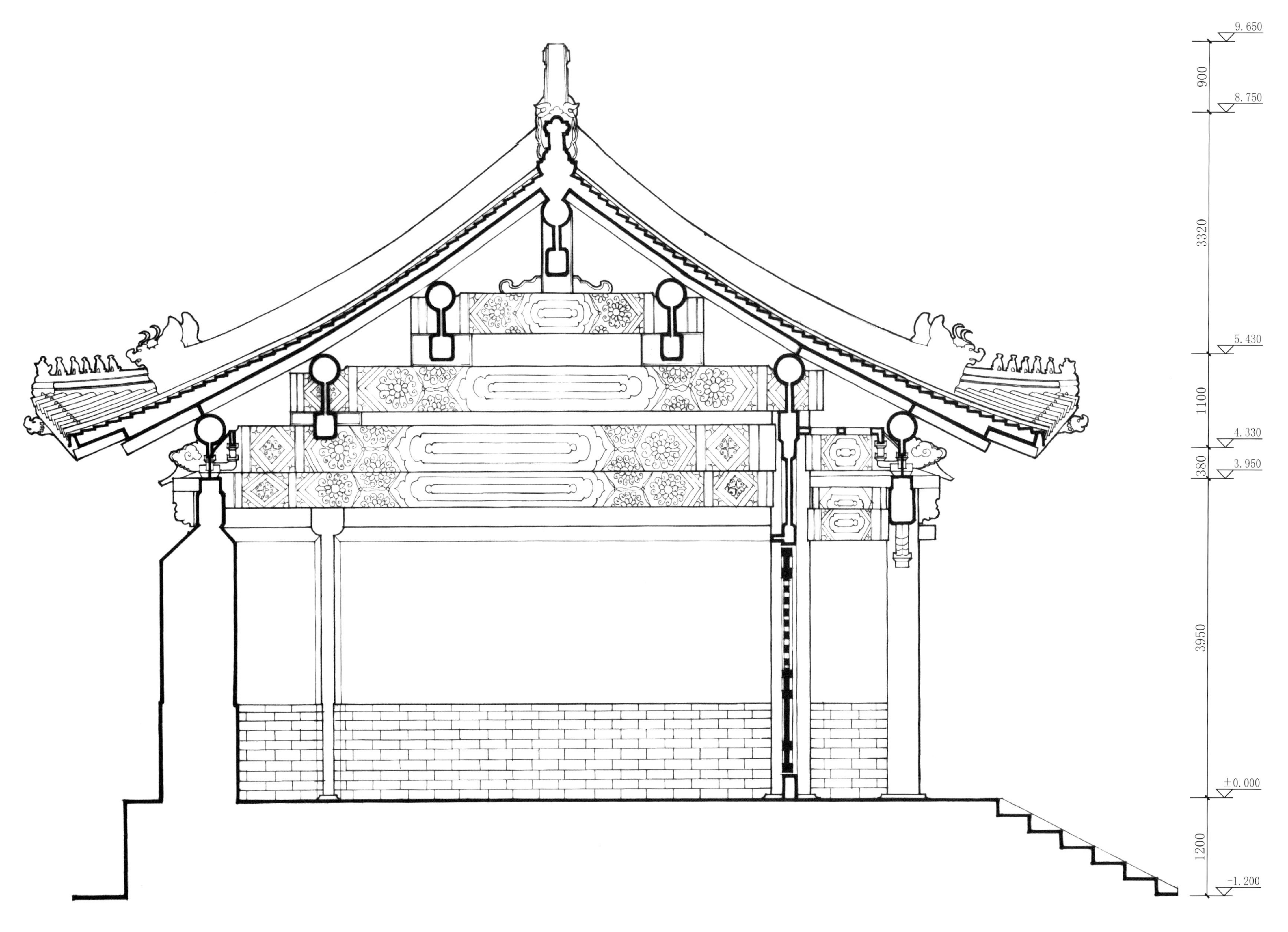

泰陵配殿明间横剖面图

Cross section of the Central Chamber within the Side Hall,Tomb of Peace

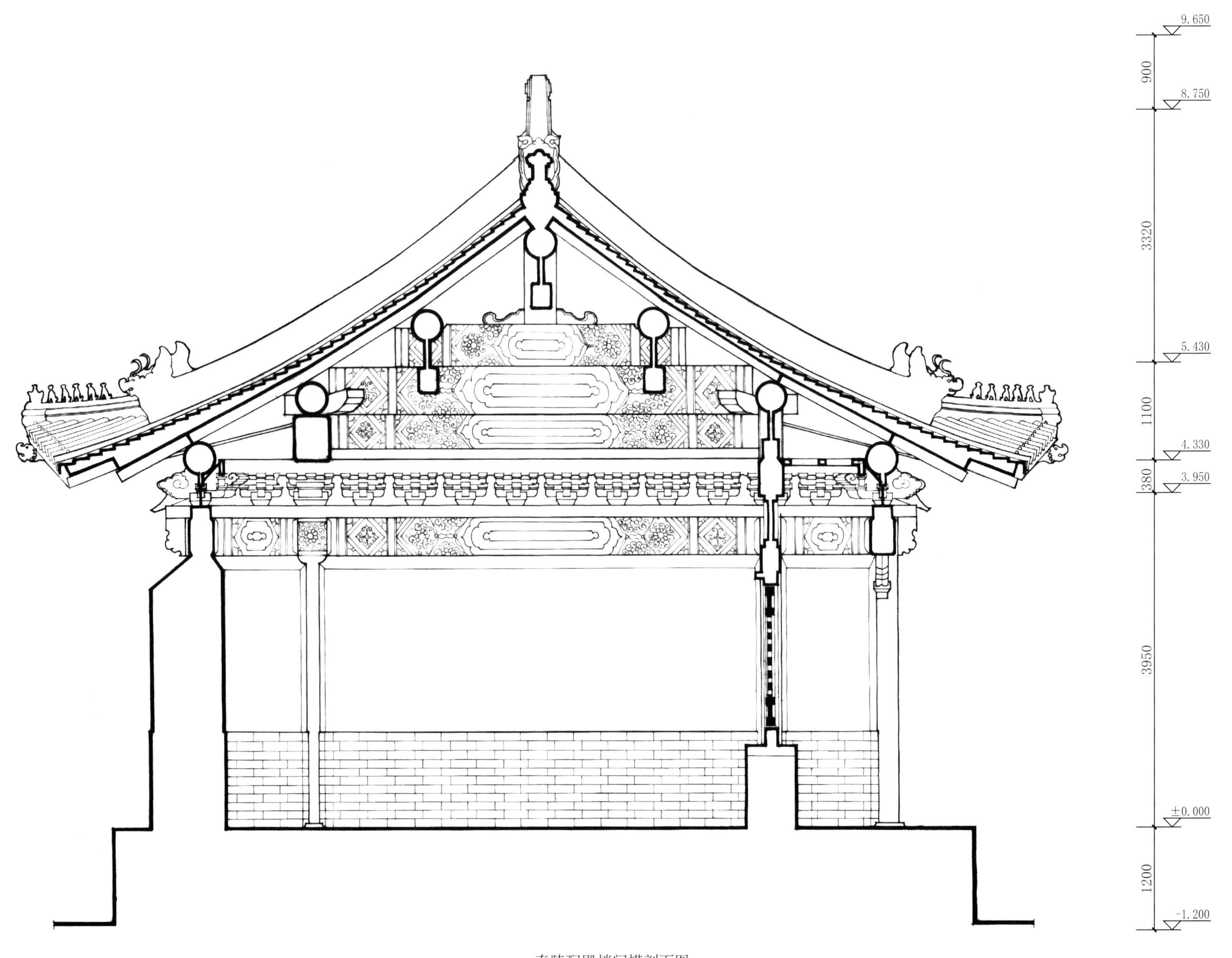

泰陵配殿梢间横剖面图
Cross section of the Second-to-last Chamber within the Side Hall,Tomb of Peace

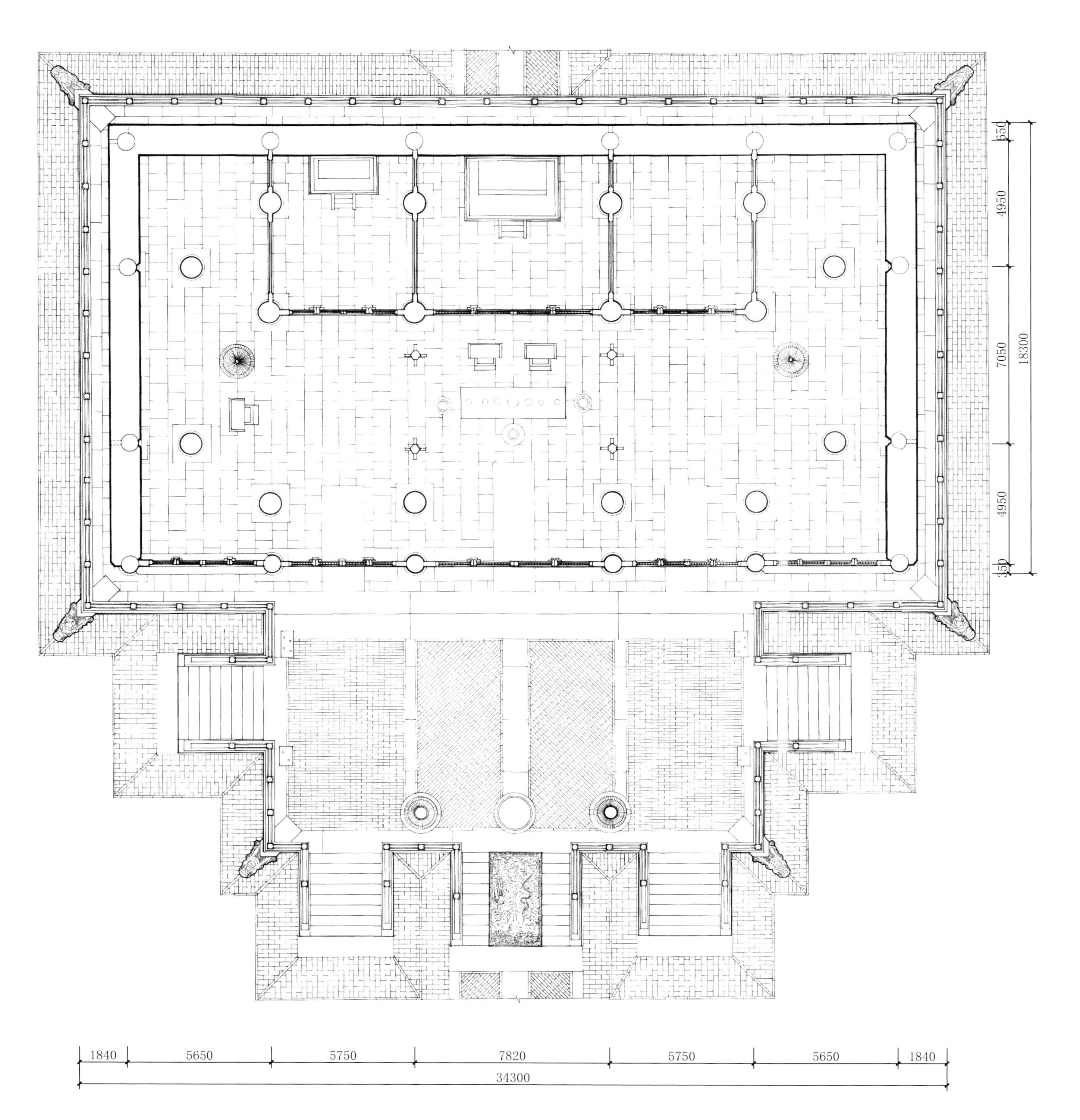

泰陵隆恩殿平面图
Plan of the Hall of Monumental Grace,Tomb of Peace

泰陵隆恩殿正立面图

Front elevation of the Hall of Monumental Grace,Tomb of Peace

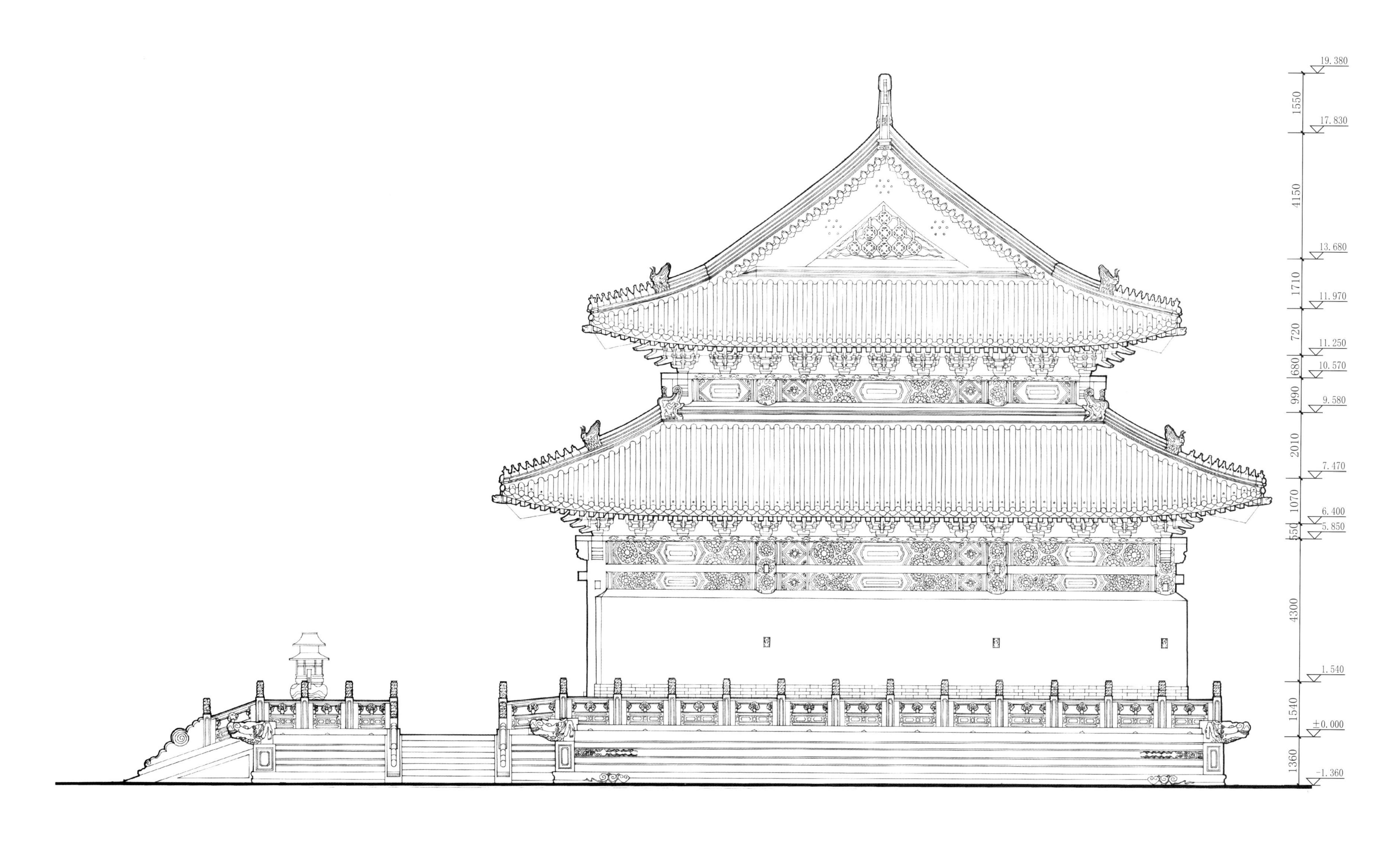

泰陵隆恩殿侧立面图

Side elevation of the Hall of Monumental Grace,Tomb of Peace

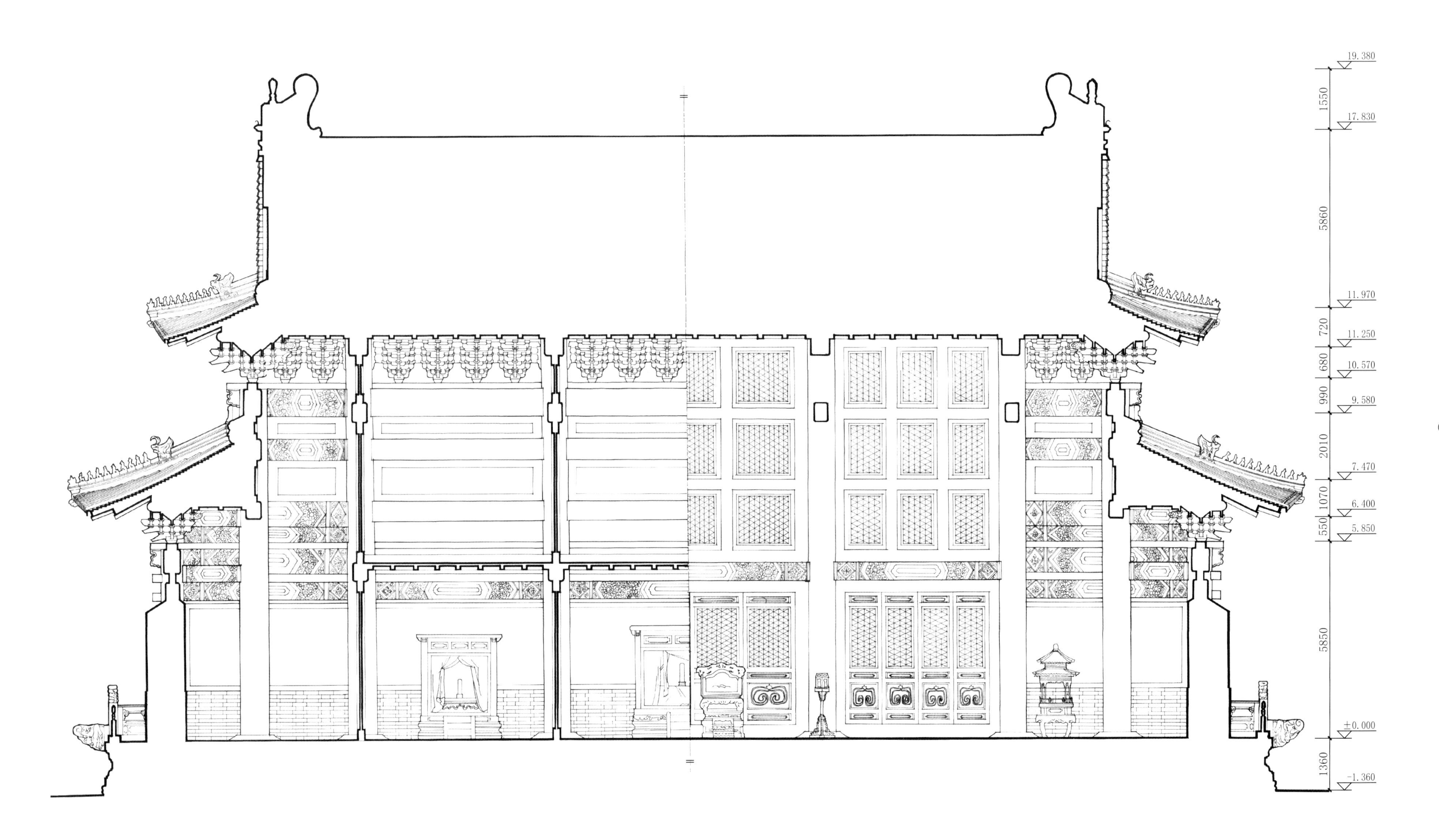

泰陵隆恩殿纵剖面图
Longitudinal section of the Hall of Monumental Grace,Tomb of Peace

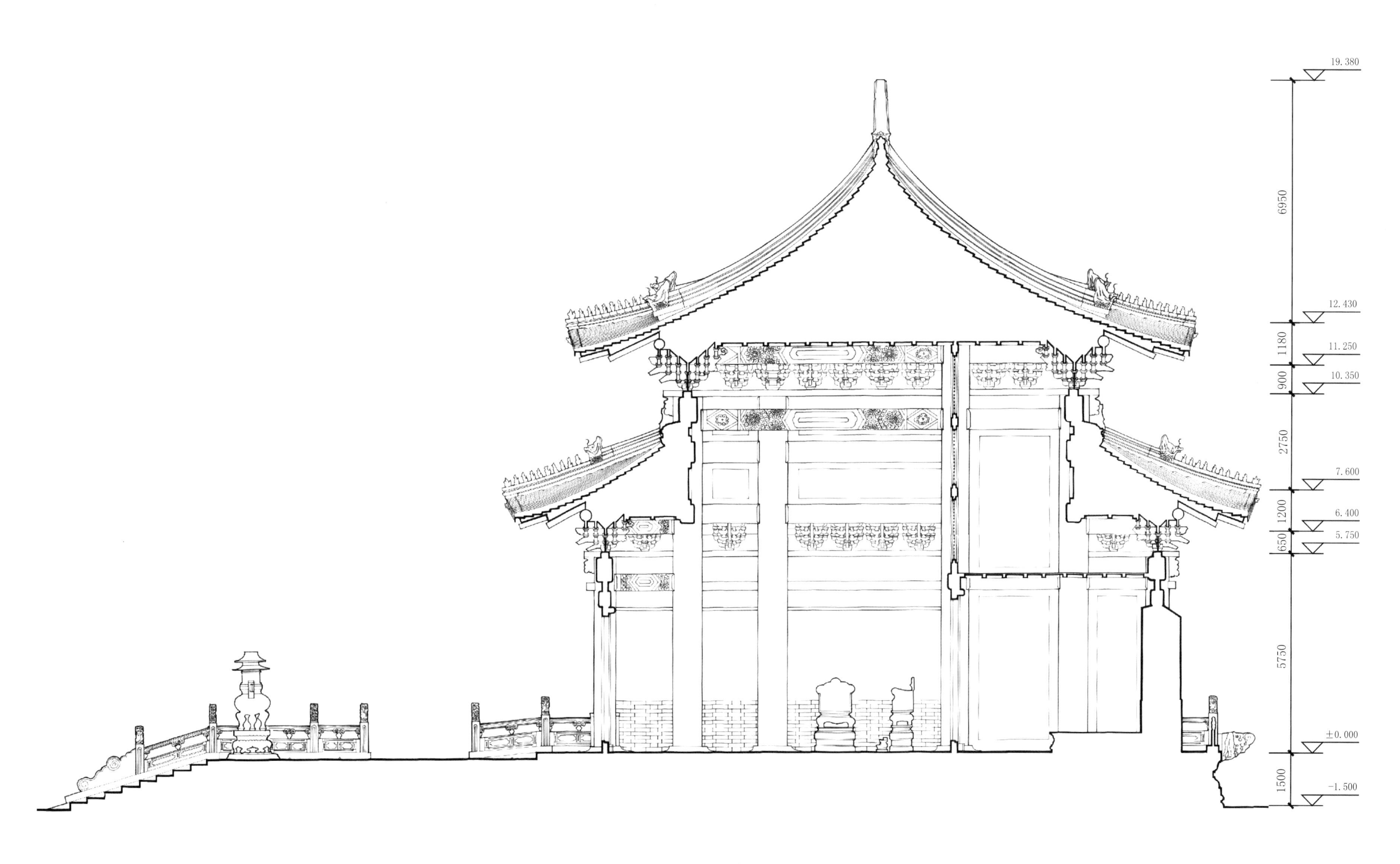

泰陵隆恩殿明间横剖面图

Cross section of the Central Chamber within the Hall of Monumental Grace,Tomb of Peace

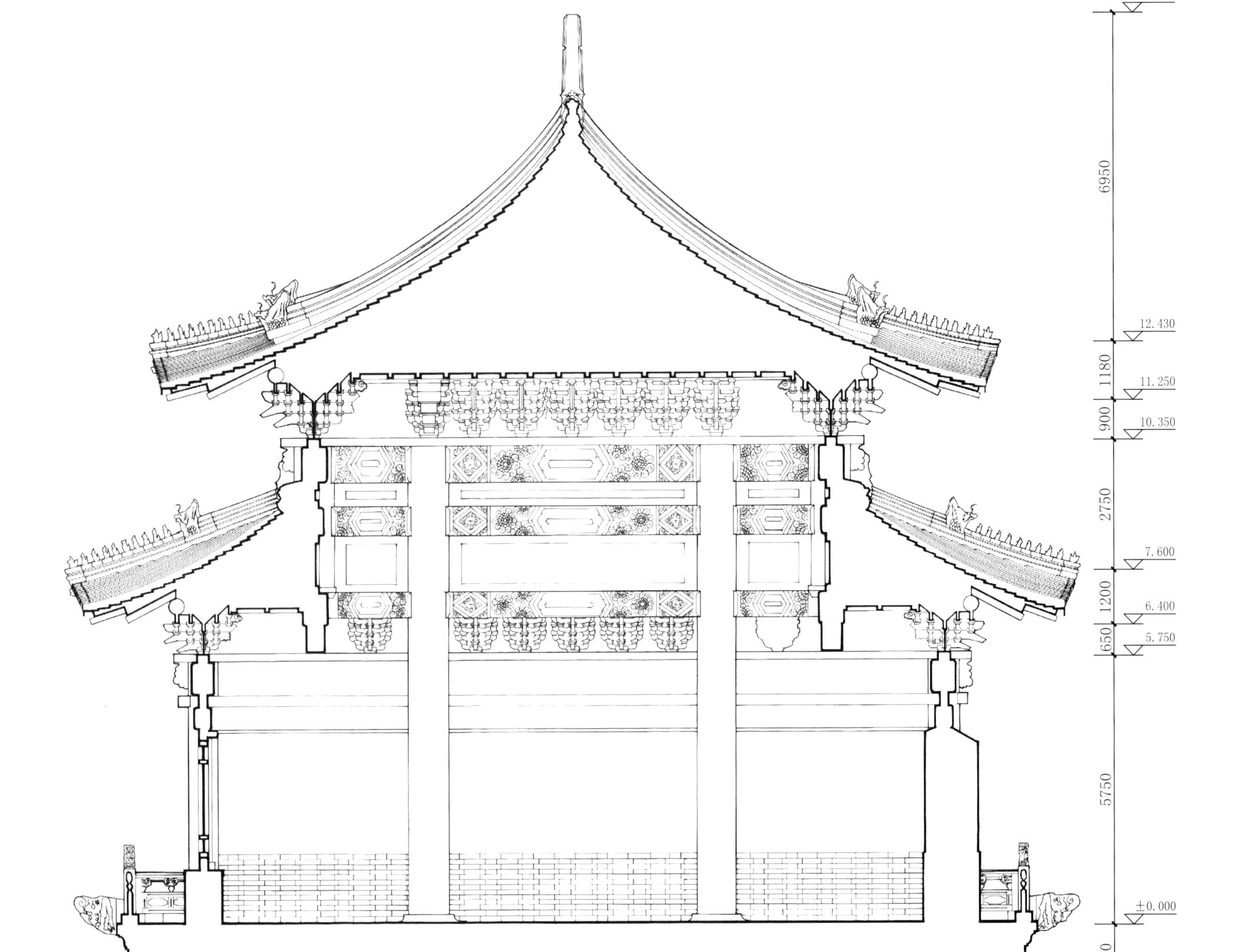

泰陵隆恩殿梢间横剖面图

Cross section of the Second-to-last Chamber within the Hall of Monumental Grace,Tomb of Peace

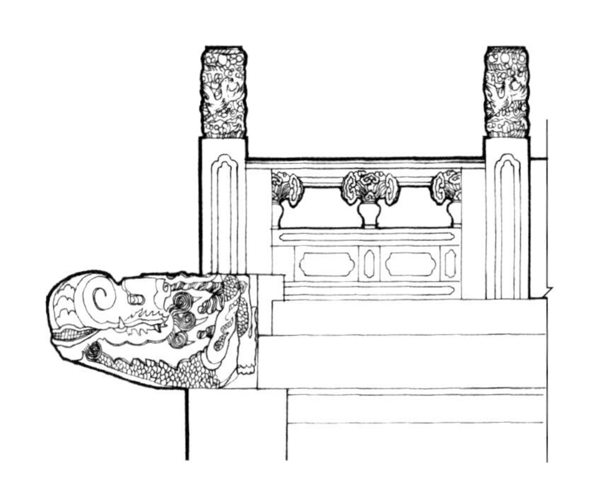

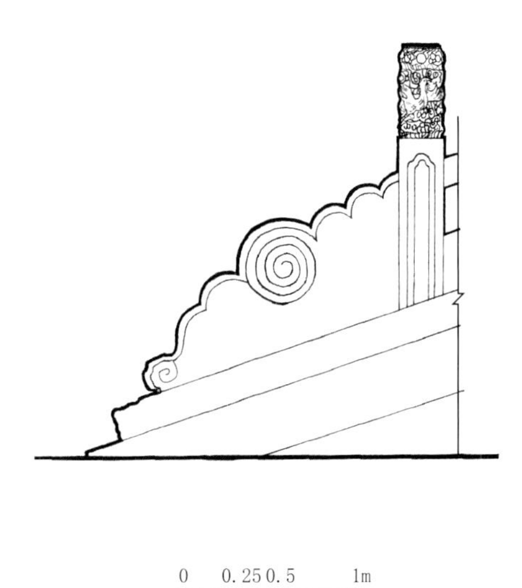

泰陵隆恩殿栏板大样图

Detailed drawing of the Railing Board of the Hall of Monumental Grace,Tomb of Peace

0 0.1 0.2 0.4m

泰陵隆恩殿香炉大样图

Detailed drawing of the Incense Burner of the Hall of Monumental Grace,Tomb of Peace

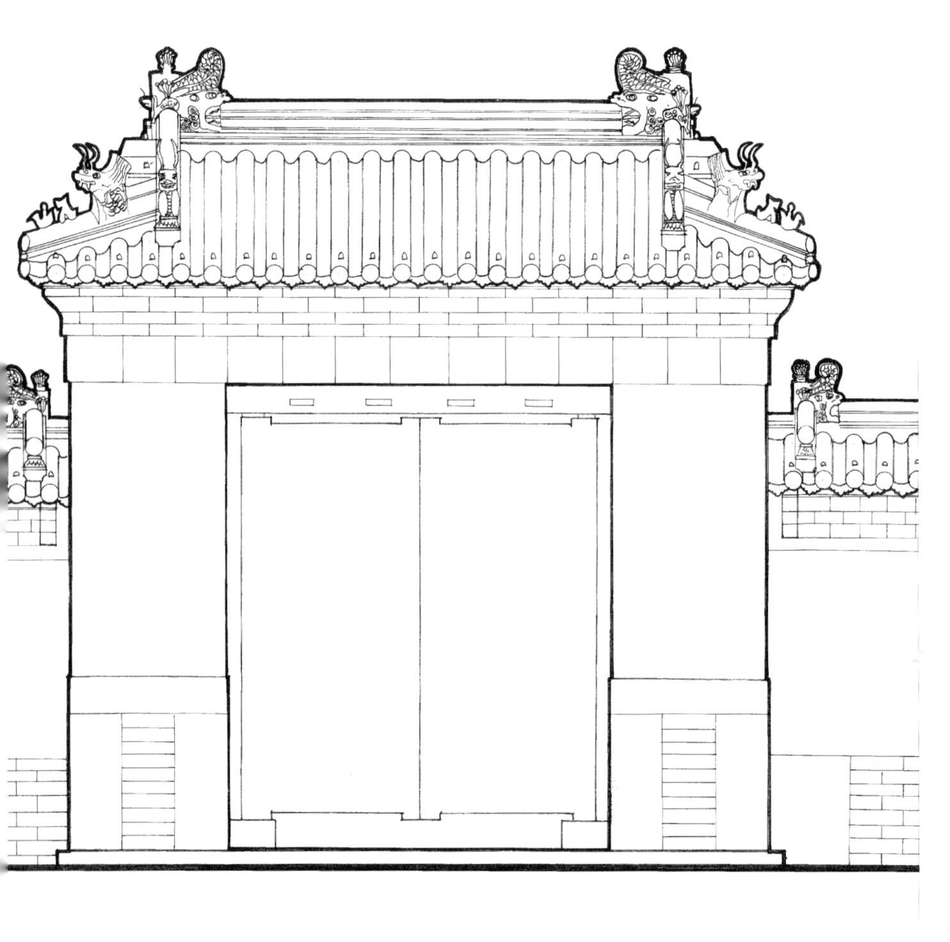

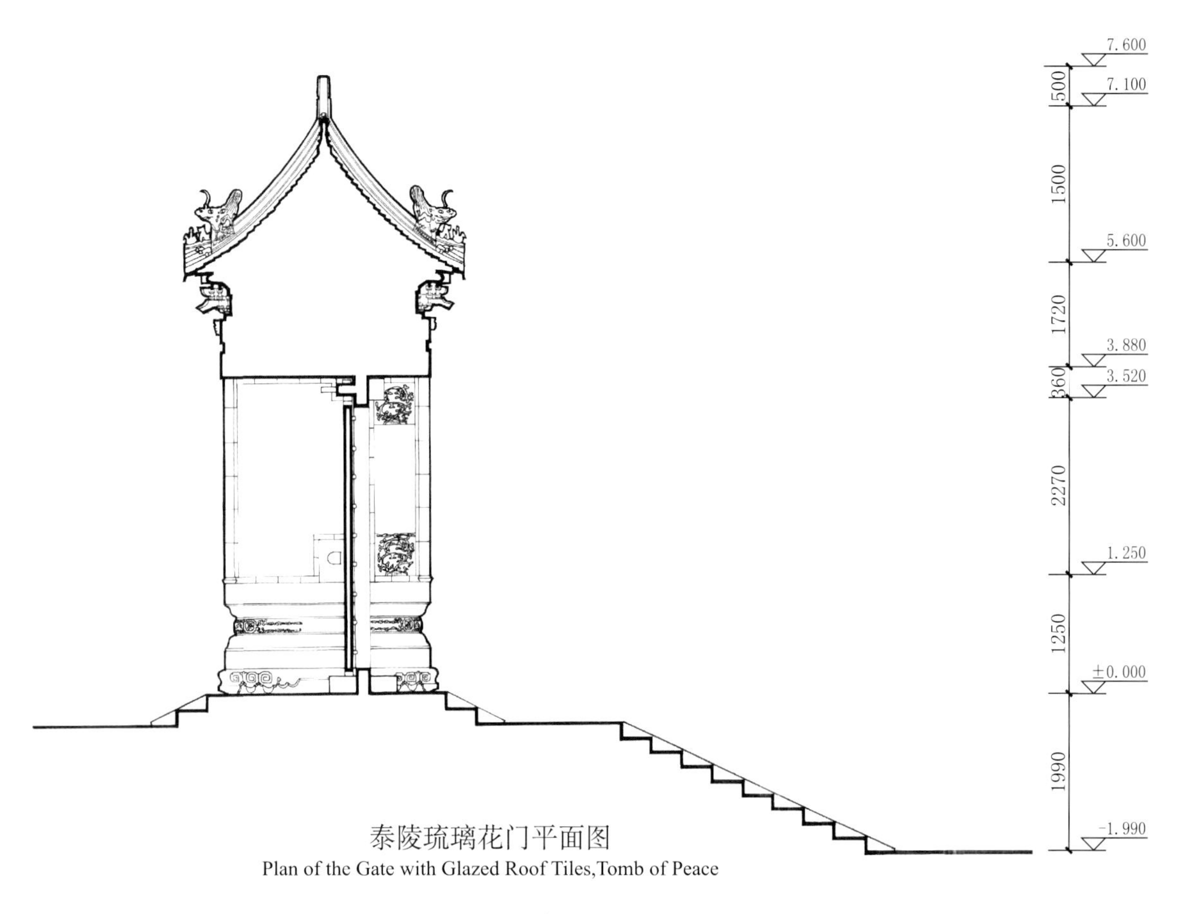

泰陵琉璃花门平面图

Plan of the Gate with Glazed Roof Tiles,Tomb of Peace

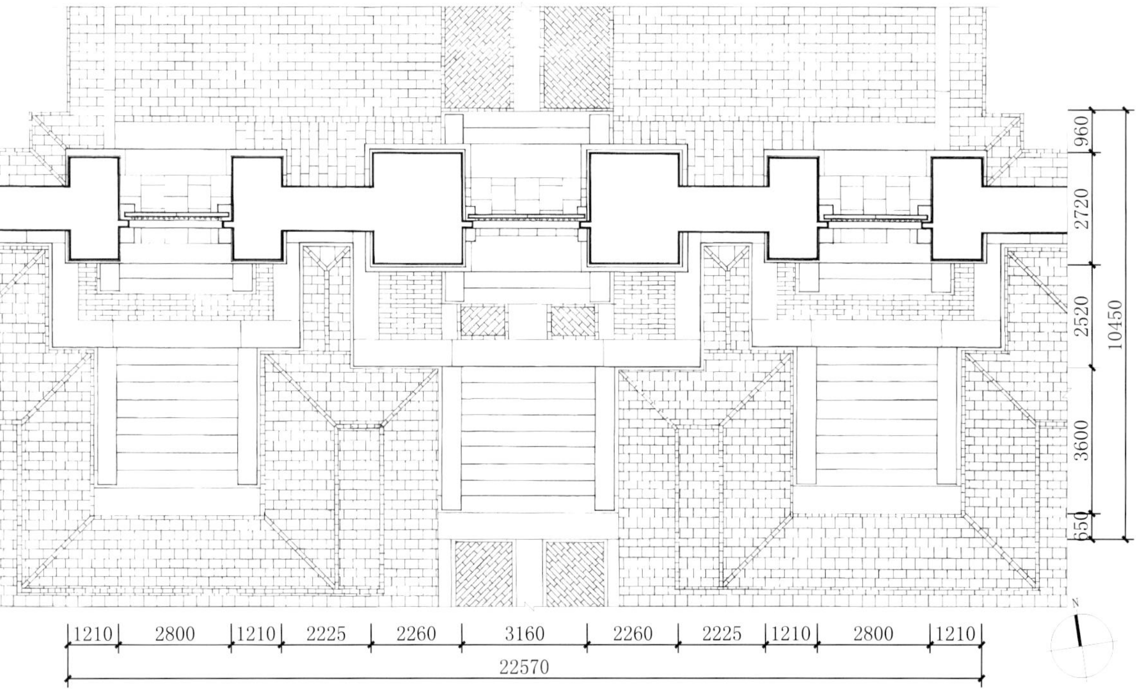

泰陵琉璃花门横剖面图

Cross section of the Gate with Glazed Roof Tiles,Tomb of Peace

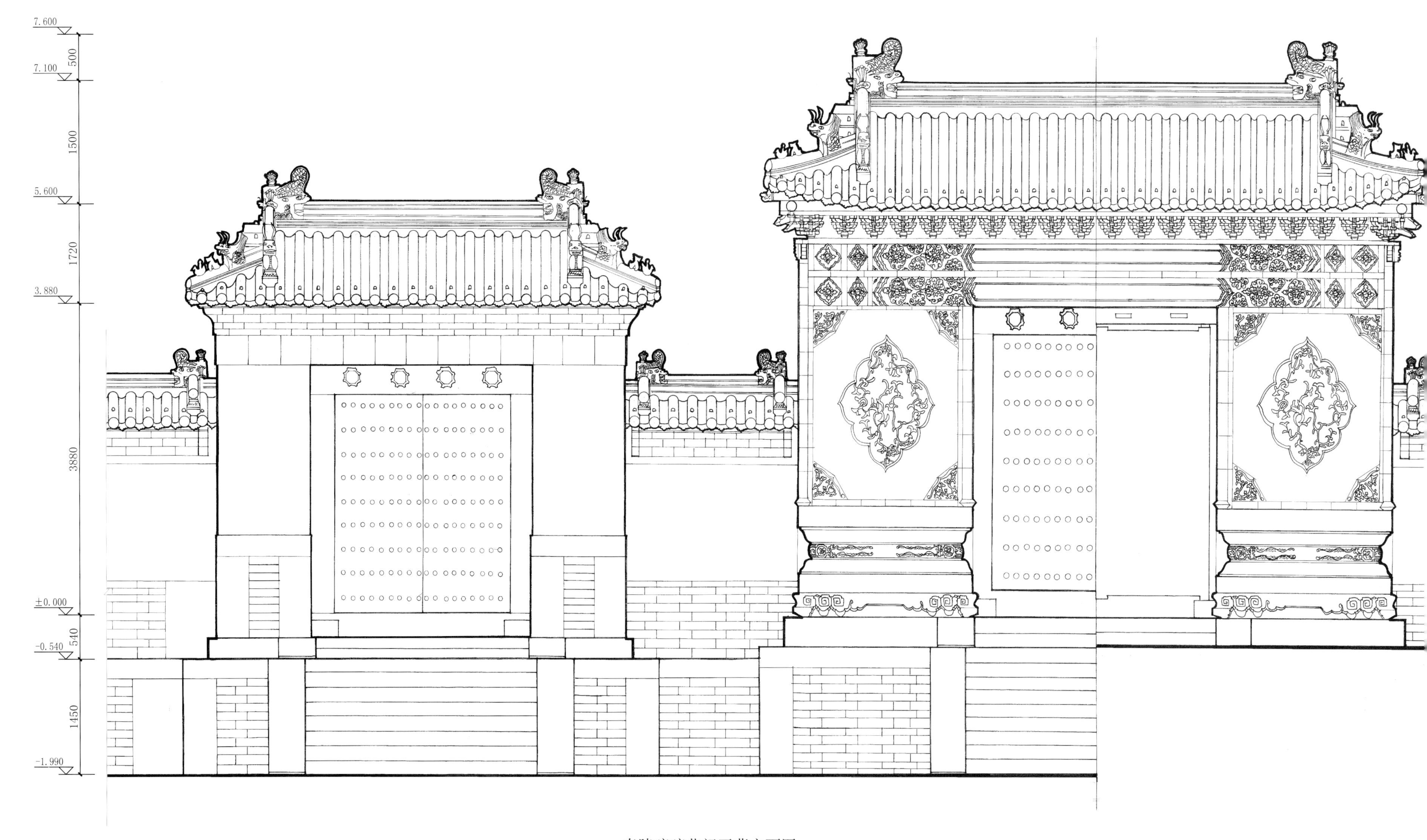

泰陵琉璃花门正背立面图

Front and rear elevations of the Gate with Glazed Roof Tiles,Tomb of Peace

泰陵石台五供正立面图

Front elevation of the Five Stone Ritual Vessels,Tomb of Peace

泰陵二柱门大样图
Detailed drawing of the Gate with Two Columns,Tomb of Peace

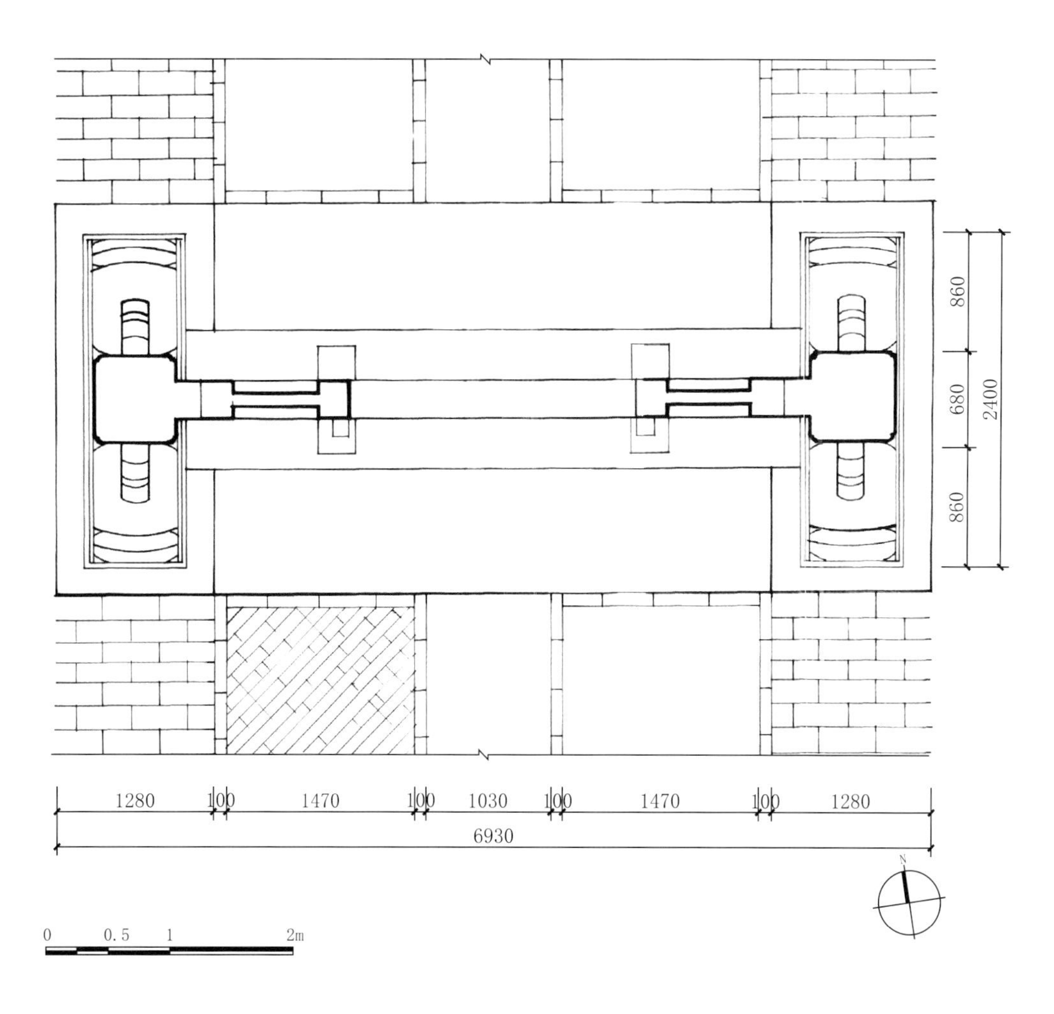

泰陵二柱门平面图
Plan of the Gate with Two Columns,Tomb of Peace

泰陵石台五供侧立面图
Side elevation of the Five Stone Ritual Vessels,Tomb of Peace

泰陵二柱门正立面图
Front elevation of the Gate with Two Columns,Tomb of Peace

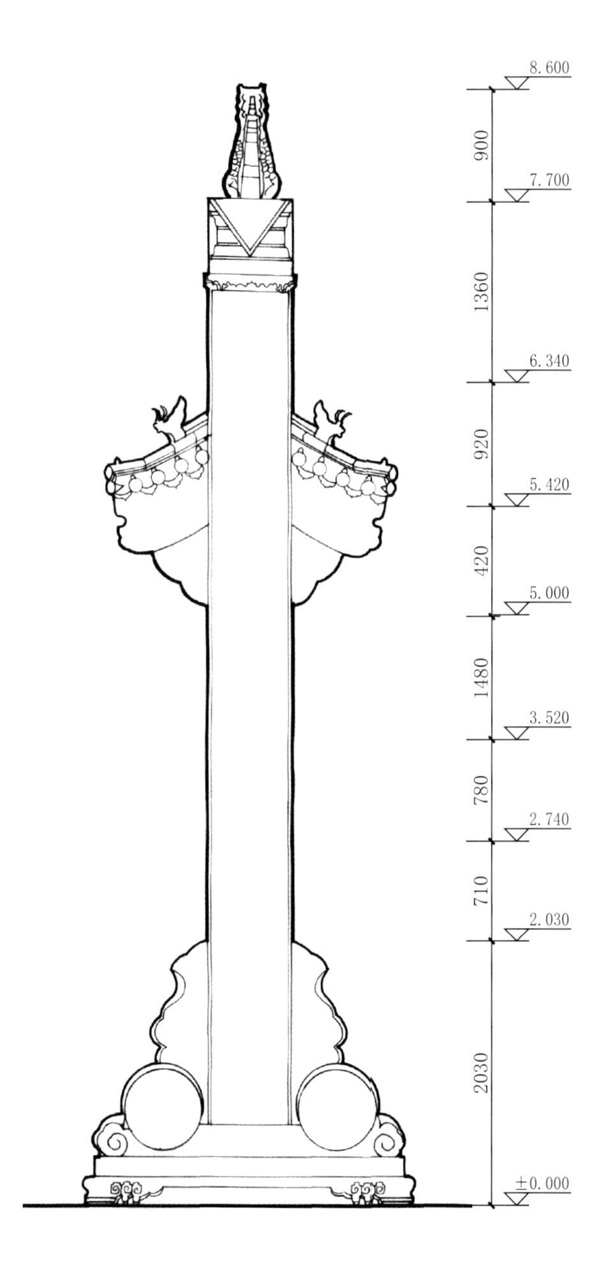

泰陵二柱门侧立面图
Side elevation of the Gate with Two Columns,Tomb of Peace

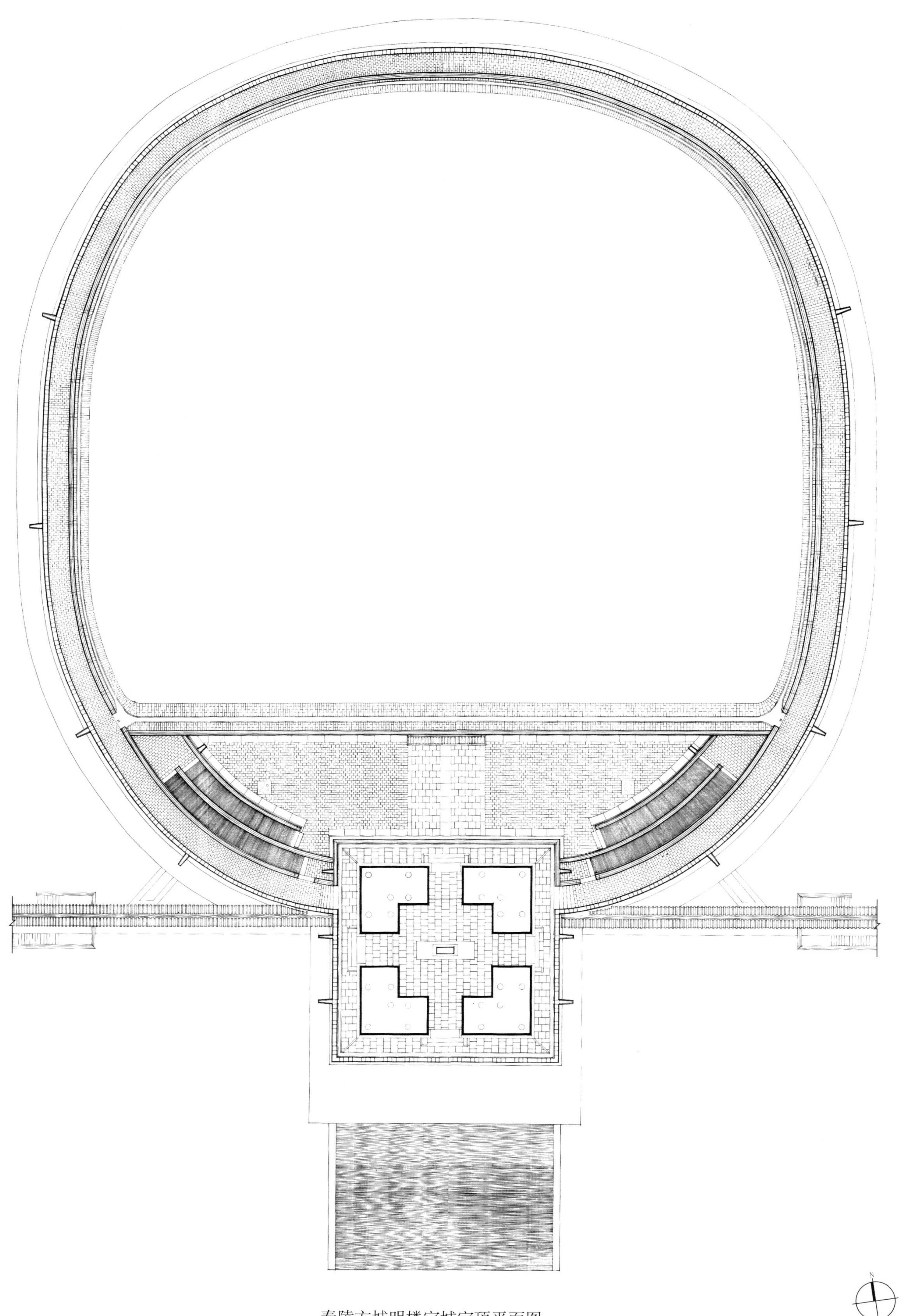

泰陵方城明楼宝城宝顶平面图
Plan of the Square Walled Terrace and the Memorial Tower as well as the Encircled Realm of Treasure and the Tumulus,Tomb of Peace

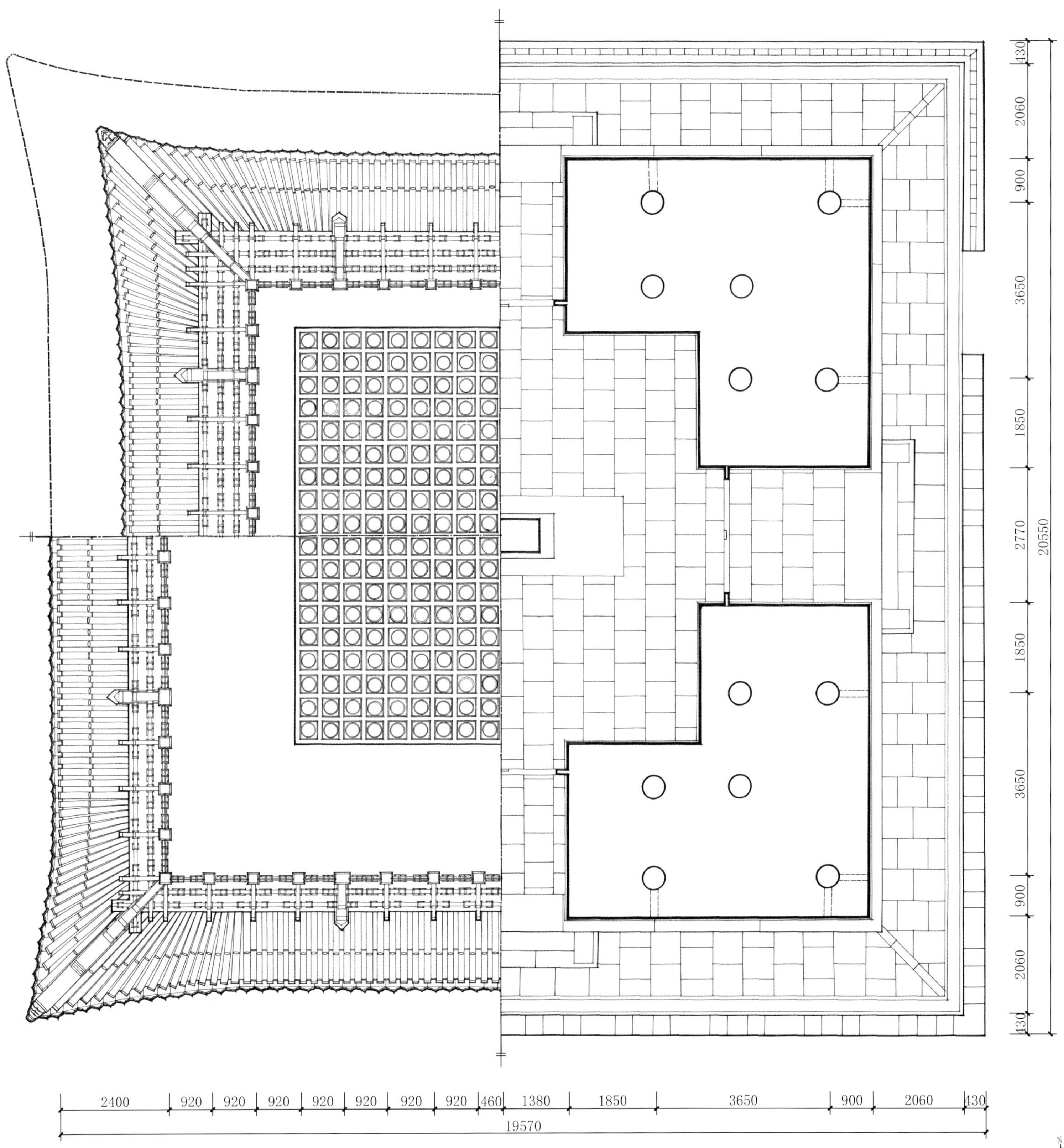

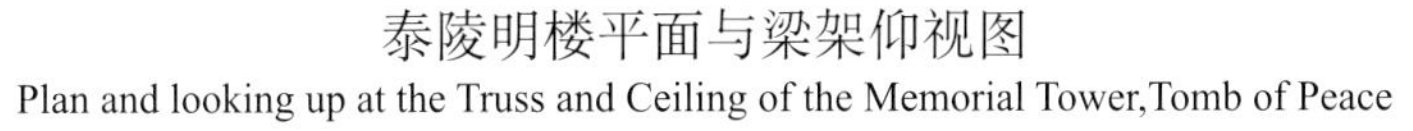

泰陵明楼平面与梁架仰视图

Plan and looking up at the Truss and Ceiling of the Memorial Tower,Tomb of Peace

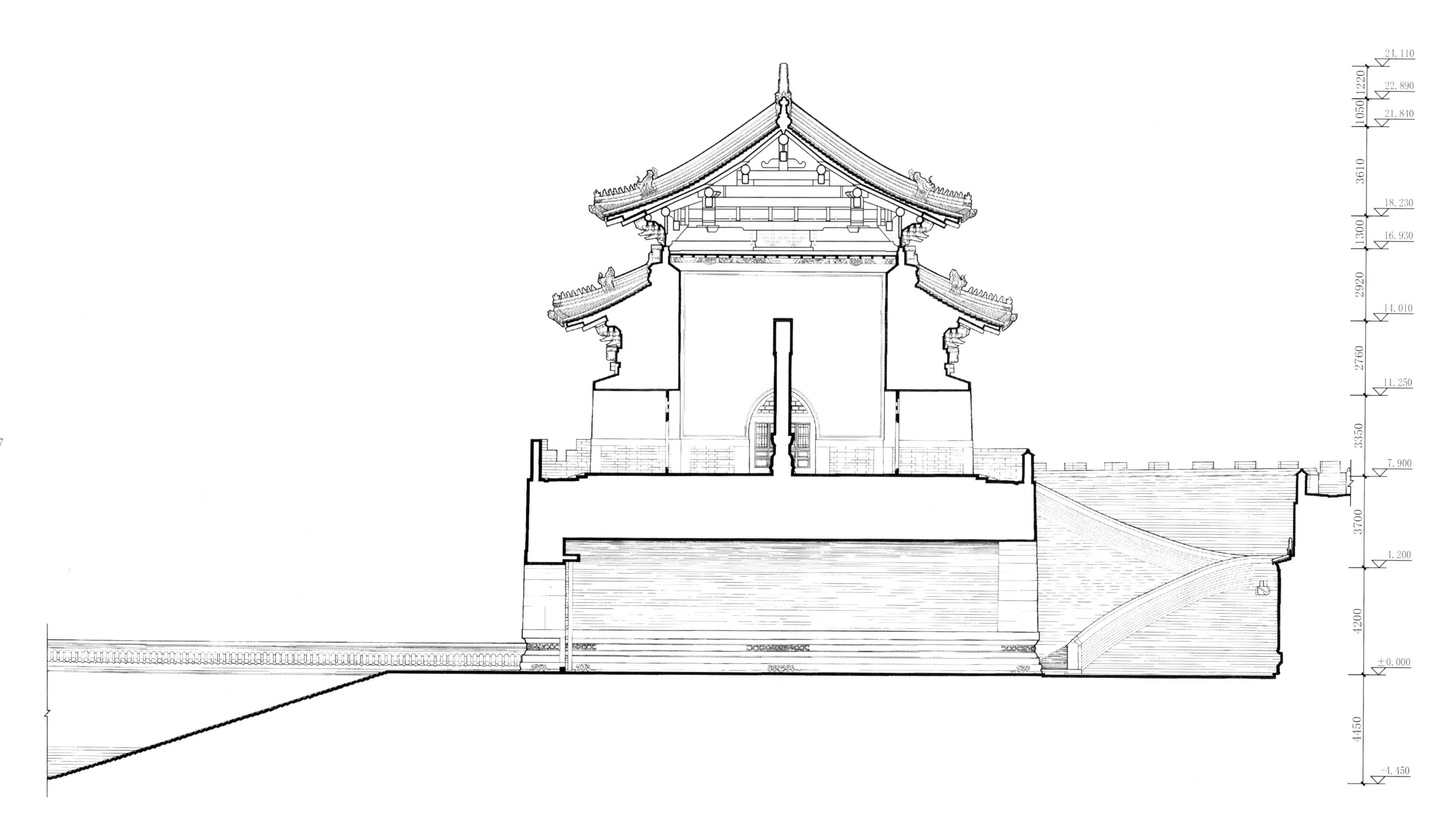

泰陵明楼明间横剖面图

Cross section of the Central Chamber within the Memorial Tower,Tomb of Peace

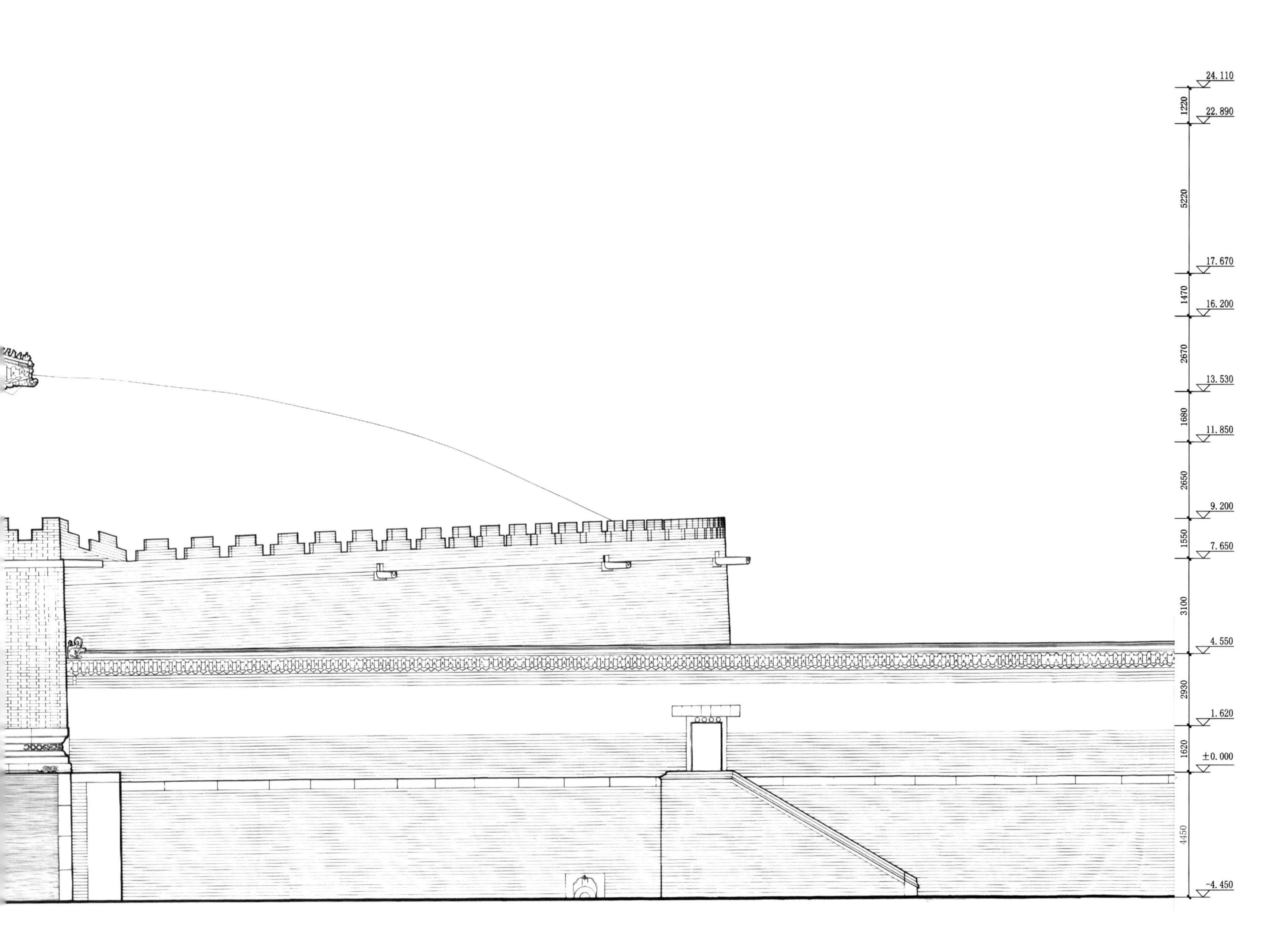

泰陵明楼宝城宝顶正立面图

Front elevation of the Memorial Tower as well as the Encircled Realm of Treasure and the Tumulus,Tomb of Peace

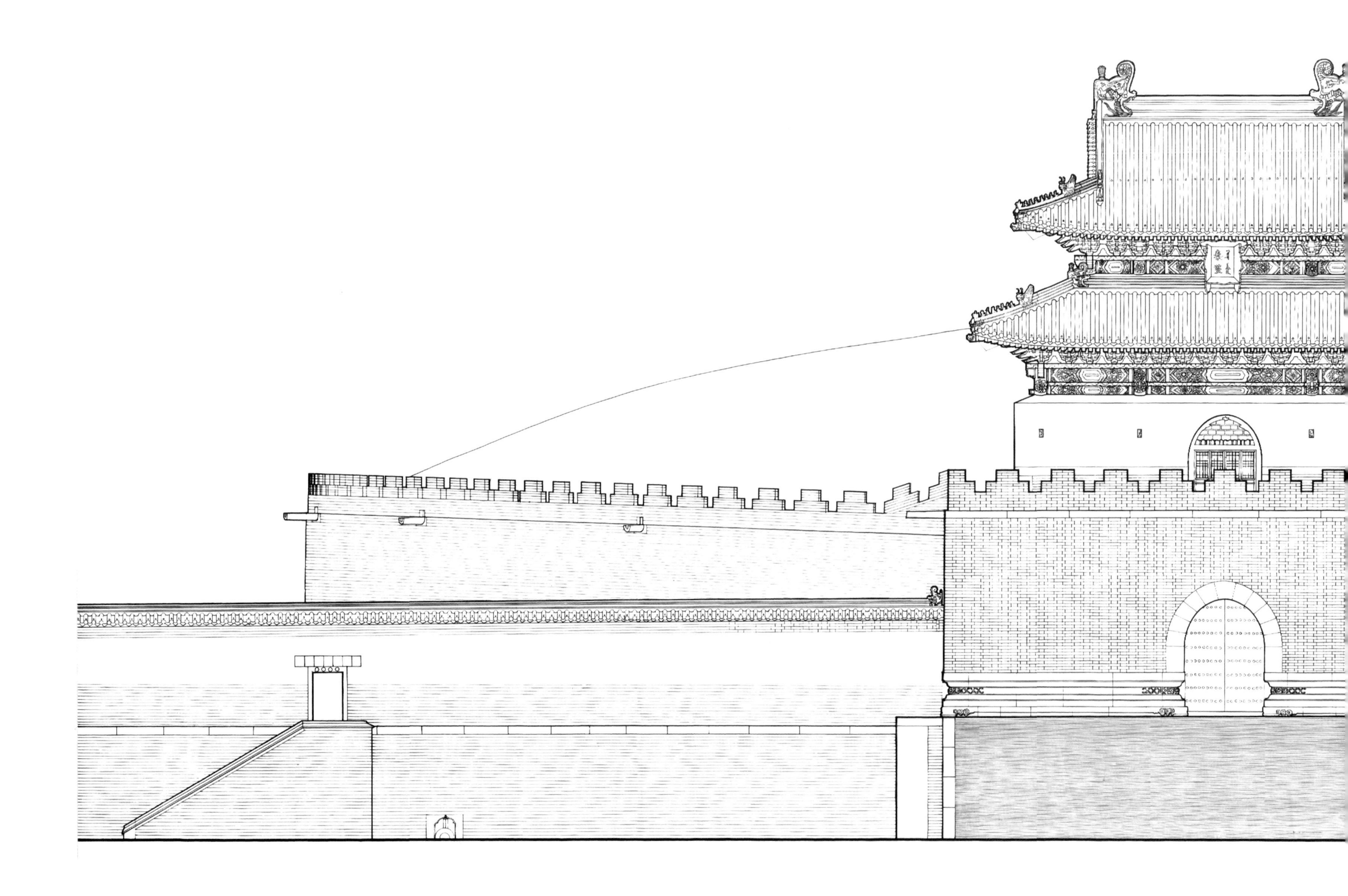

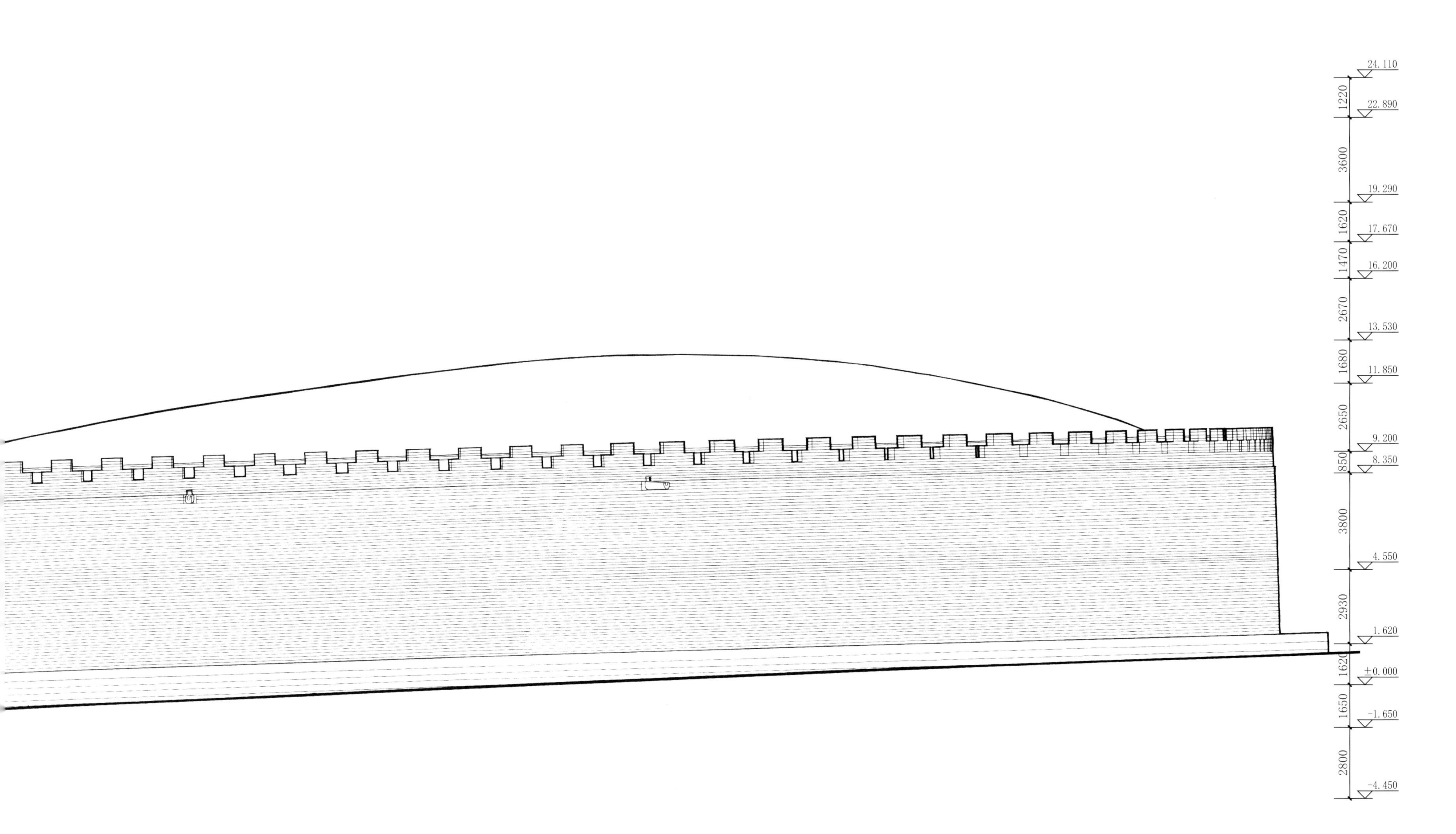

泰陵明楼宝城宝顶侧立面图

Side elevation of the Memorial Tower as well as the Encircled Realm of Treasure and the Tumulus,Tomb of Peace

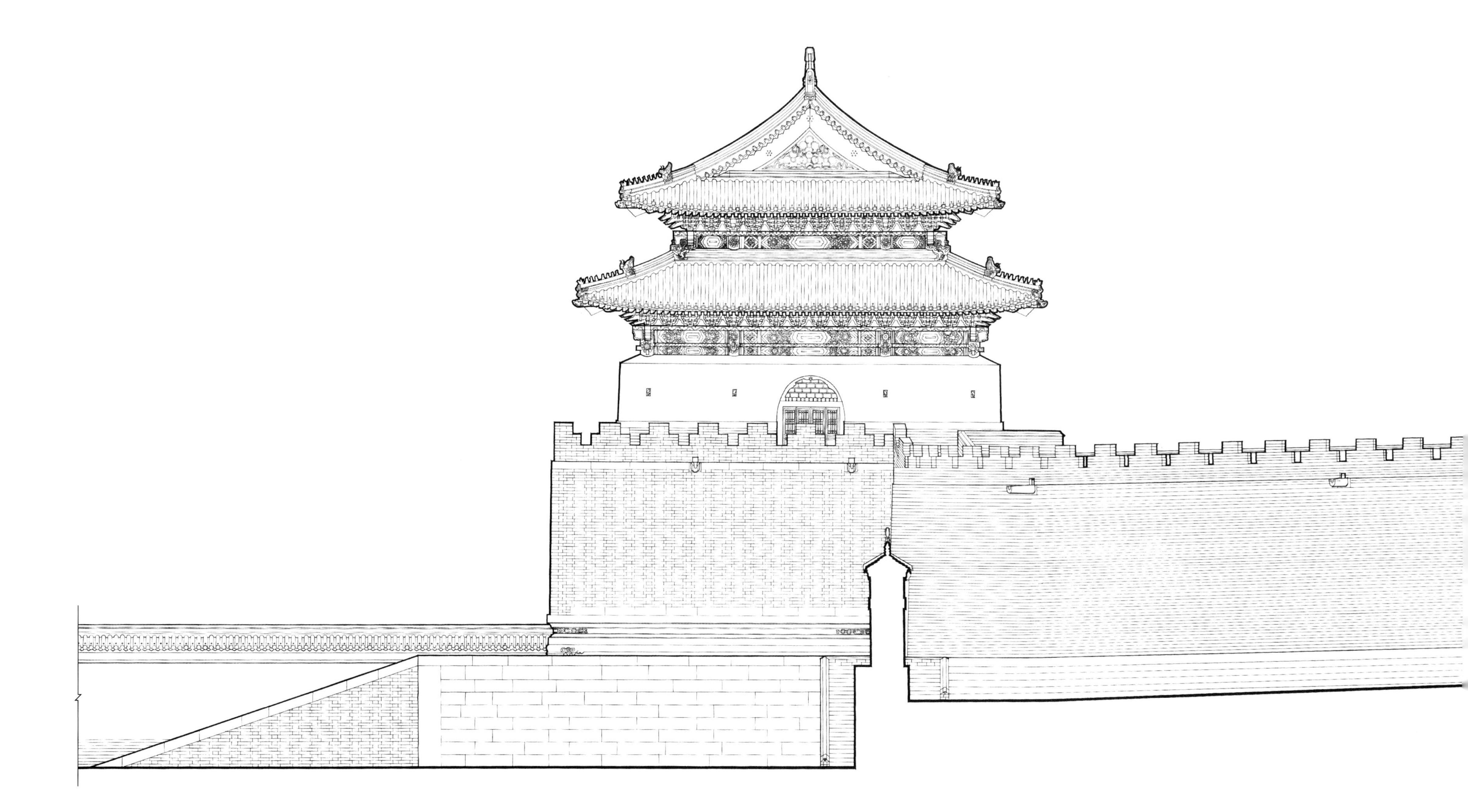

泰东陵
East Tomb of Peace (Tai Dongling)

1 石拱桥 Arched Stone Bridge
2 神厨库门 Gate in the Culinary Courtyard for Sacrifices
3 神厨 Kitchen in the Culinary Courtyard for Sacrifices
4 神库 Repository in the Culinary Courtyard for Sacrifices
5 宰牲亭 Ritual Abattoir in the Culinary Courtyard for Sacrifices
6 朝房 Reception Hall for Court Officials
7 值房 Guard House
8 隆恩门 Gate of Monumental Grace
9 焚帛炉 Sacrificial Burner
10 配殿 Side Hall
11 隆恩殿 Hall of Monumental Grace
12 琉璃花门 Gate with Glazed Roof Tiles
13 石台五供 Five Stone Ritual Vessels
14 方城明楼 Square Walled Terrace and Memorial Tower
15 宝城宝顶 Encircled Realm of Treasure and Tumulus

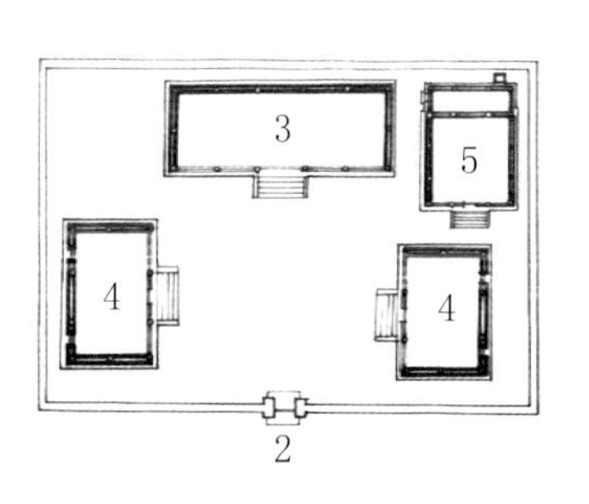

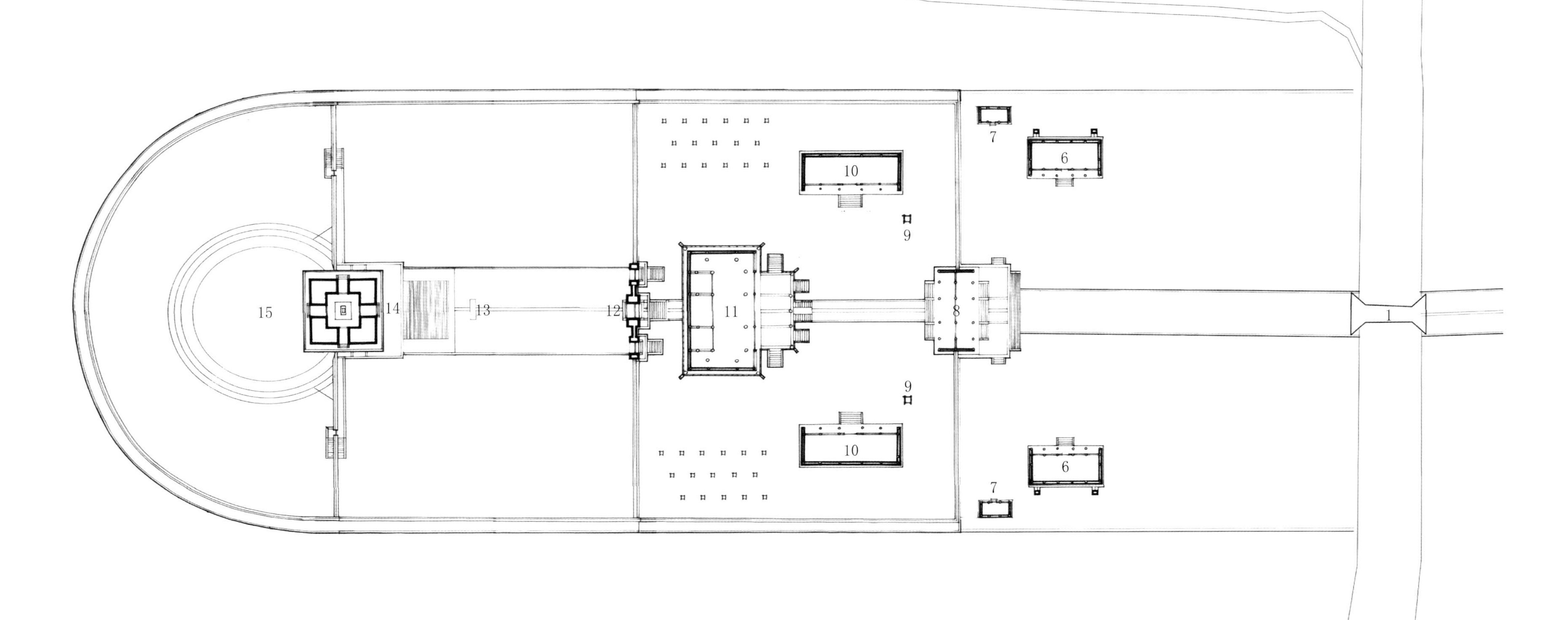

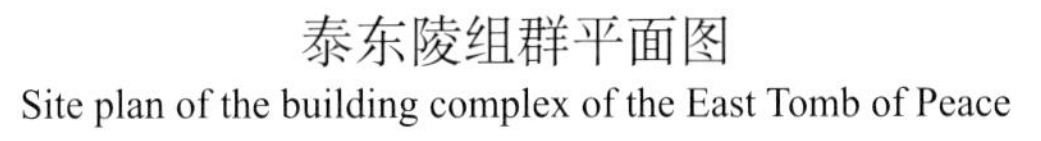
泰东陵组群平面图
Site plan of the building complex of the East Tomb of Peace

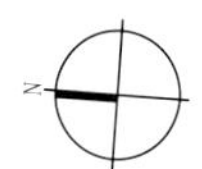

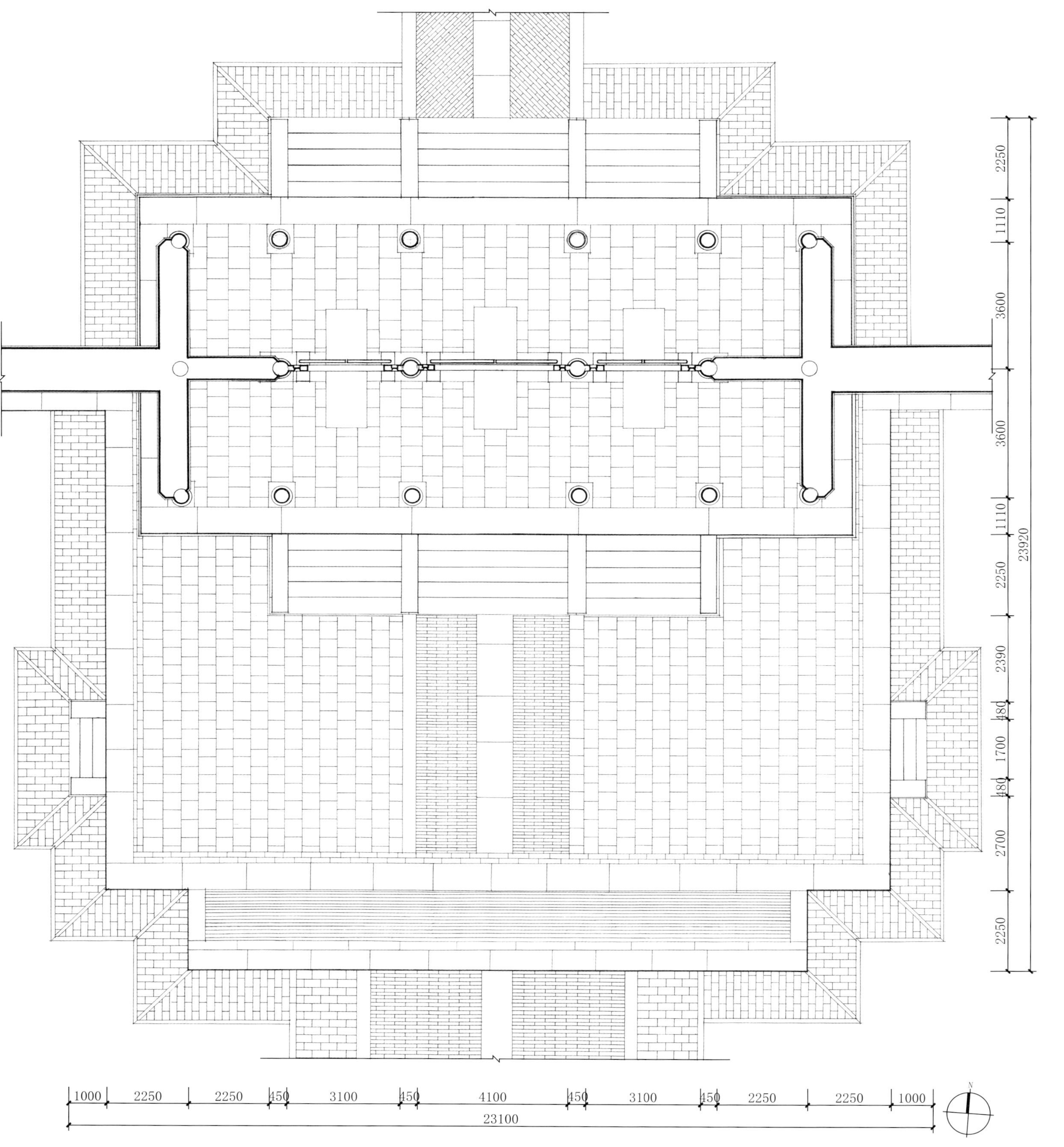

泰东陵隆恩门平面图
Plan of the Gate of Monumental Grace,East Tomb of Peace

泰东陵隆恩门正立面图

Front elevation of the Gate of Monumental Grace,East Tomb of Peace

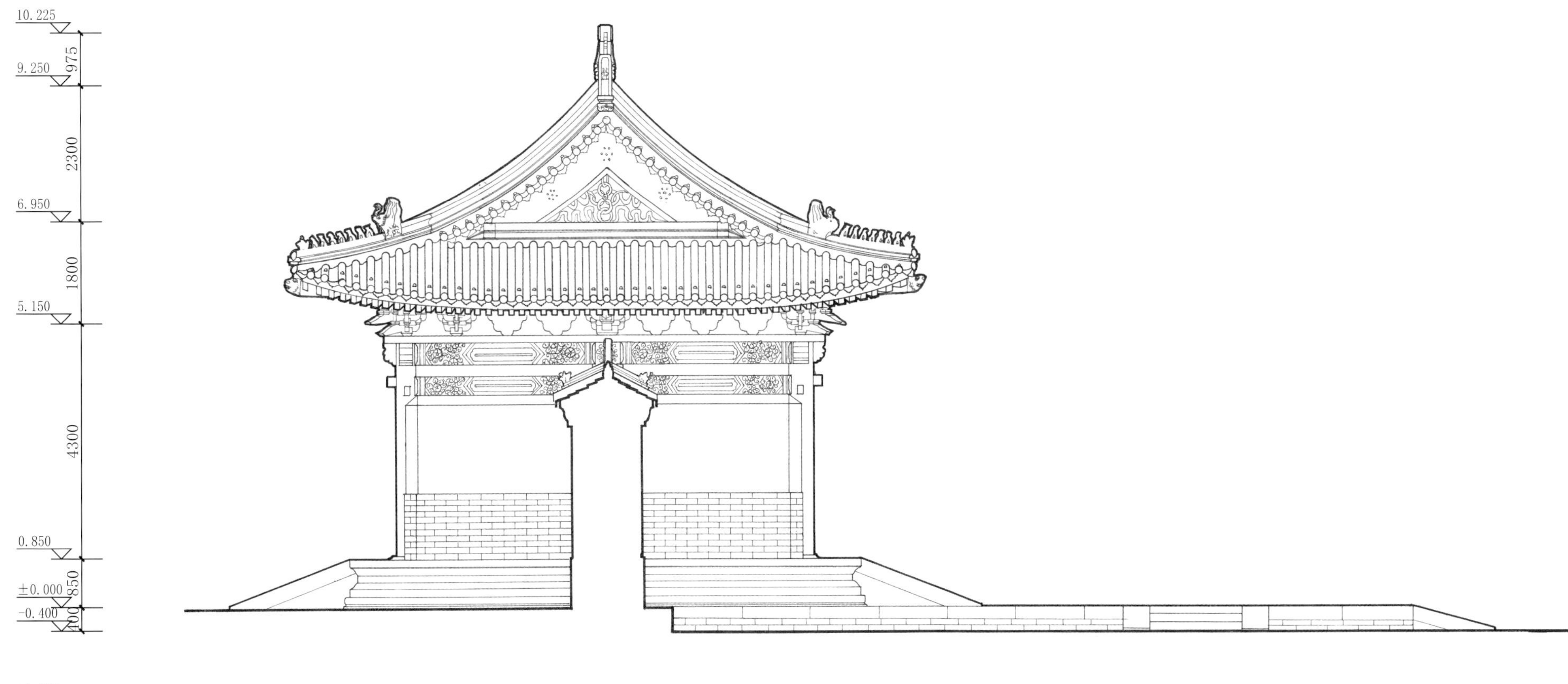

泰东陵隆恩门侧立面图

Side elevation of the Gate of Monumental Grace,East Tomb of Peace

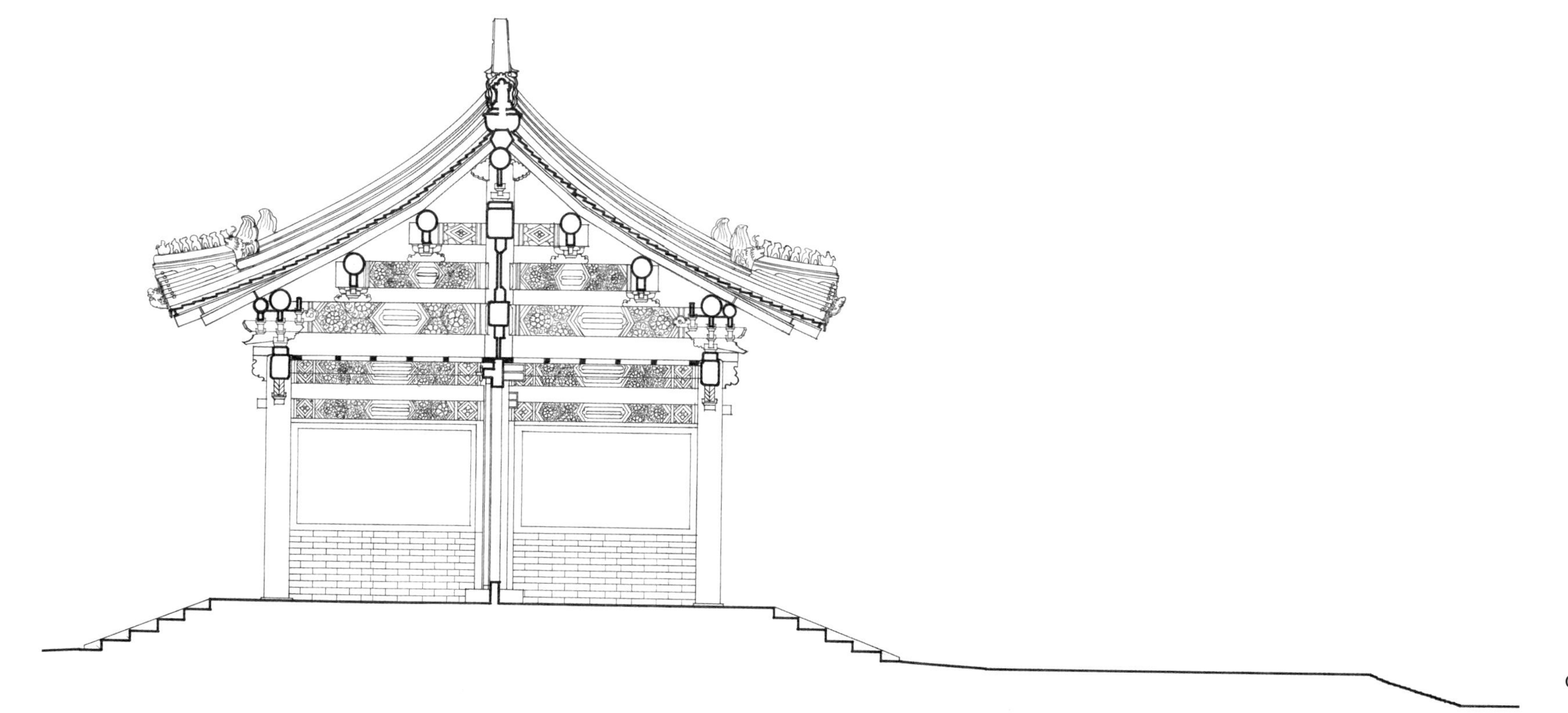

泰东陵隆恩门明间横剖面图

Cross section of the Central Chamber within the Gate of Monumental Grace,East Tomb of Peace

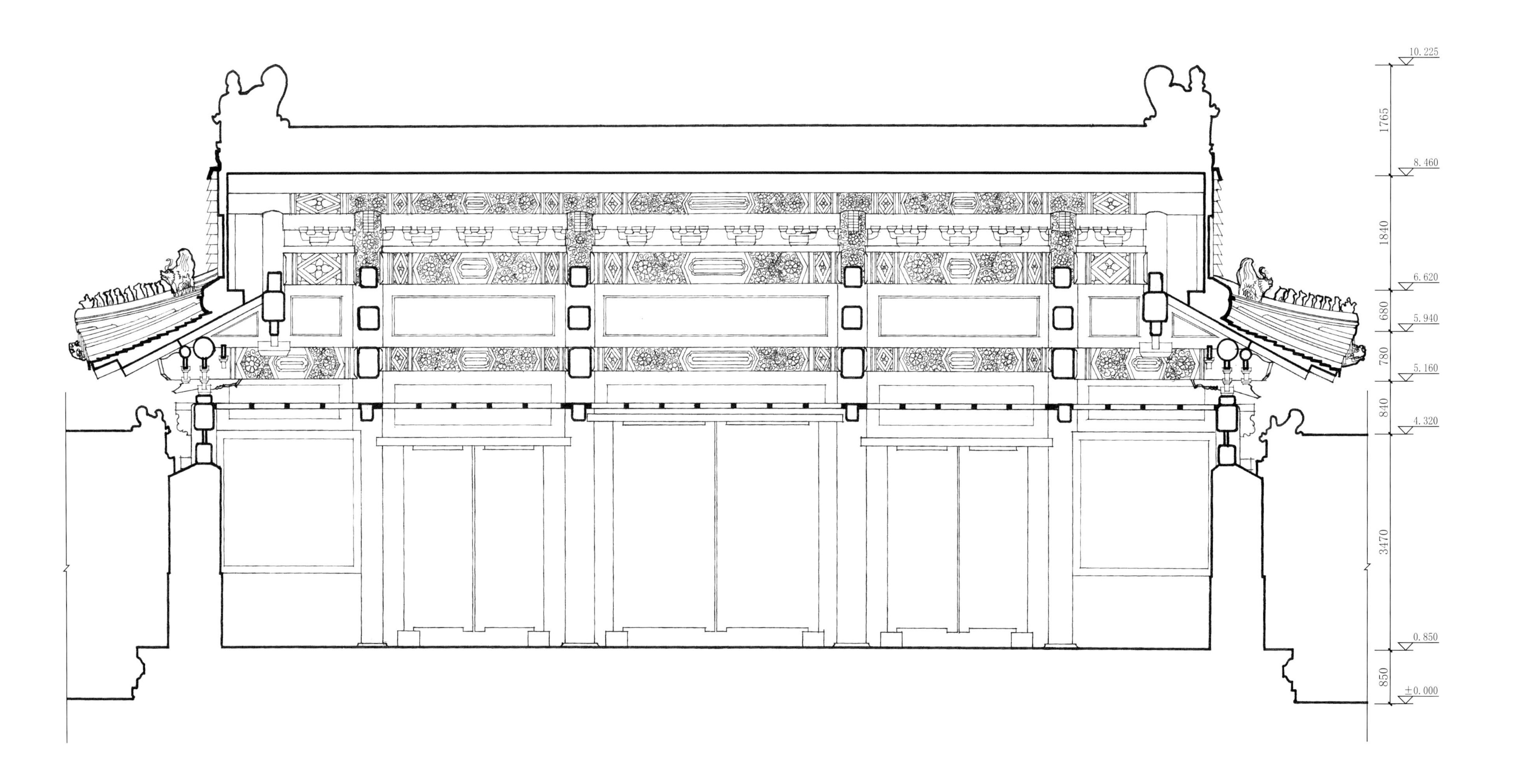

泰东陵隆恩门纵剖面图
Longitudinal section of the Gate of Monumental Grace,East Tomb of Peace

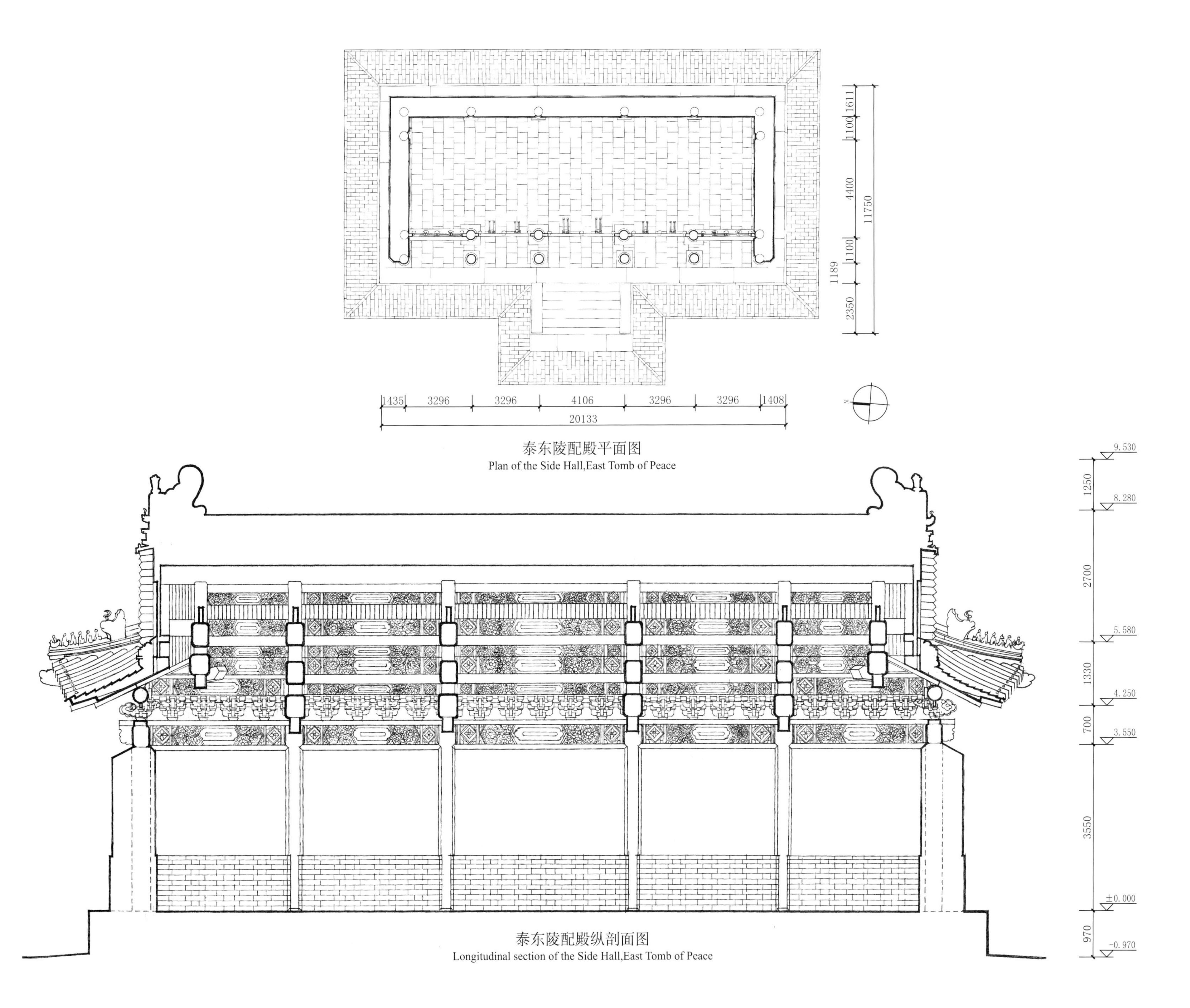

泰东陵配殿平面图

Plan of the Side Hall,East Tomb of Peace

泰东陵配殿纵剖面图

Longitudinal section of the Side Hall,East Tomb of Peace

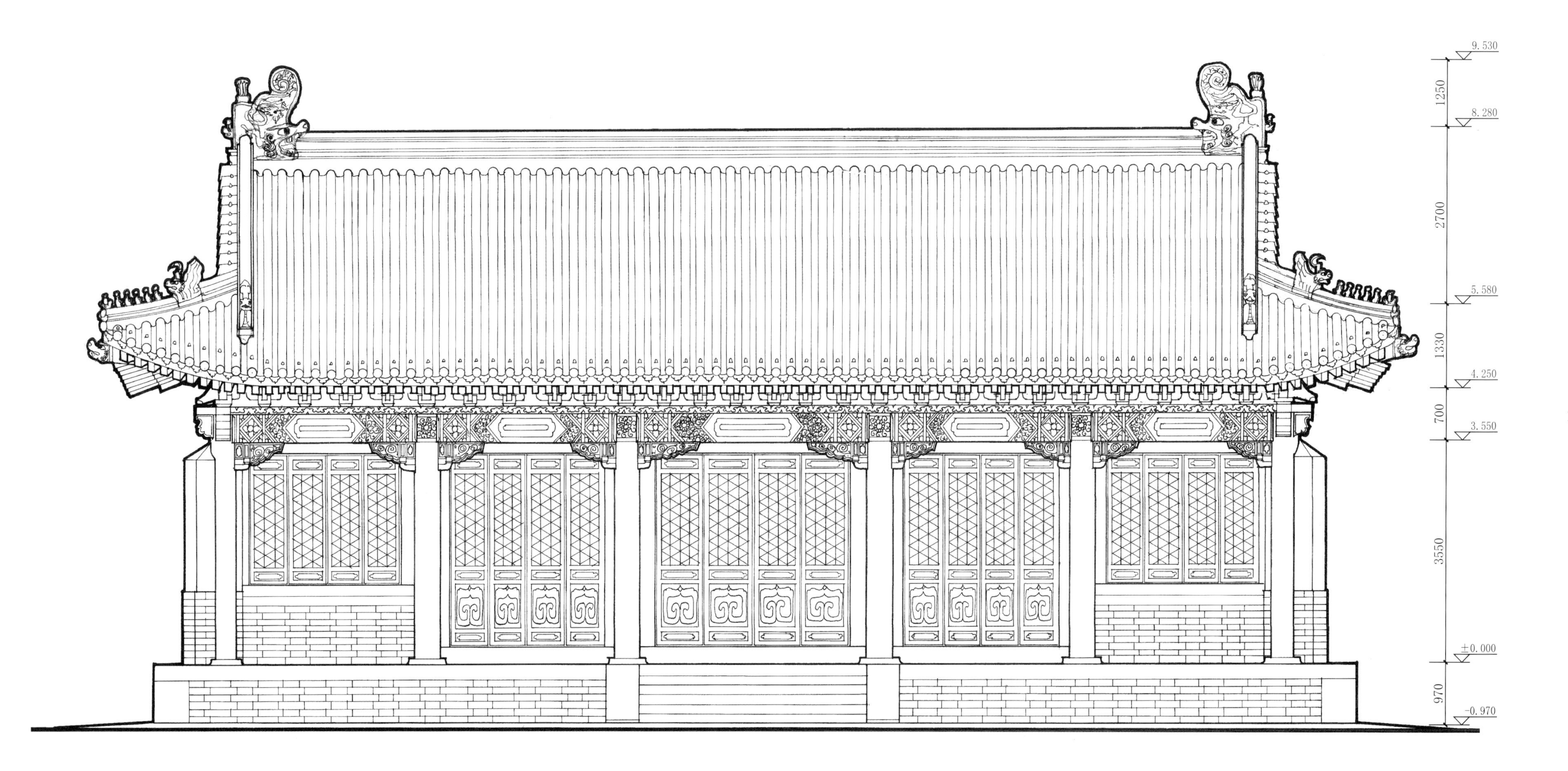

0 0.5 1 2m

泰东陵配殿正立面图

Front elevation of the Side Hall,East Tomb of Peace

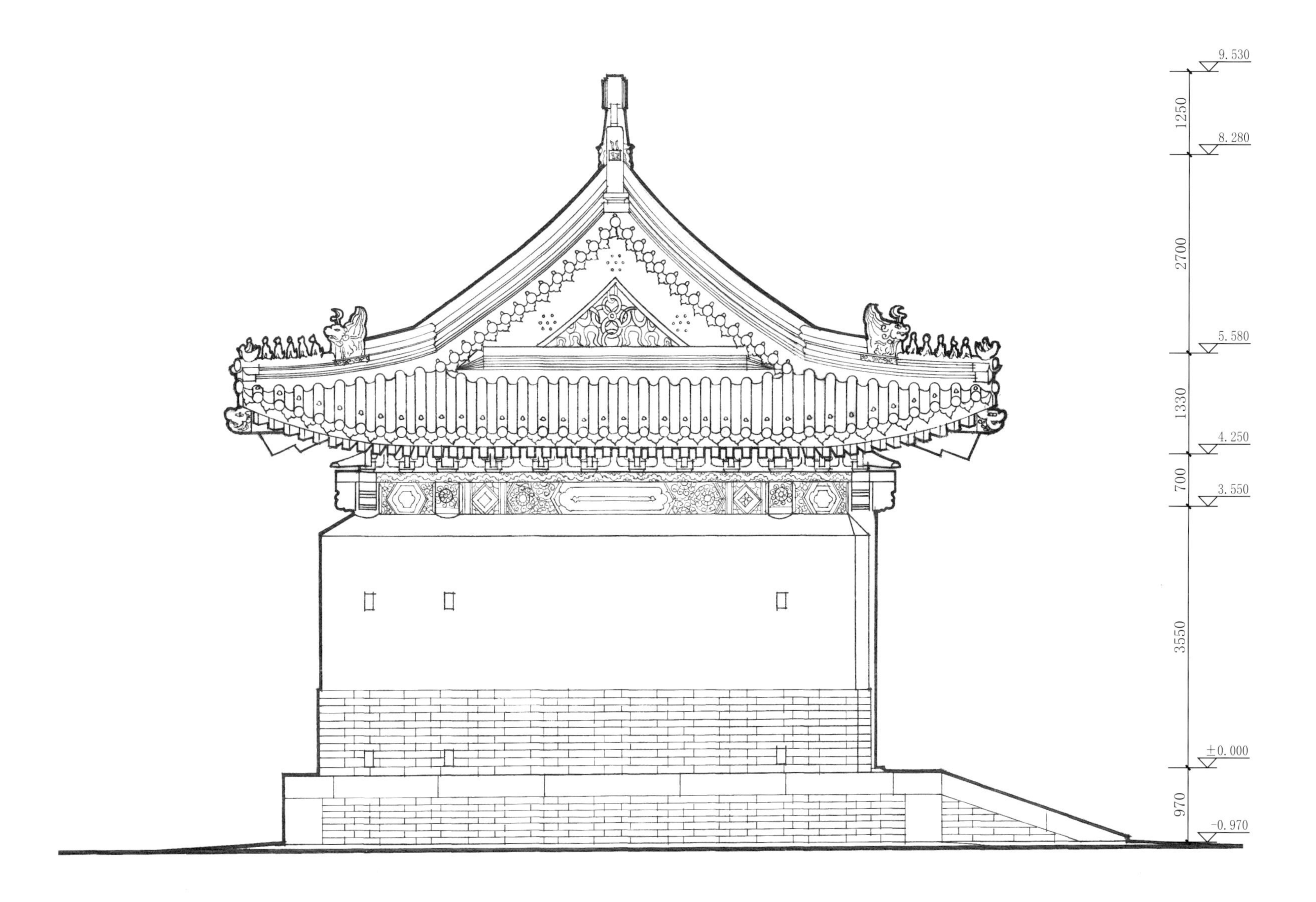

泰东陵配殿侧立面图
Side elevation of the Side Hall,East Tomb of Peace

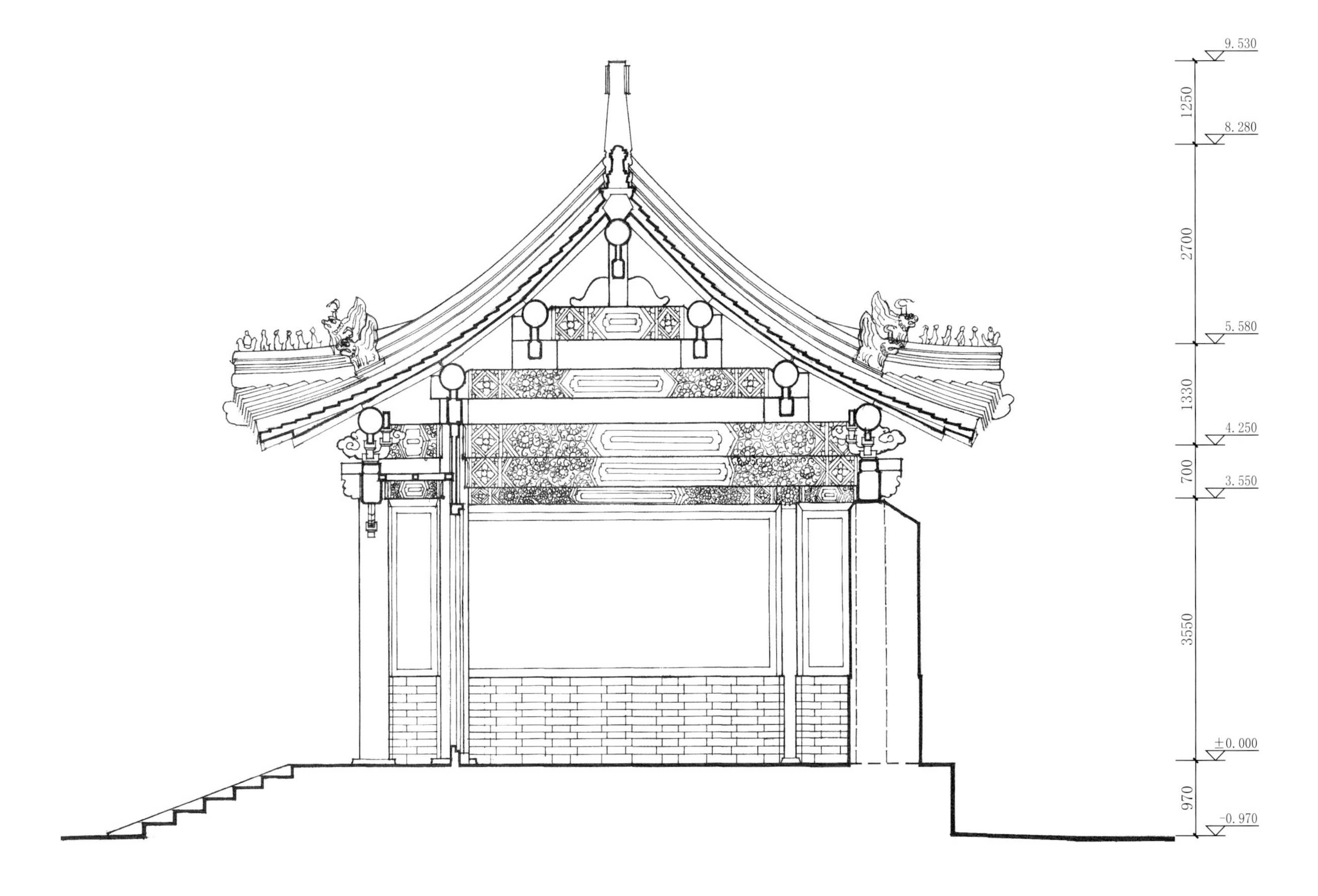

泰东陵配殿明间横剖面图

Cross section of the Central Chamber within the Side Hall,East Tomb of Peace

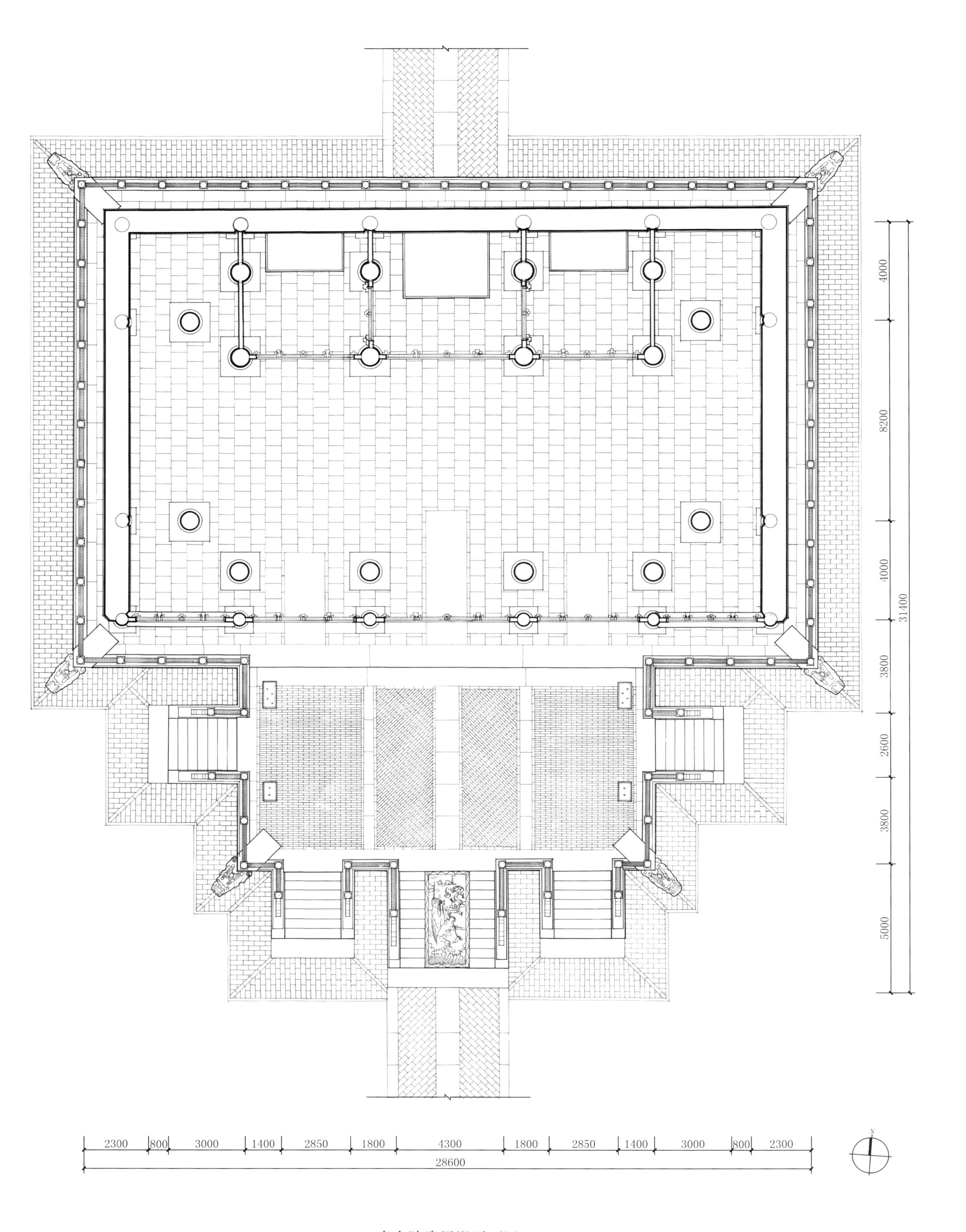

泰东陵隆恩殿平面图
Plan of the Hall of Monumental Grace,East Tomb of Peace

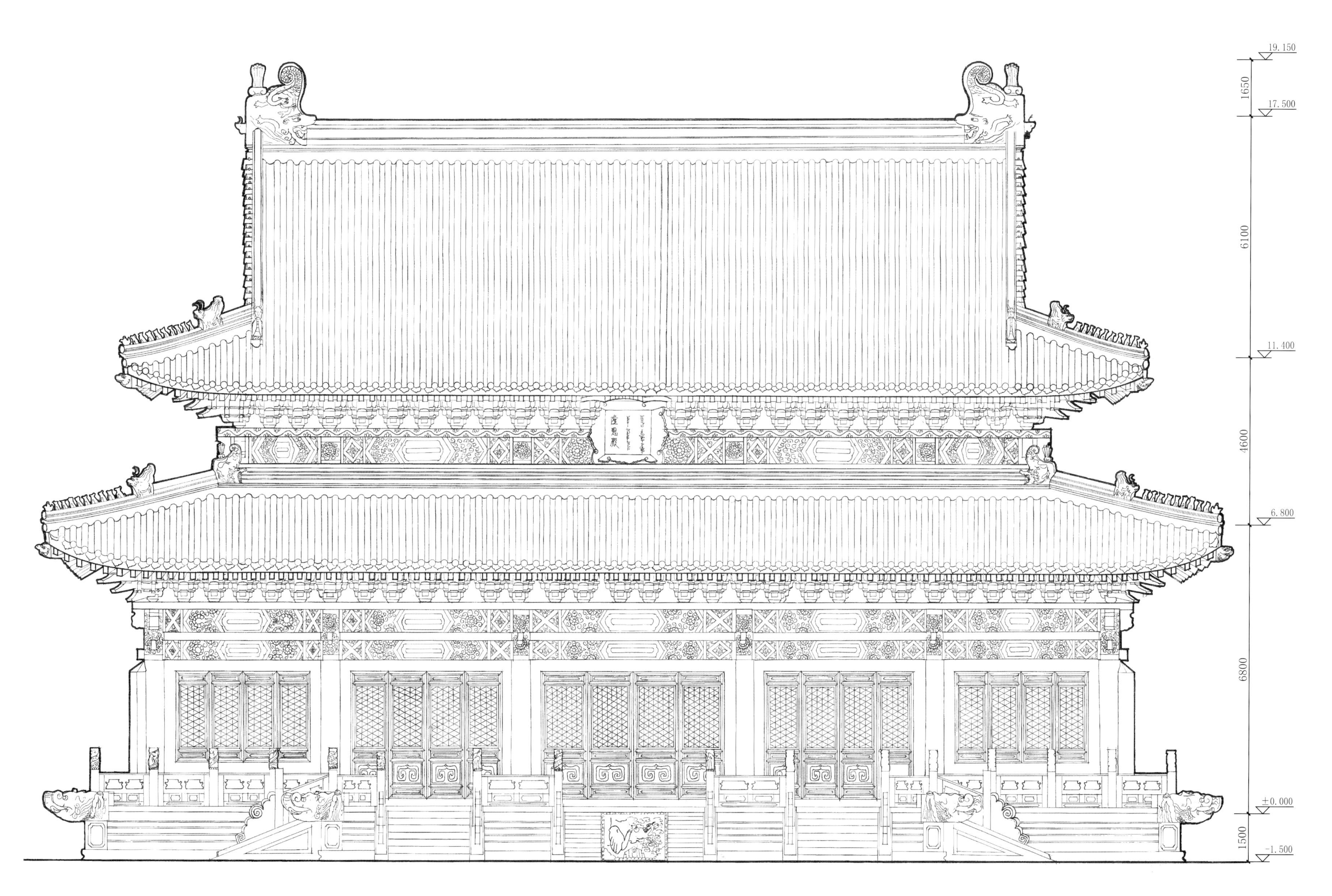

泰东陵隆恩殿正立面图
Front elevation of the Hall of Monumental Grace,East Tomb of Peace

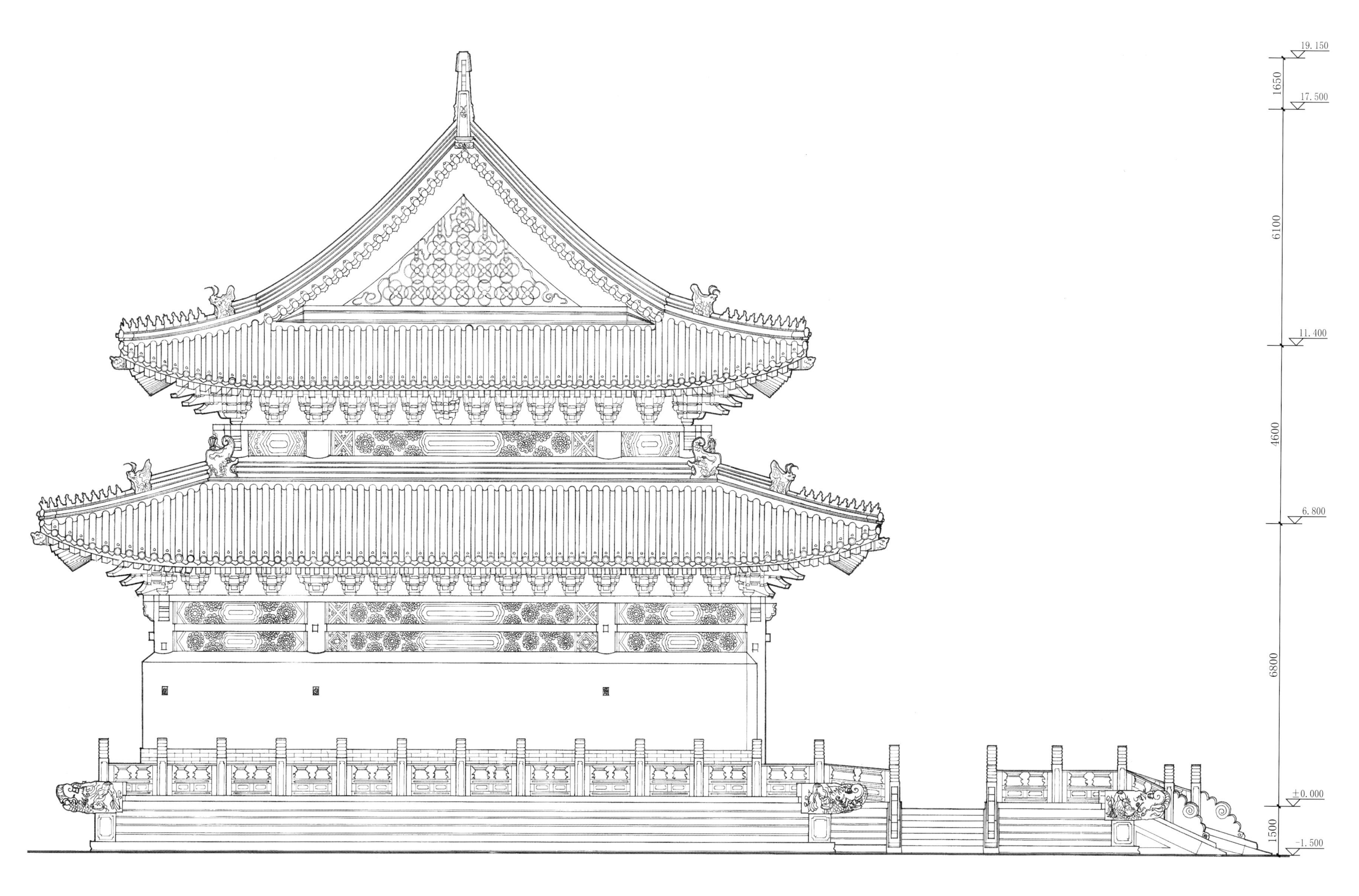

泰东陵隆恩殿侧立面图
Side elevation of the Hall of Monumental Grace,East Tomb of Peace

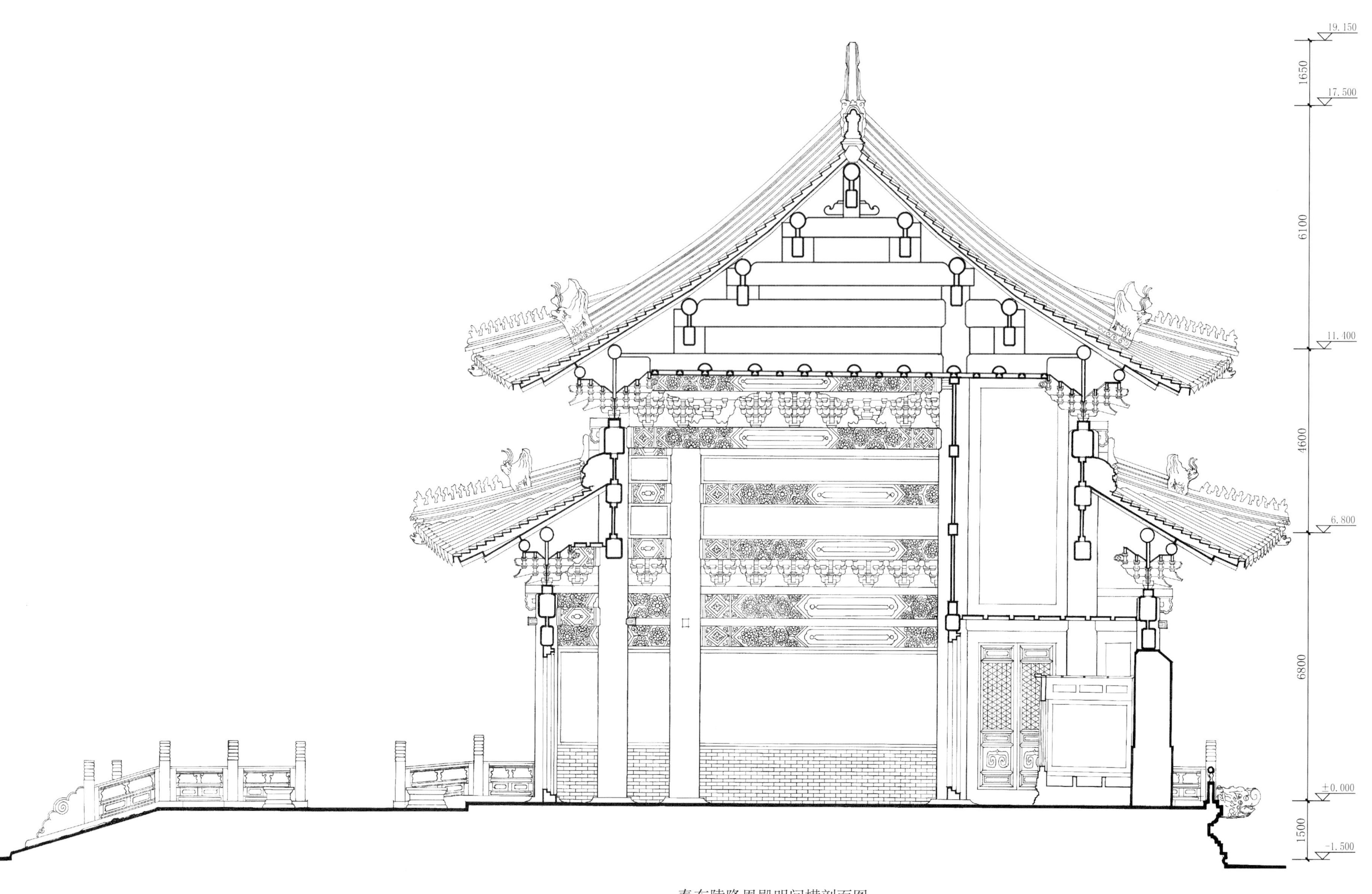

泰东陵隆恩殿明间横剖面图

Cross section of the Central Chamber within the Hall of Monumental Grace,East Tomb of Peace

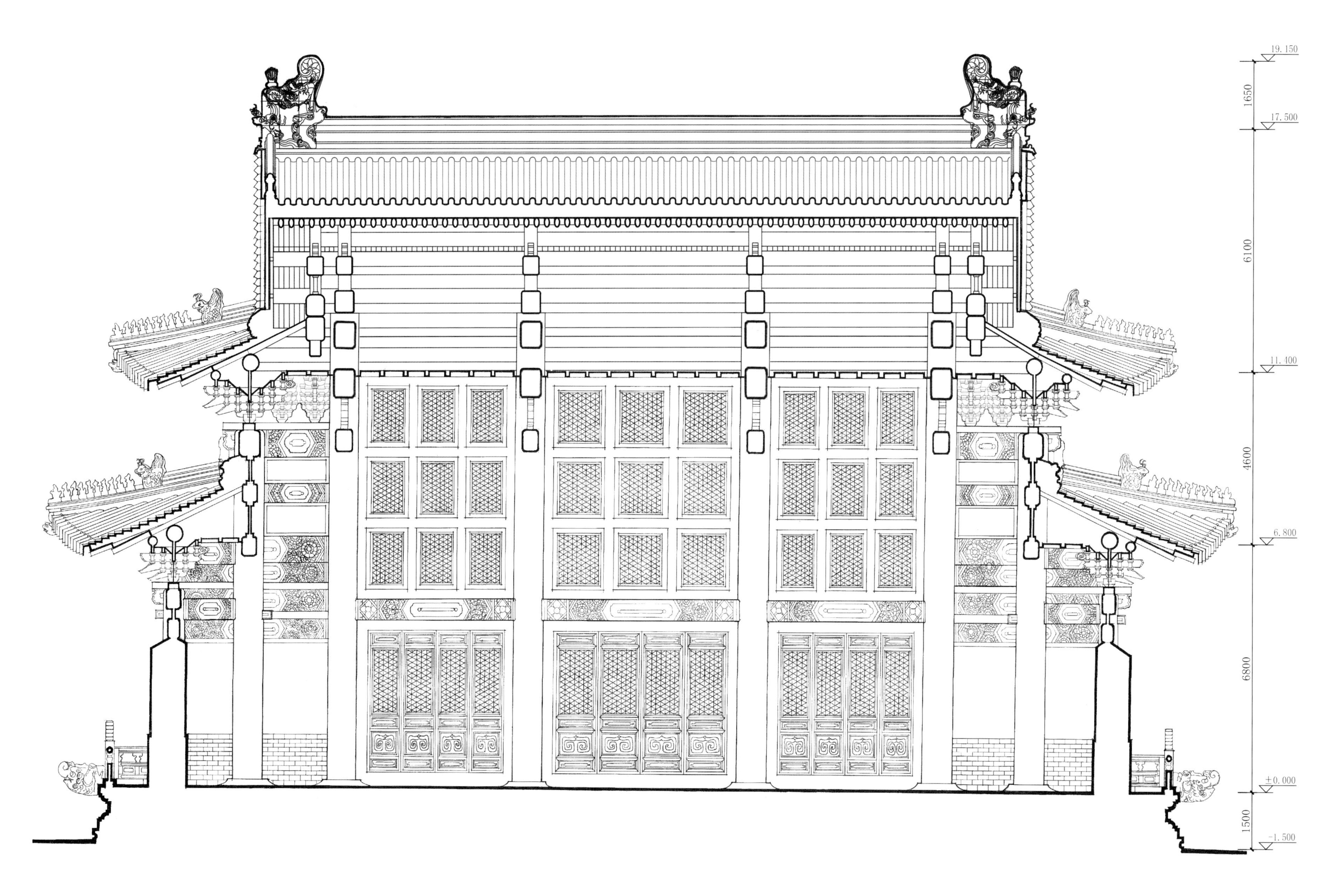

泰东陵隆恩殿纵剖面图

Longitudinal section of the Hall of Monumental Grace,East Tomb of Peace

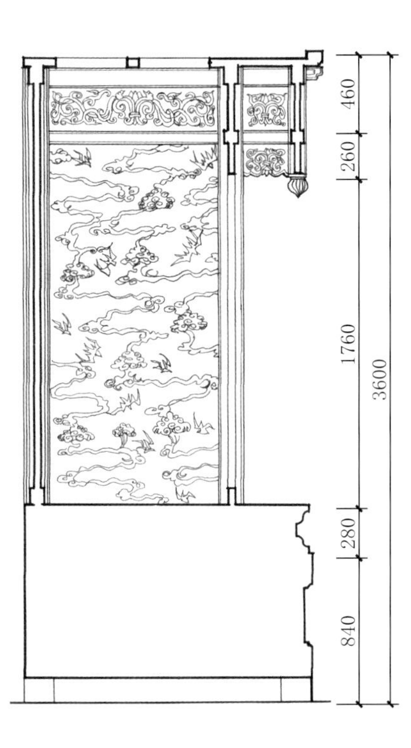

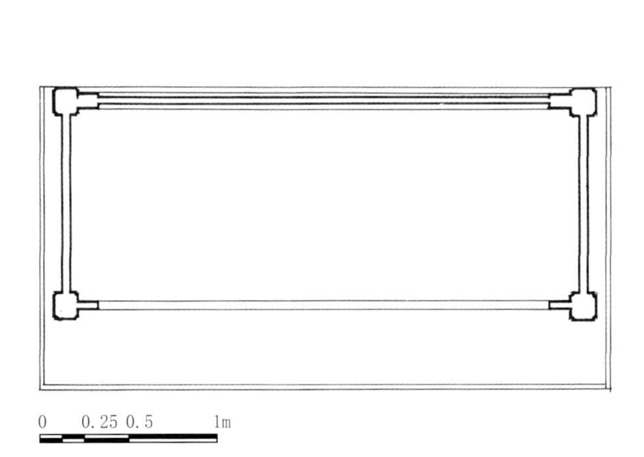

0 0.25 0.5 1m

泰东陵隆恩殿佛龛宝床大样图（一）

Detailed drawing of the Buddhist cabinet and the Precious bed of the Hall of Monumental Grace,East Tomb of Peace(1)

0 0.25 0.5 1m

泰东陵隆恩殿佛龛宝床大样图（二）

Detailed drawing of the Buddhist cabinet and the Precious bed of the Hall of Monumental Grace,East Tomb of Peace(2)

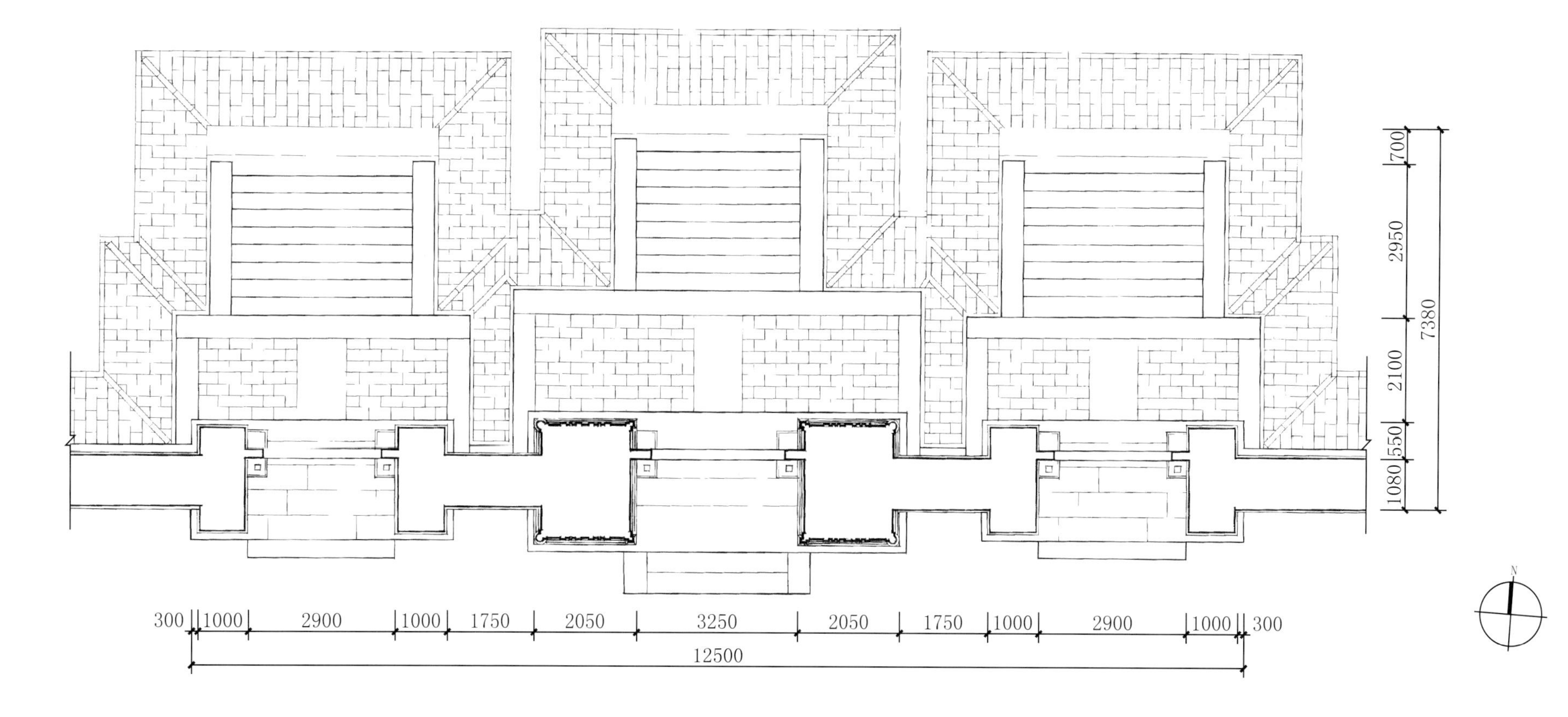

泰东陵琉璃花门平面图

Plan of the Gate with Glazed Roof Tiles,East Tomb of Peace

泰东陵琉璃花门正立面图
Front elevation of the Gate with Glazed Roof Tiles,East Tomb of Peace

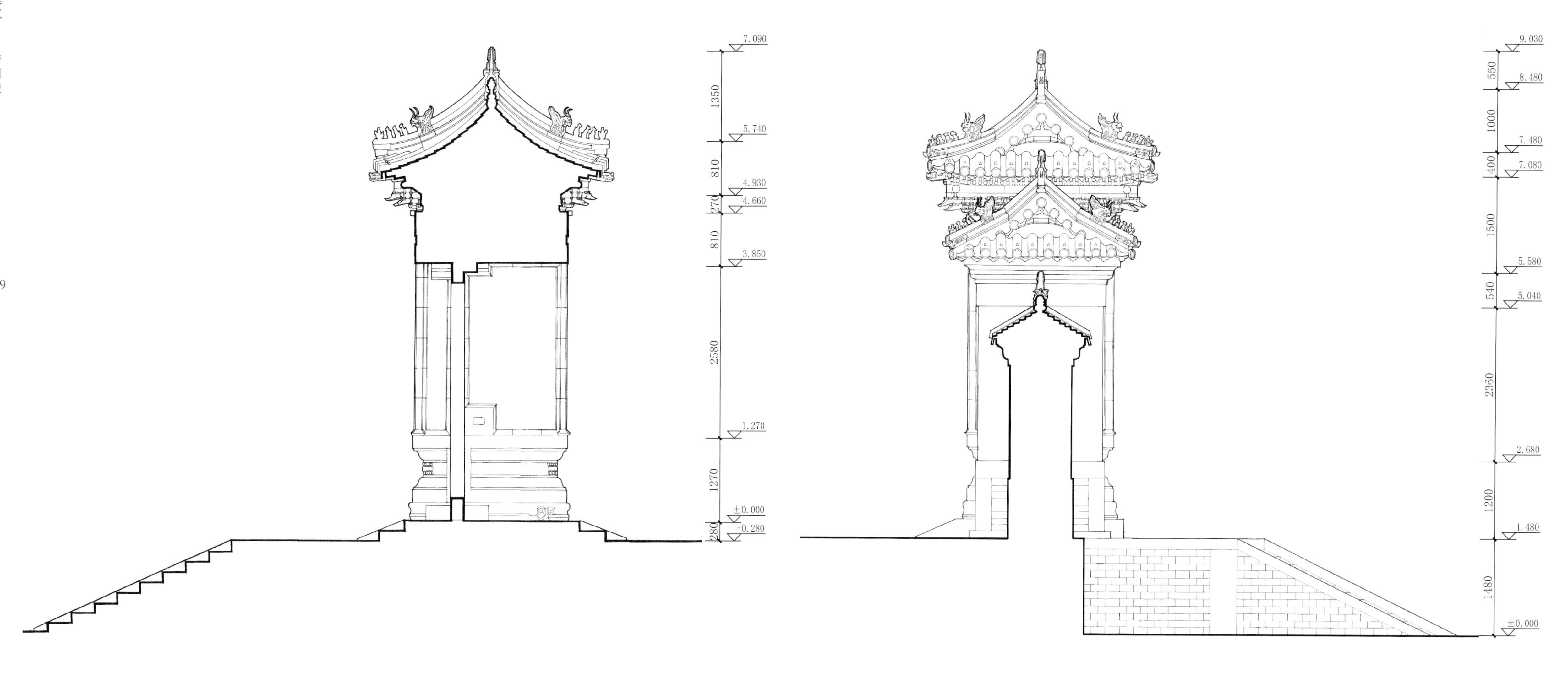

泰东陵琉璃花门横剖面图
Cross section of the Gate with Glazed Roof Tiles,East Tomb of Peace

泰东陵琉璃花门侧立面图
Side elevation of the Gate with Glazed Roof Tiles,East Tomb of Peace

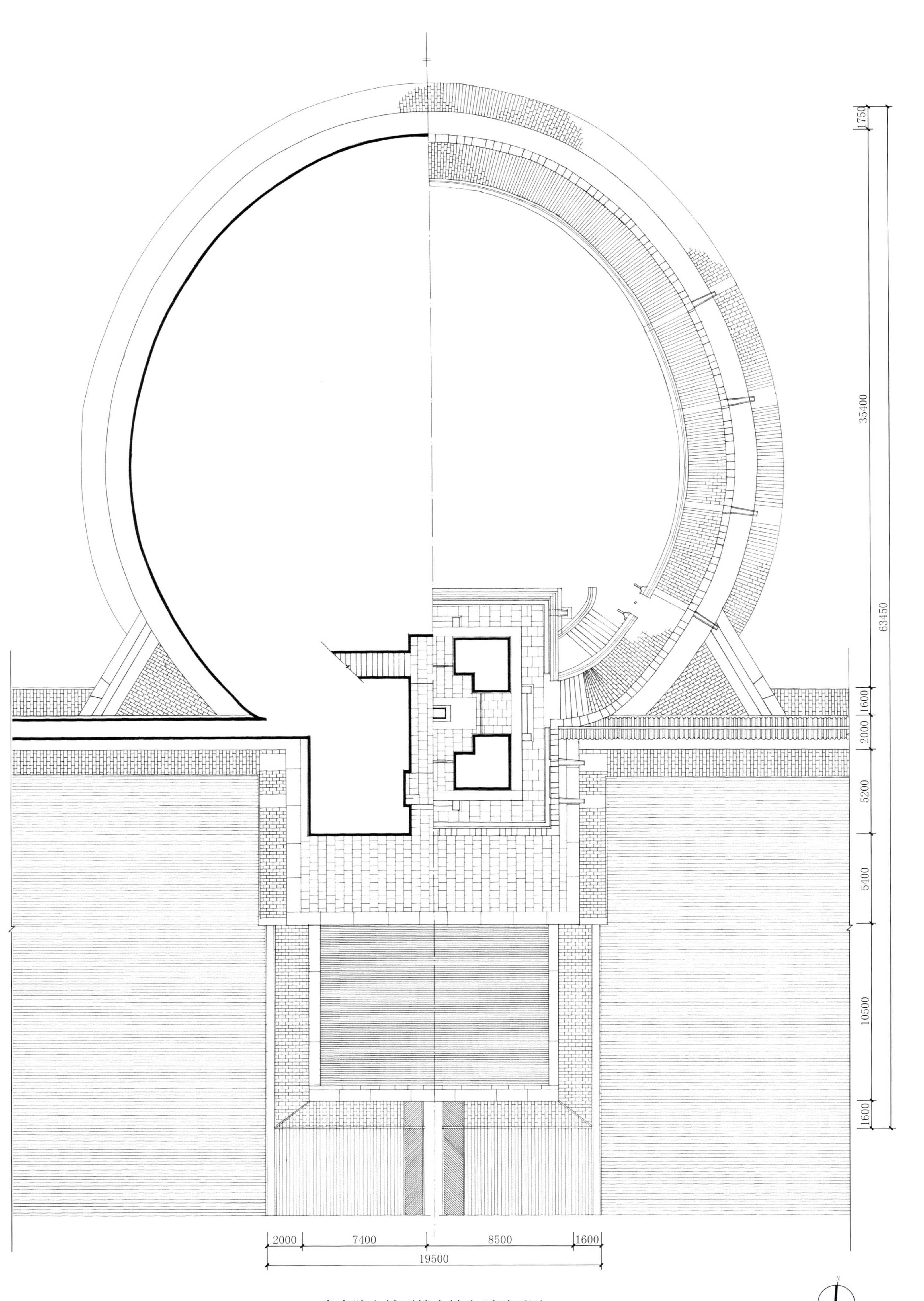

泰东陵方城明楼宝城宝顶平面图

Plan of the Square Walled Terrace and the Memorial Tower as well as the Encircled Realm of Treasure and the Tumulus,East Tomb of Peace

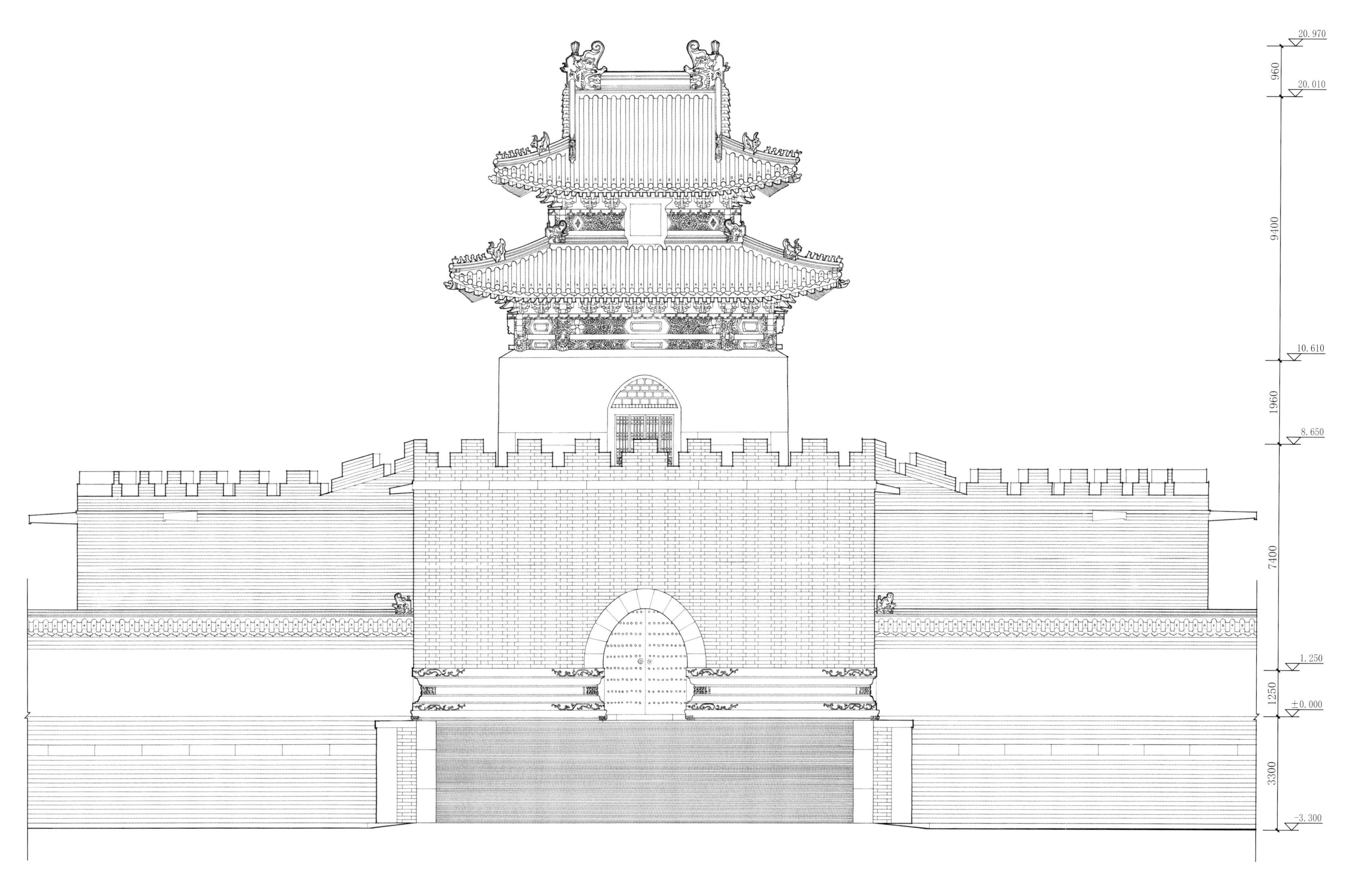

泰东陵方城明楼宝城宝顶正立面图

Front elevation of the Square Walled Terrace and the Memorial Tower as well as the Encircled Realm of Treasure and the Tumulus,East Tomb of Peace

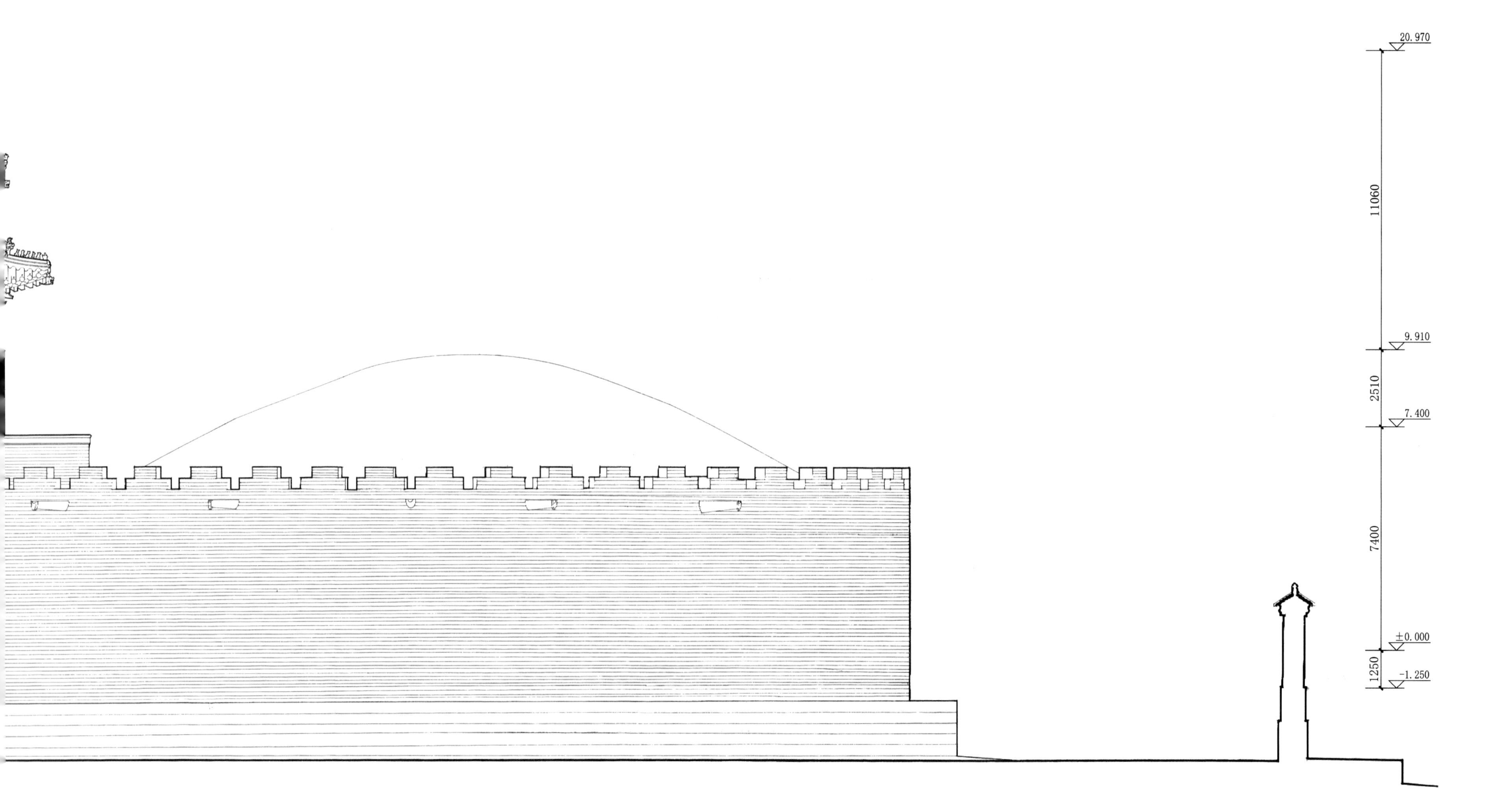
20.970
11060
9.910
2510
7.400
7400
±0.000
1250
-1.250

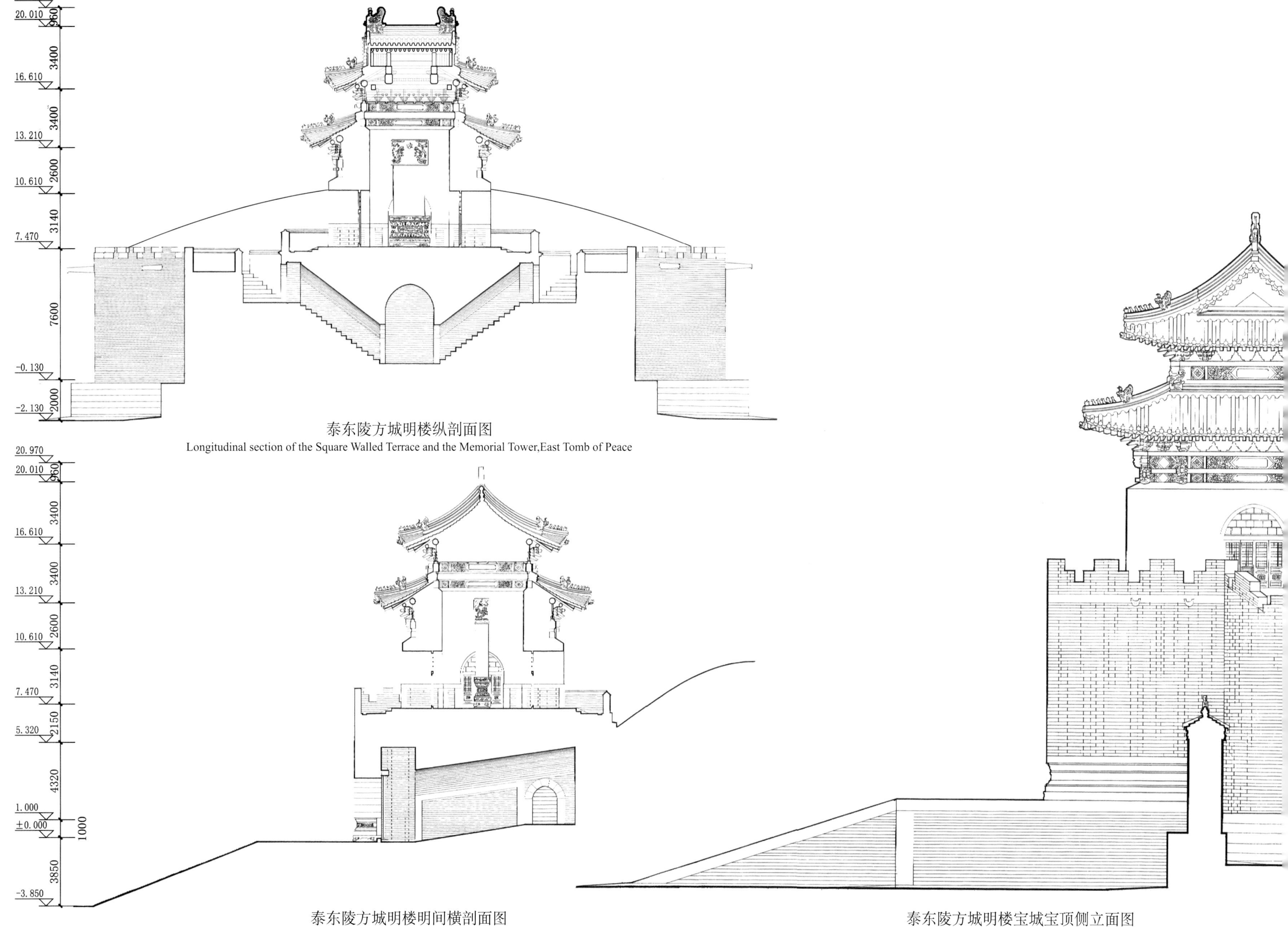

泰东陵方城明楼纵剖面图
Longitudinal section of the Square Walled Terrace and the Memorial Tower,East Tomb of Peace

泰东陵方城明楼明间横剖面图
Cross section of the Central Chamber within the Square Walled Terrace and the Memorial Tower,East Tomb of Peace

泰东陵方城明楼宝城宝顶侧立面图
Side elevation of the Square Walled Terrace and the Memorial Tower as well as the Encircled Realm of Treasure a
the Tumulus,East Tomb of Peace

泰陵妃园寝

Imperial Consorts' Tombs affiliated with the Tomb of Peace (Tailing Feiyuanqin)

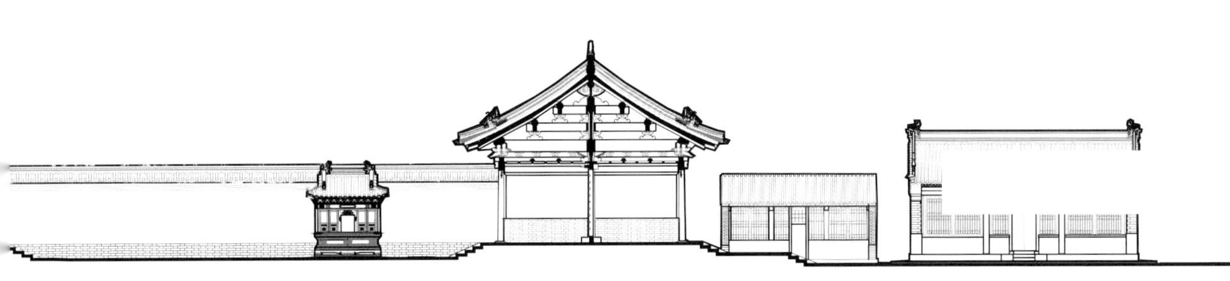

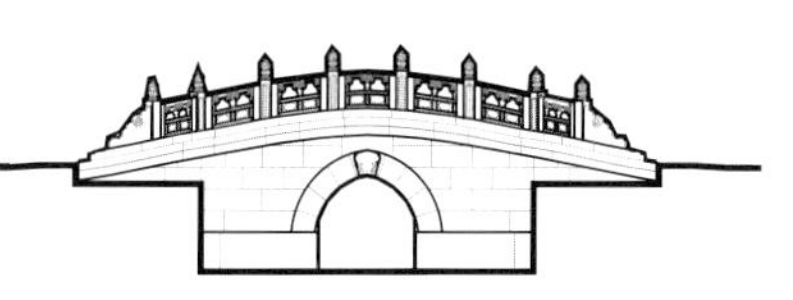

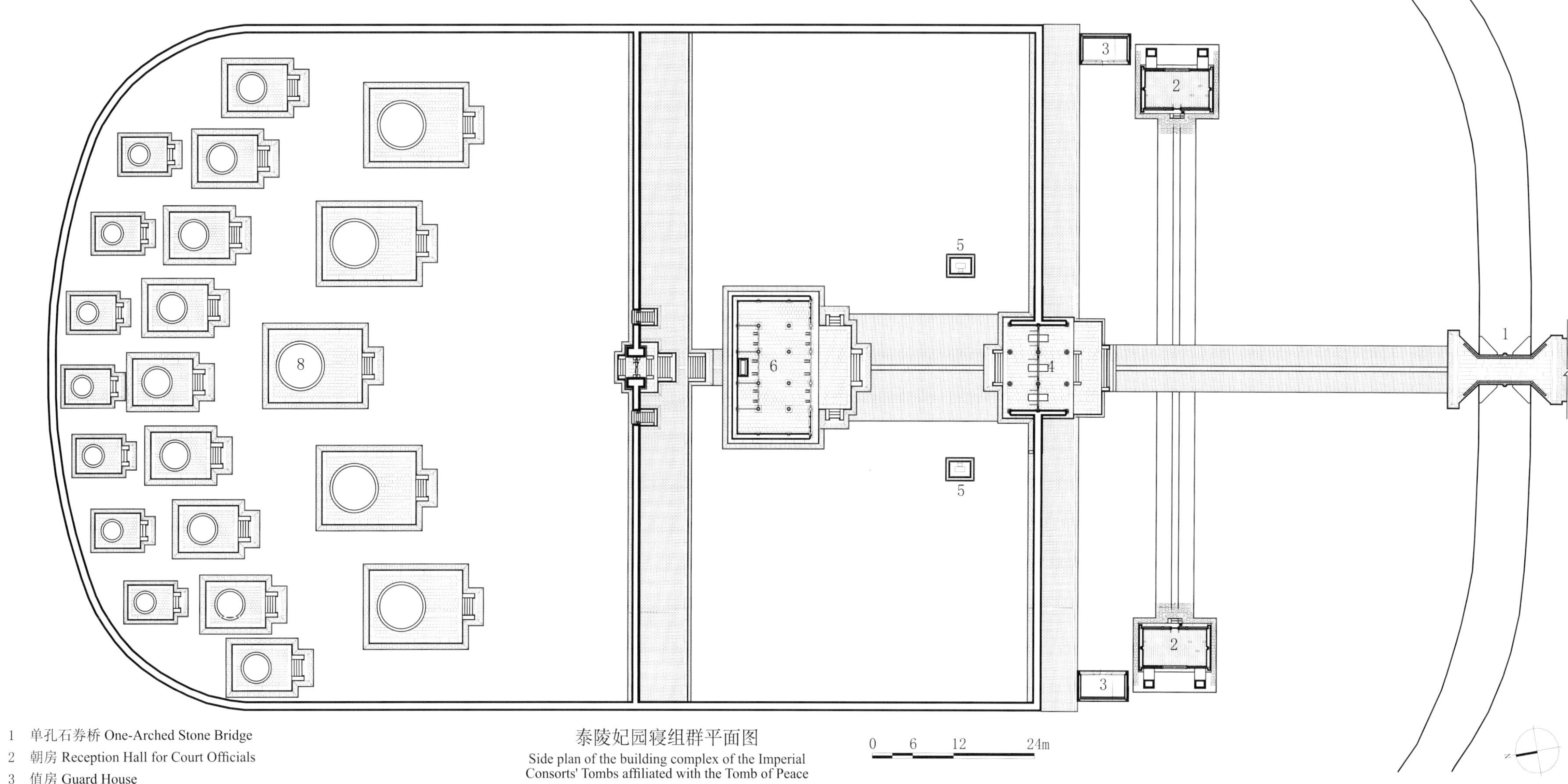

1 单孔石券桥 One-Arched Stone Bridge
2 朝房 Reception Hall for Court Officials
3 值房 Guard House
4 宫门 Main Gate
5 焚帛炉 Sacrificial Burner
6 享殿 Hall of Ritual Sacrifice
7 琉璃花门 Gate with Glazed Roof Tiles
8 宝顶 Tumulus

泰陵妃园寝组群平面图
Side plan of the building complex of the Imperial Consorts' Tombs affiliated with the Tomb of Peace

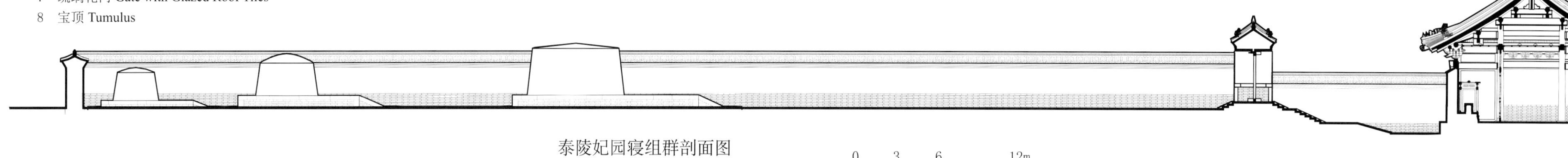

泰陵妃园寝组群剖面图
Site section of the building complex of the Imperial Consorts'Tombs affiliated with the Tomb of Peace

昌　陵

Tomb of Good Fortune (Changling)

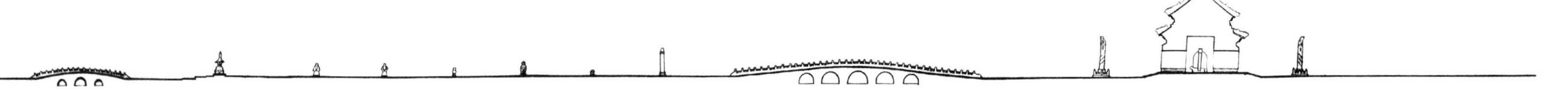

1　华表 Ceremonial Column
2　圣德神功碑亭 Pavilion for the Stela of Sage Virtue and Divine Merit
3　五孔石券桥 Five-Arched Stone Bridge
4　望柱 Ornamental Column
5　石狮 Lion among the Stone Statues
6　石象 Elephant among the Stone Statues
7　石马 Horse among the Stone Statues
8　武官 Martial Official among the Stone Statues
9　文官 Literary Official among the Stone Statues
10　龙凤门 Dragon and Phoenix Gate
11　三孔石券桥 Three-Arched Stone Bridge
12　三路三孔桥 Three-Way Three-Arch Bridges
13　下马牌 Stela Marking the Place for Dismounting from One's Horse
14　神道碑亭 Pavilion for the Stela on the Spirit Way
15　石平桥 Flat Stone Bridge
16　神厨库门 Gate in the Culinary Courtyard for Sacrifices
17　神厨 Kitchen in the Culinary Courtyard for Sacrifices
18　神库 Repository in the Culinary Courtyard for Sacrifices
19　宰牲亭 Ritual Abattoir in the Culinary Courtyard for Sacrifices
20　井亭 Pavilion for the Well
21　朝房 Reception Hall for Court Officials
22　值房 Guard House
23　隆恩门 Gate of Monumental Grace
24　焚帛炉 Sacrificial Burner
25　配殿 Side Hall
26　隆恩殿 Hall of Monumental Grace
27　香炉 Incense Burner
28　琉璃花门 Gate with Glazed Roof Tiles
29　二柱门 Gate with Two Columns
30　石台五供 Five Stone Ritual Vessels
31　方城明楼 Square Walled Terrace and Memorial Tower
32　宝城宝顶 Encircled Realm of Treasure and Tumulus

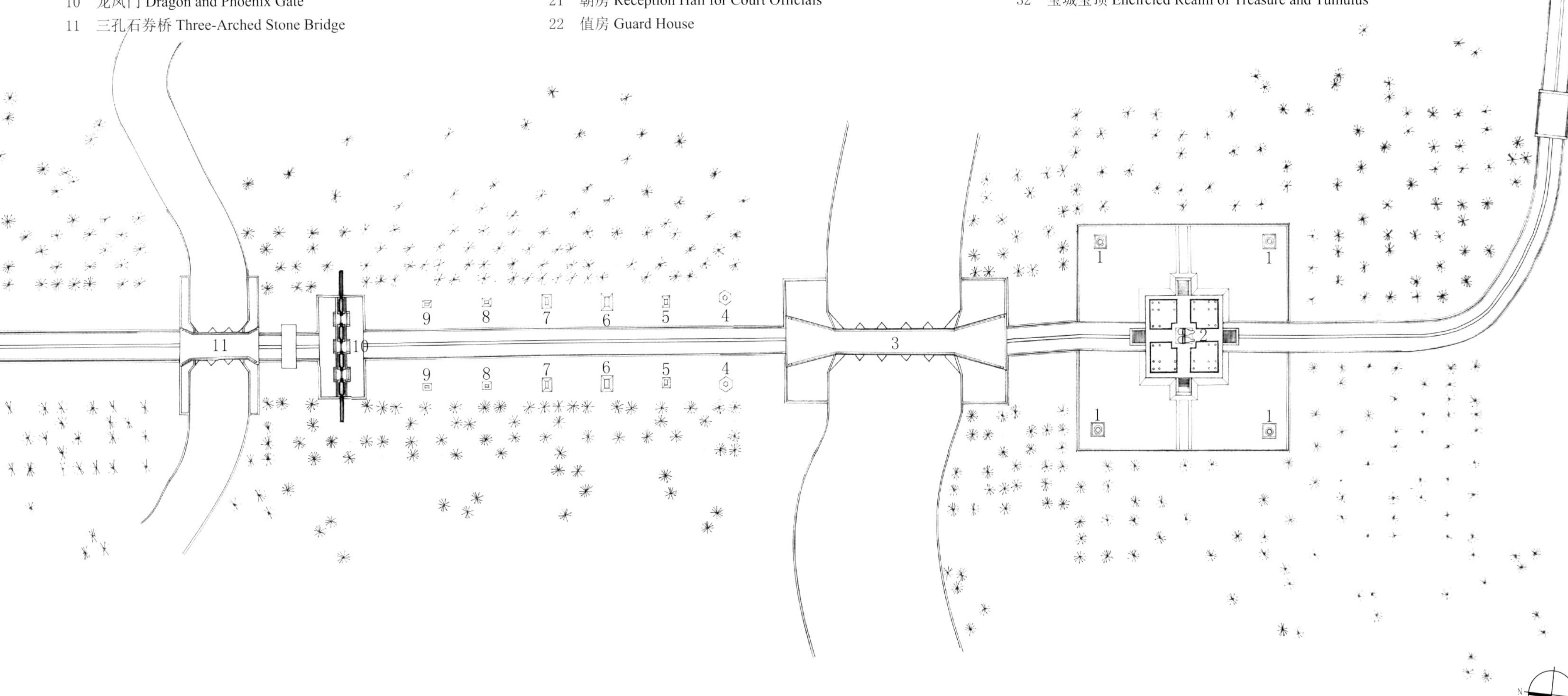

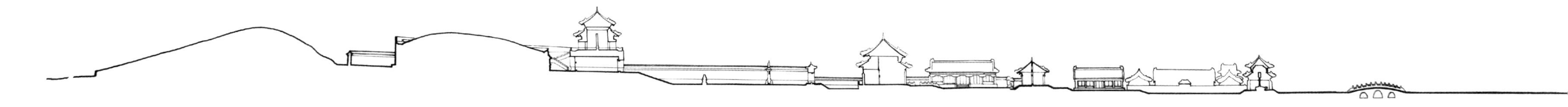

昌陵组群剖面图
Site section of the building complex of the Tomb of Good Fortune

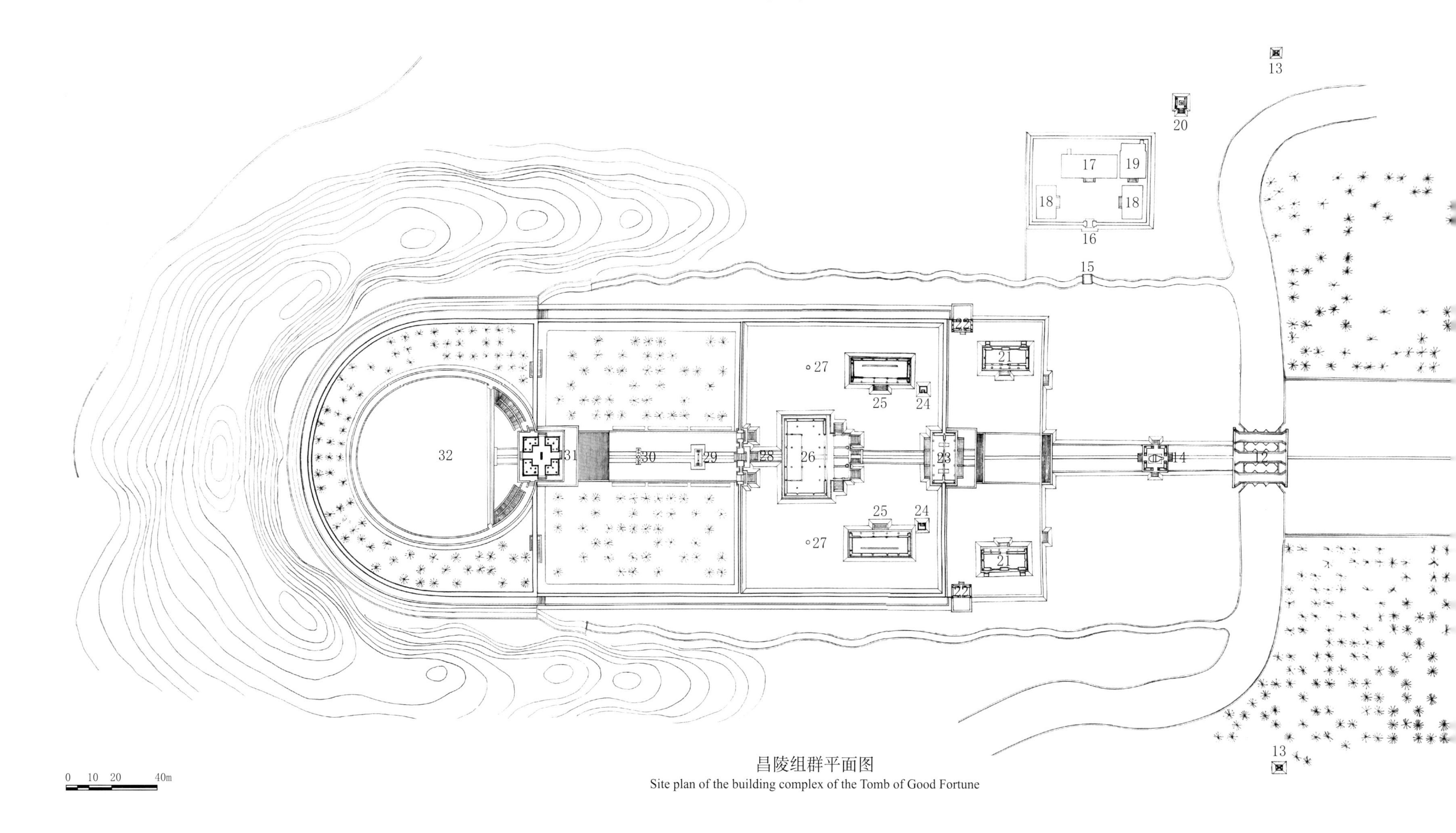

昌陵组群平面图
Site plan of the building complex of the Tomb of Good Fortune

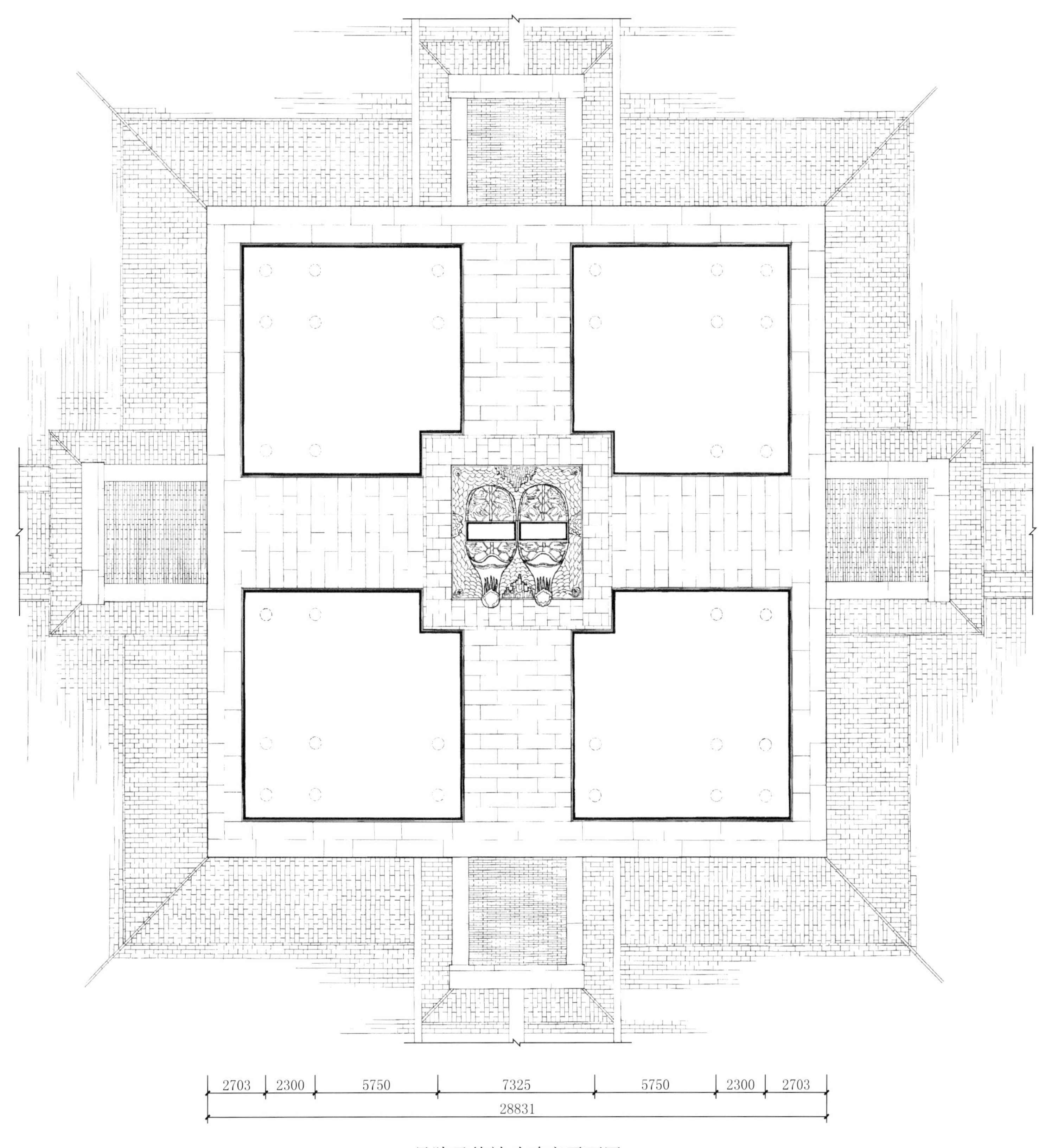

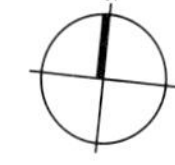

昌陵圣德神功碑亭平面图

Plan of the Pavilion for the Stela of Divine Merit and Sagely Virtue,Tomb of Good Fortune

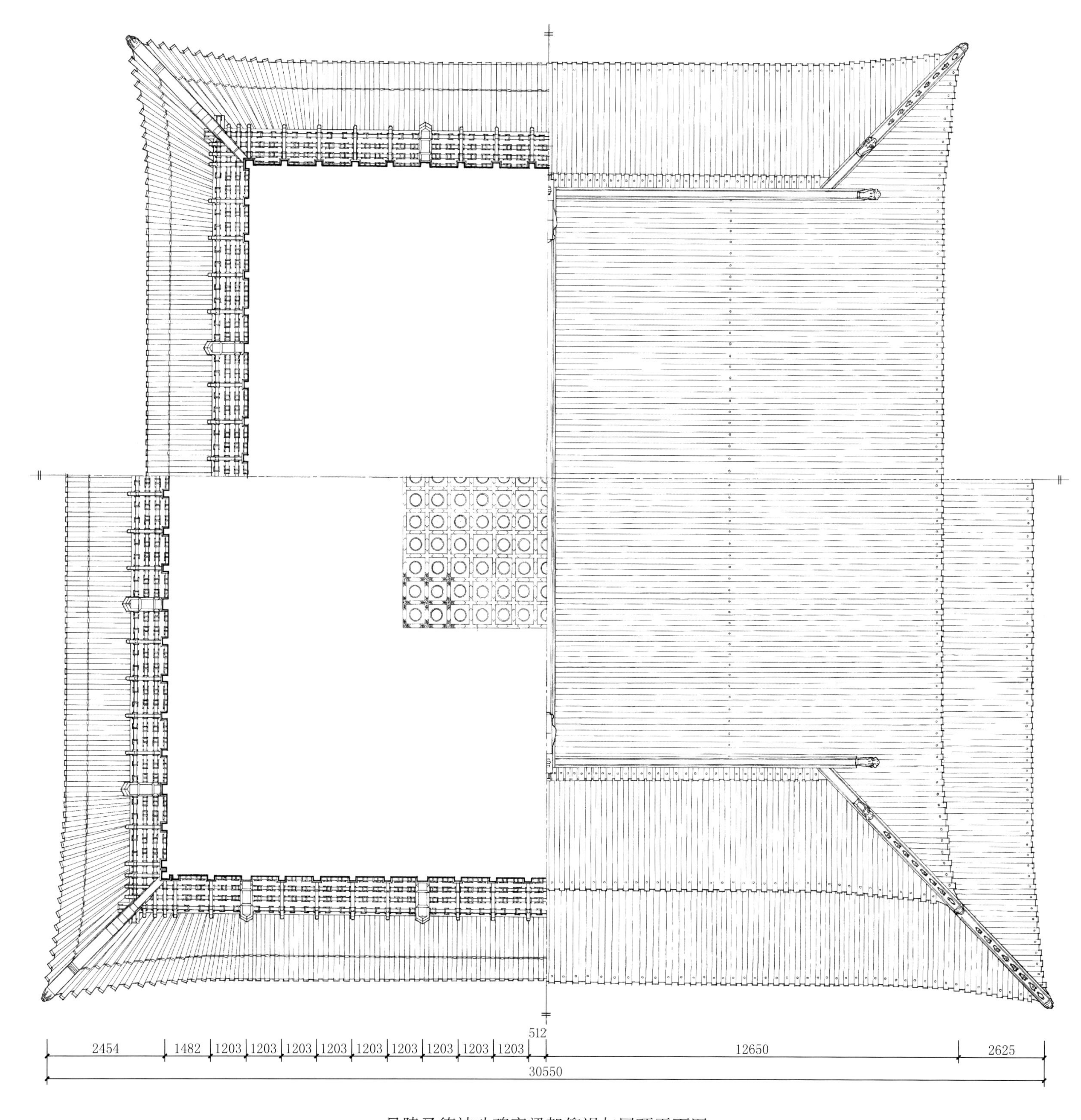

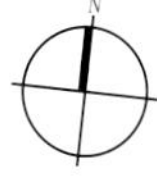

昌陵圣德神功碑亭梁架仰视与屋顶平面图

Roof plan and looking up at the Truss and Ceiling of the Pavilion for the Stela of Divine Merit and Sagely Virtue,Tomb of Good Fortune

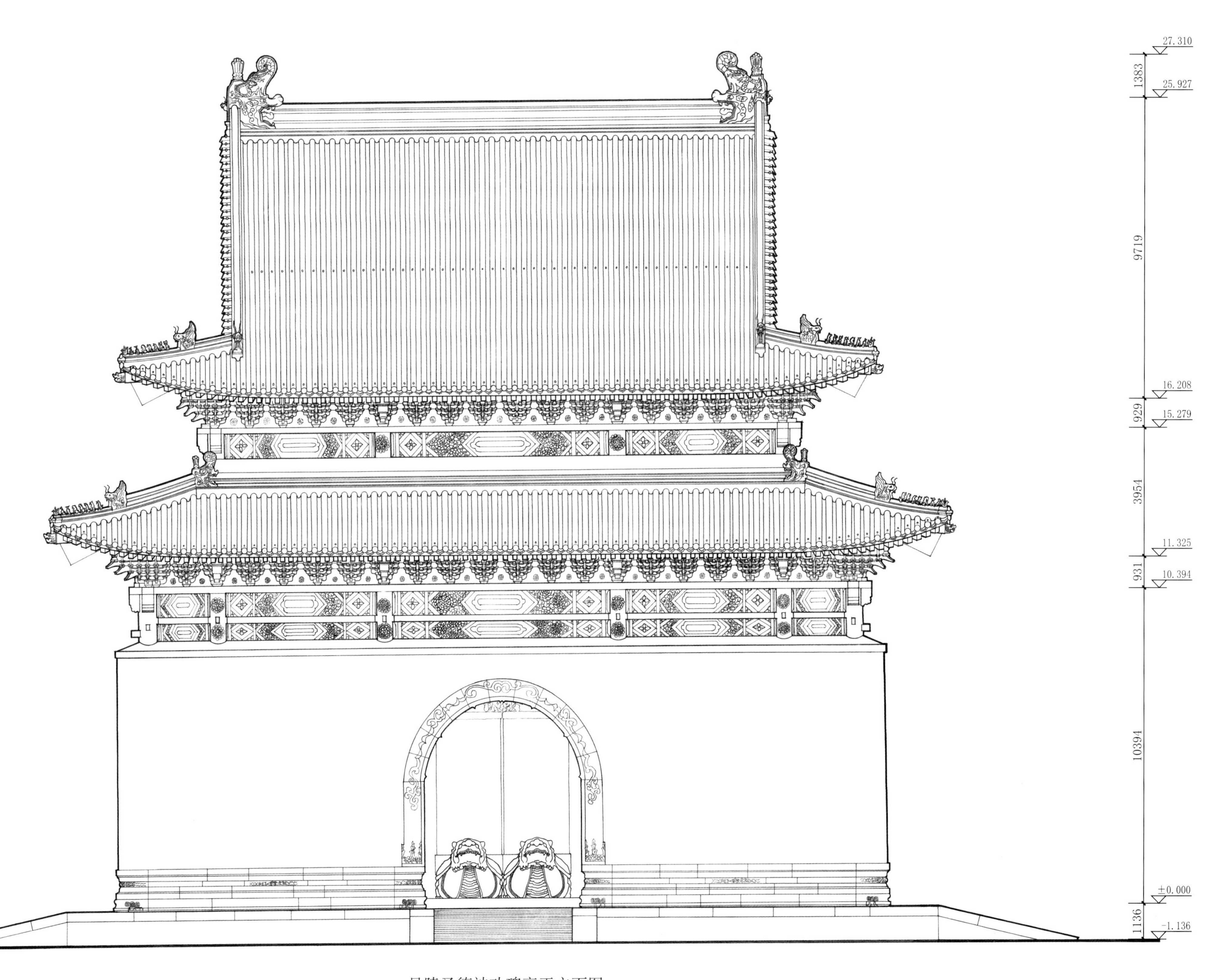

昌陵圣德神功碑亭正立面图

Front elevation of the Pavilion for the Stela of Divine Merit and Sagely Virtue,Tomb of Good Fortune

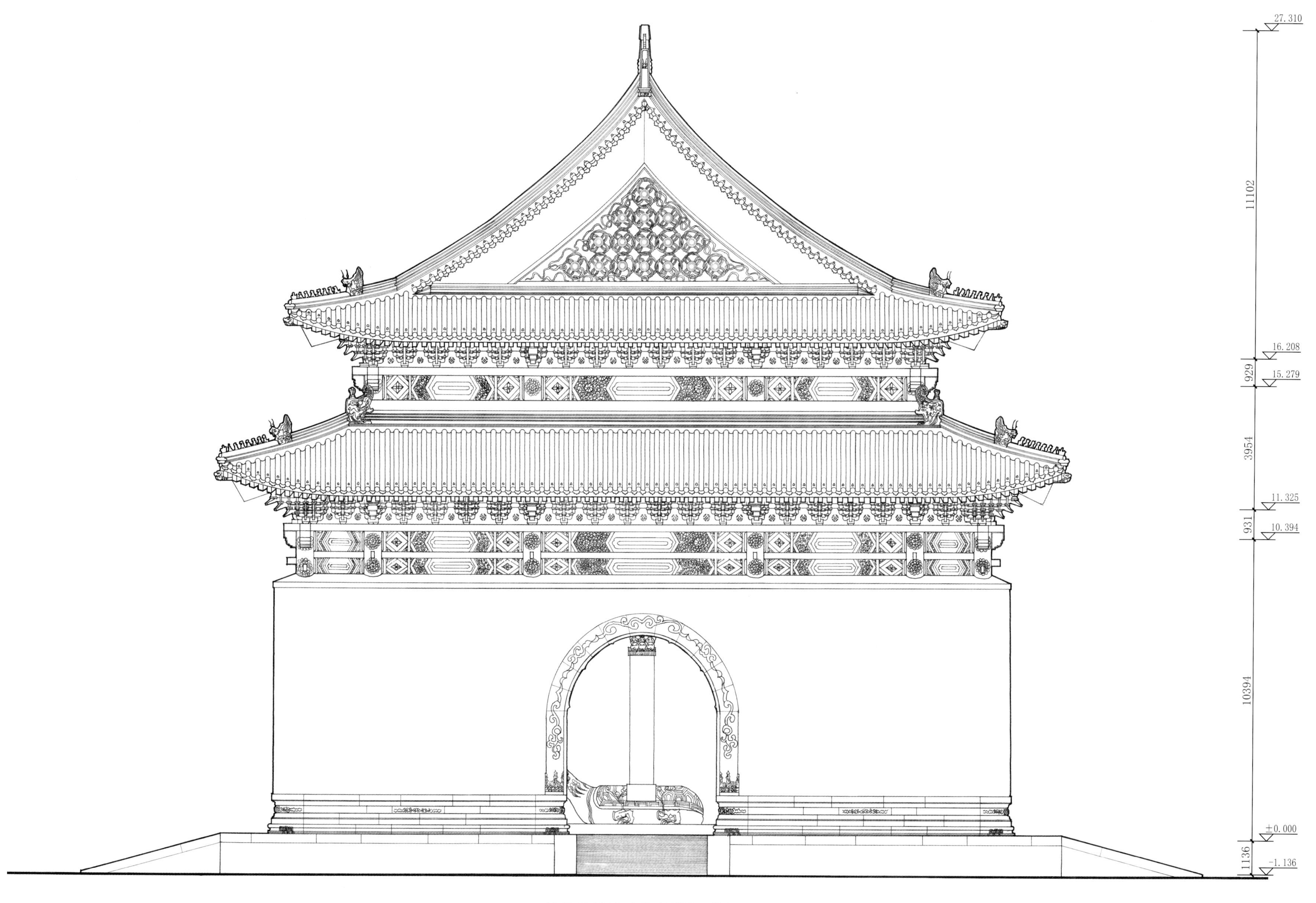

昌陵圣德神功碑亭侧立面图

Side elevation of the Pavilion for the Stela of Divine Merit and Sagely Virtue,Tomb of Good Fortune

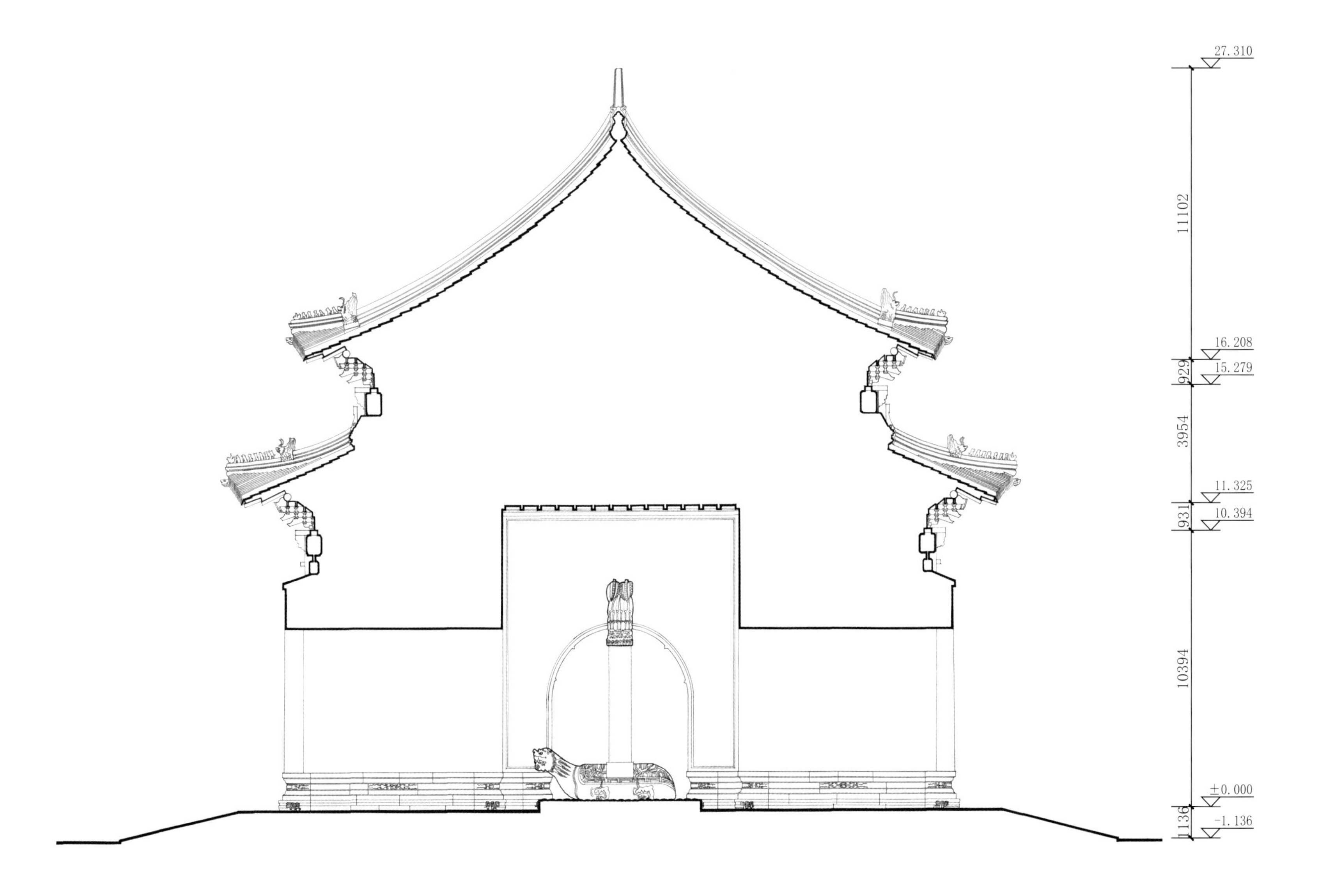

昌陵圣德神功碑亭明间横剖面图

Cross section of the Central Chamber within the Pavilion for the Stela of Divine Merit and Sagely Virtue,Tomb of Good Fortune

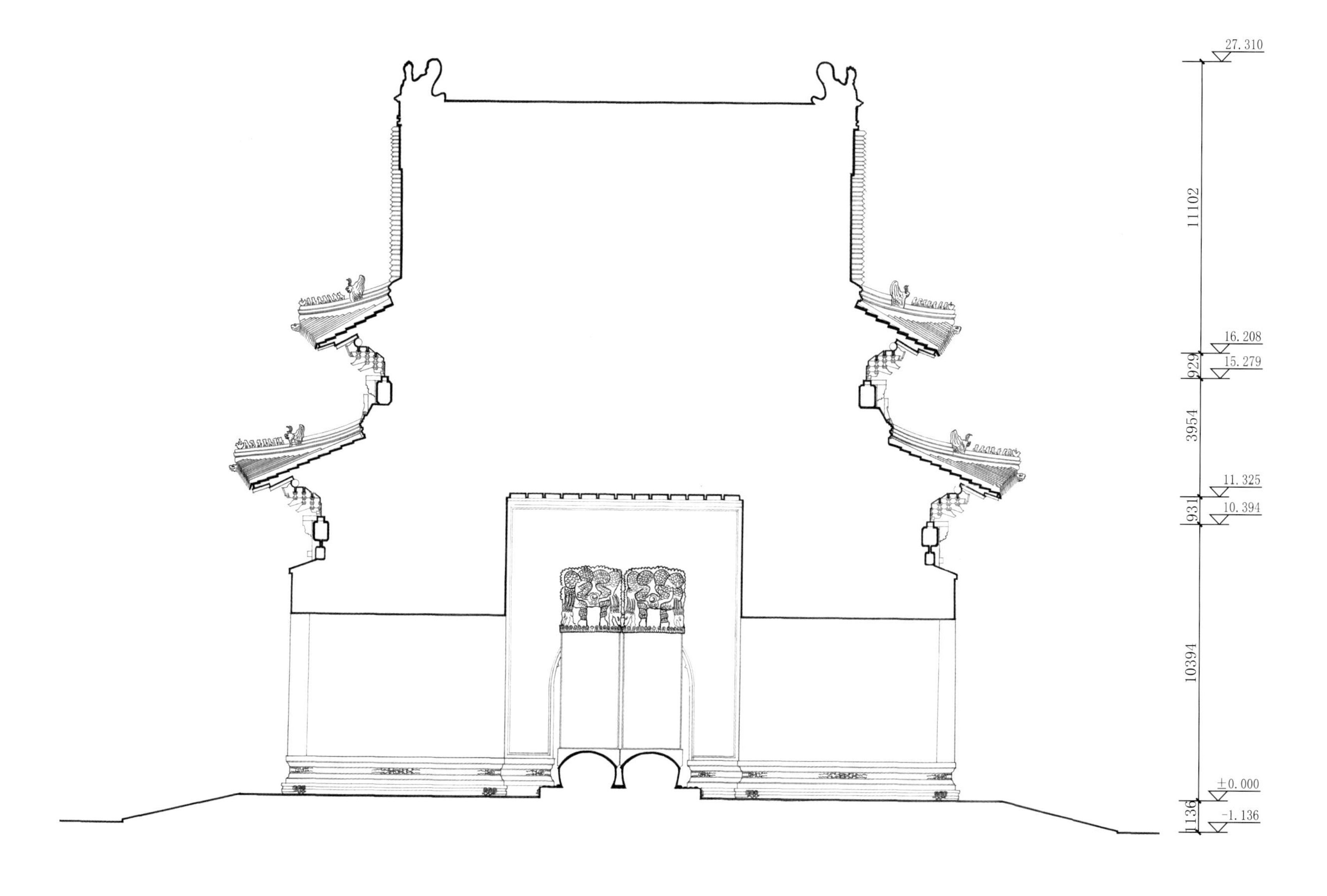

昌陵圣德神功碑亭纵剖面图

Longitudinal section of the Pavilion for the Stela of Divine Merit and Sagely Virtue,Tomb of Good Fortune

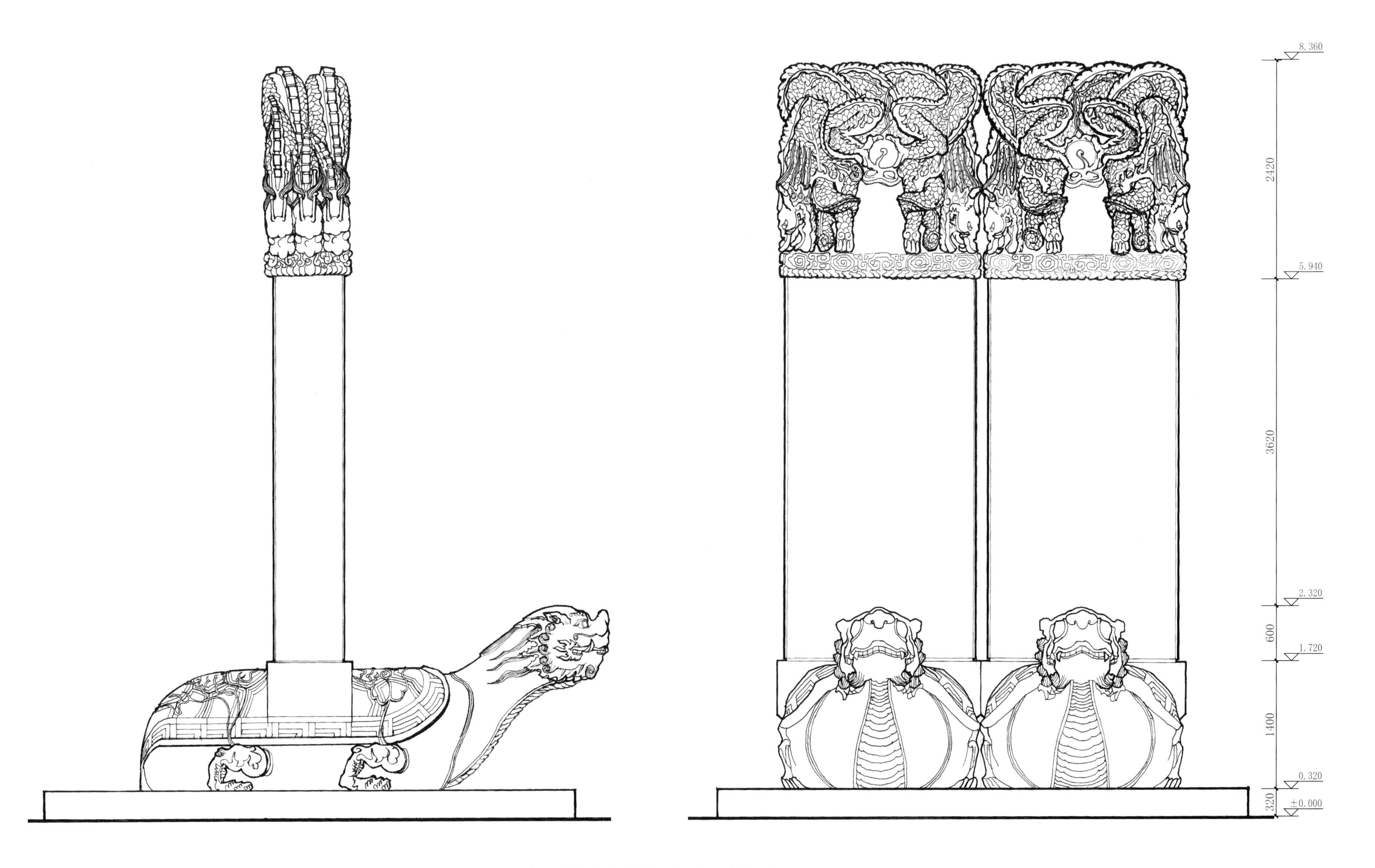

昌陵圣德神功碑亭龙蝠碑正立、侧立面图

Front elevation and side elevation of the Longfu Tablet of the Pavilion for the Stela of Divine Merit and Sagely Virtue,Tomb of Good Fortune

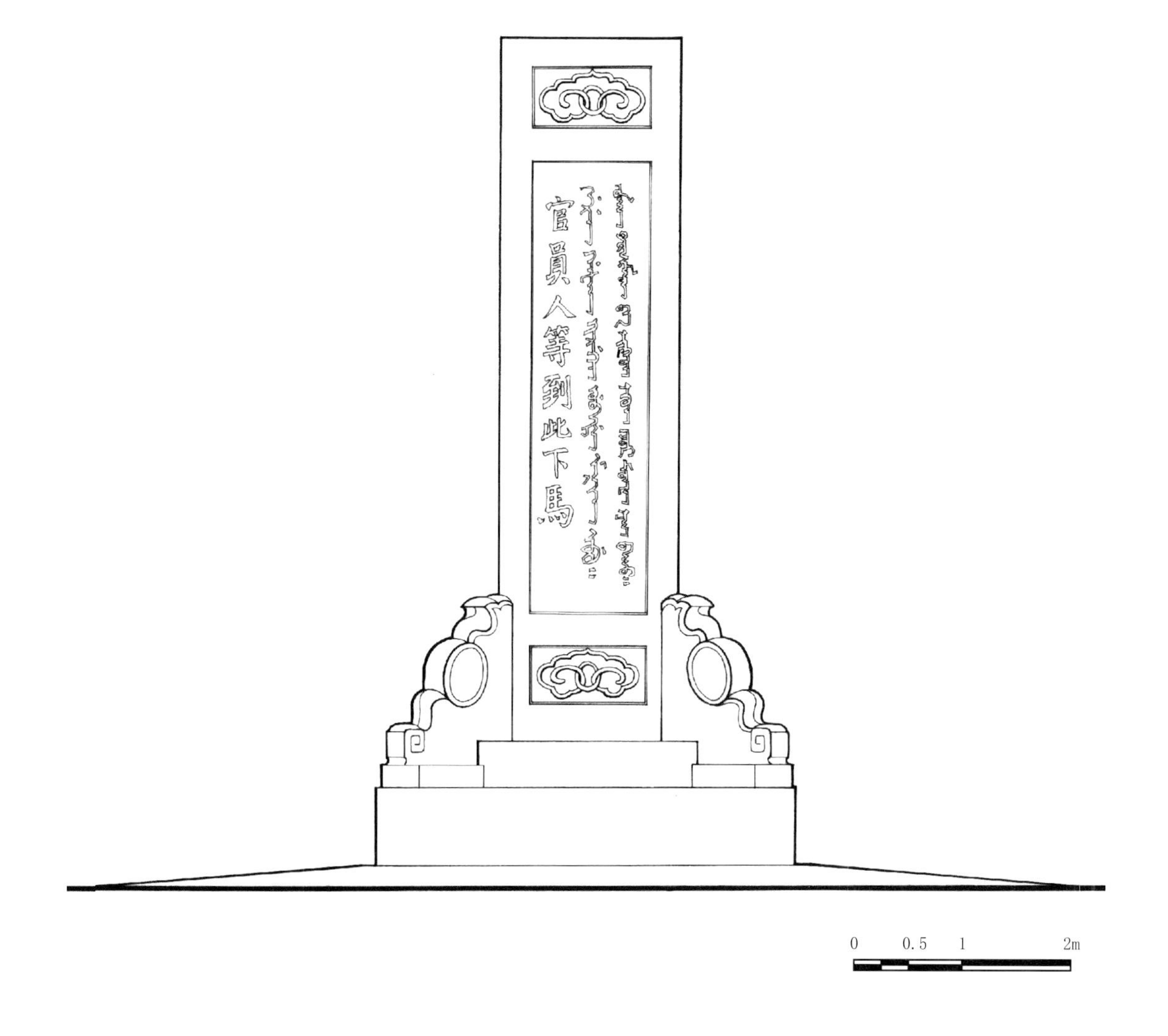

昌陵下马牌立面图
Elevation of the Stela marking the place for dismounting from one's horse,Tomb of Good Fortune

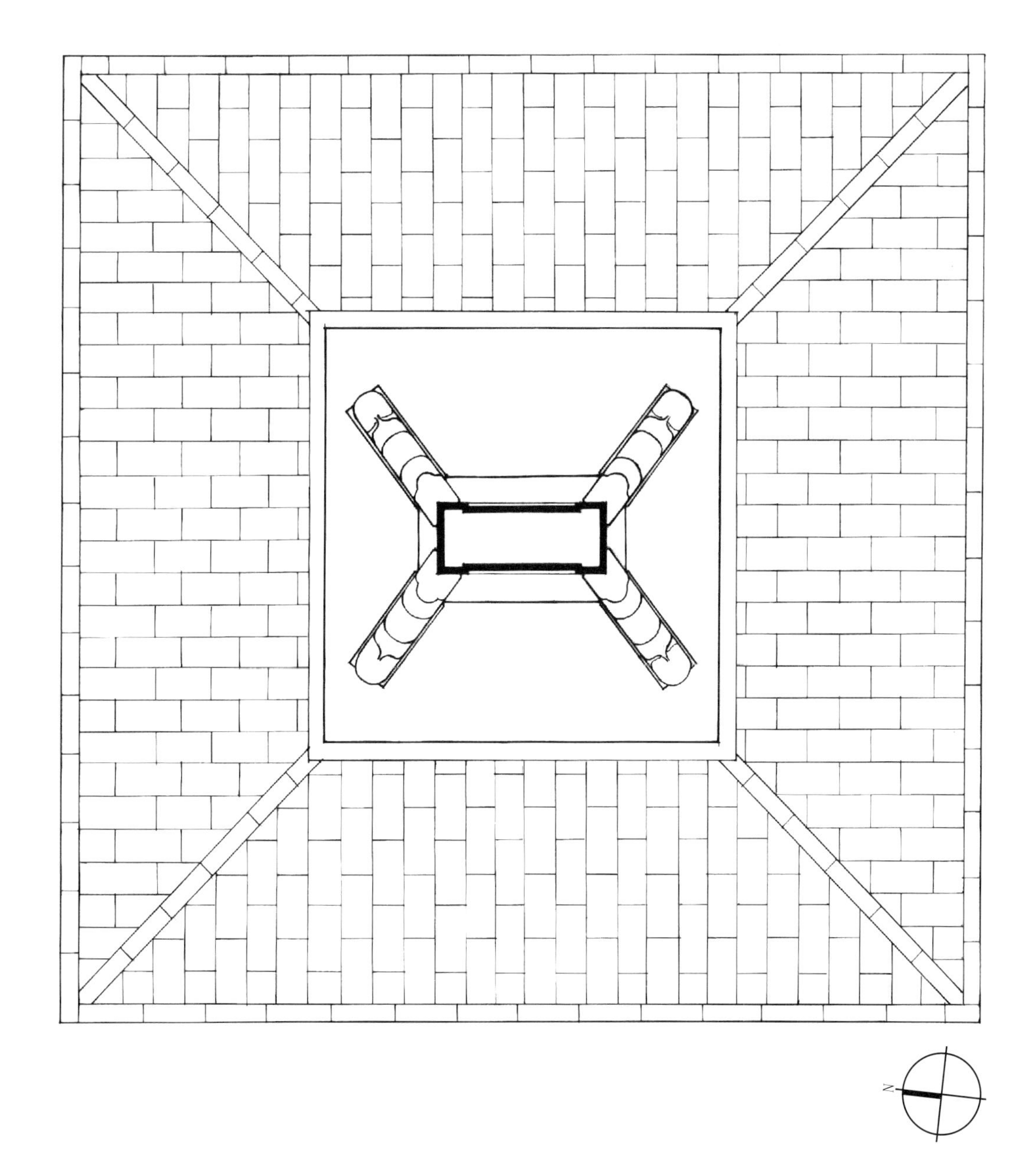

昌陵下马牌平面图
Plan of the Stela marking the place for dismounting from one's horse,Tomb of Good Fortune

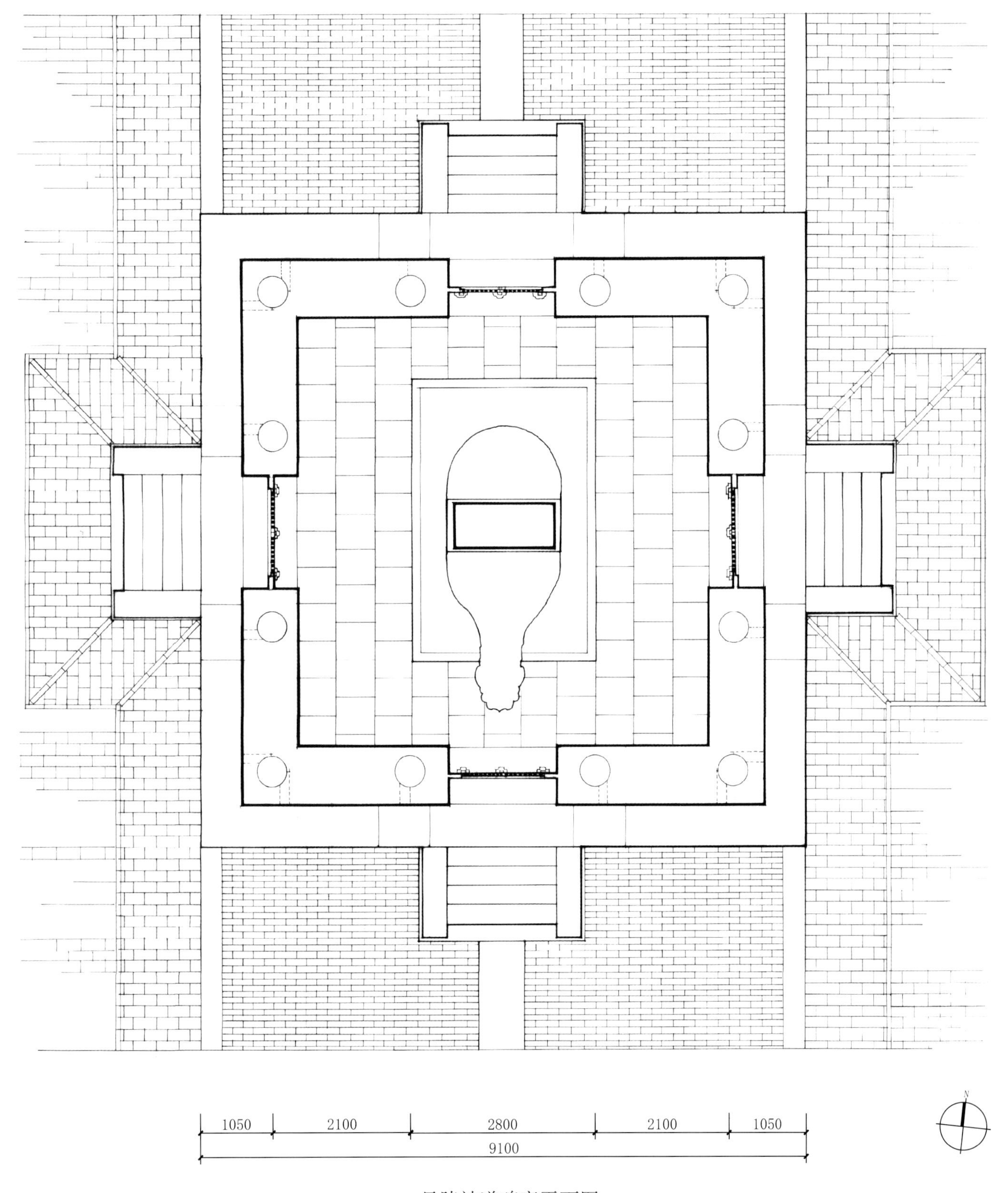

昌陵神道碑亭平面图

Plan of the Pavilion for the Stela on the Spirit Way,Tomb of Good Fortune

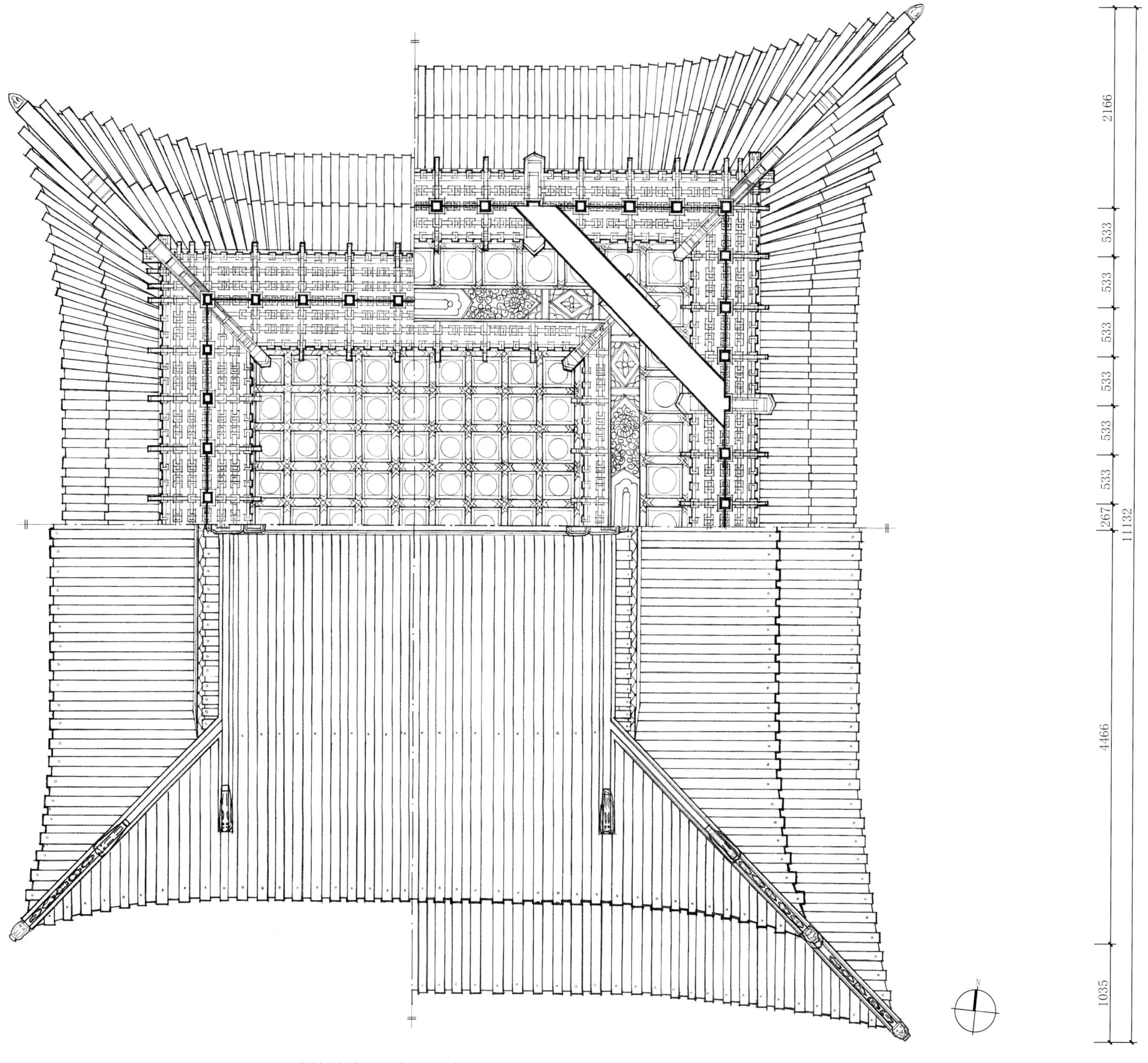

昌陵神道碑亭梁架仰视与屋顶平面图

Roof plan and looking up at the Truss and Ceiling of the Pavilion for the Stela on the Spirit Way,Tomb of Good Fortune

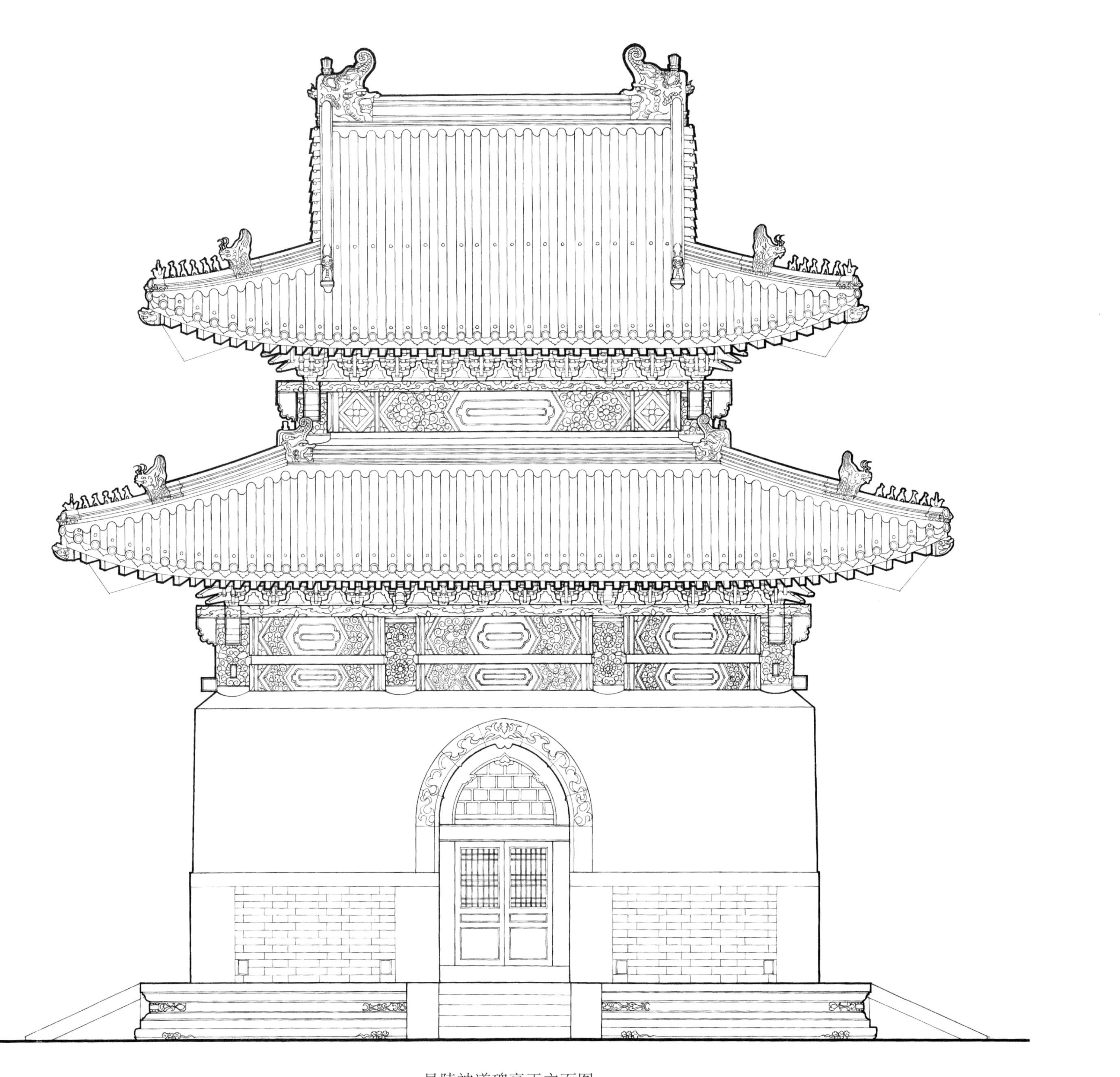

昌陵神道碑亭正立面图
Front elevation of the Pavilion for the Stela on the Spirit Way,Tomb of Good Fortune

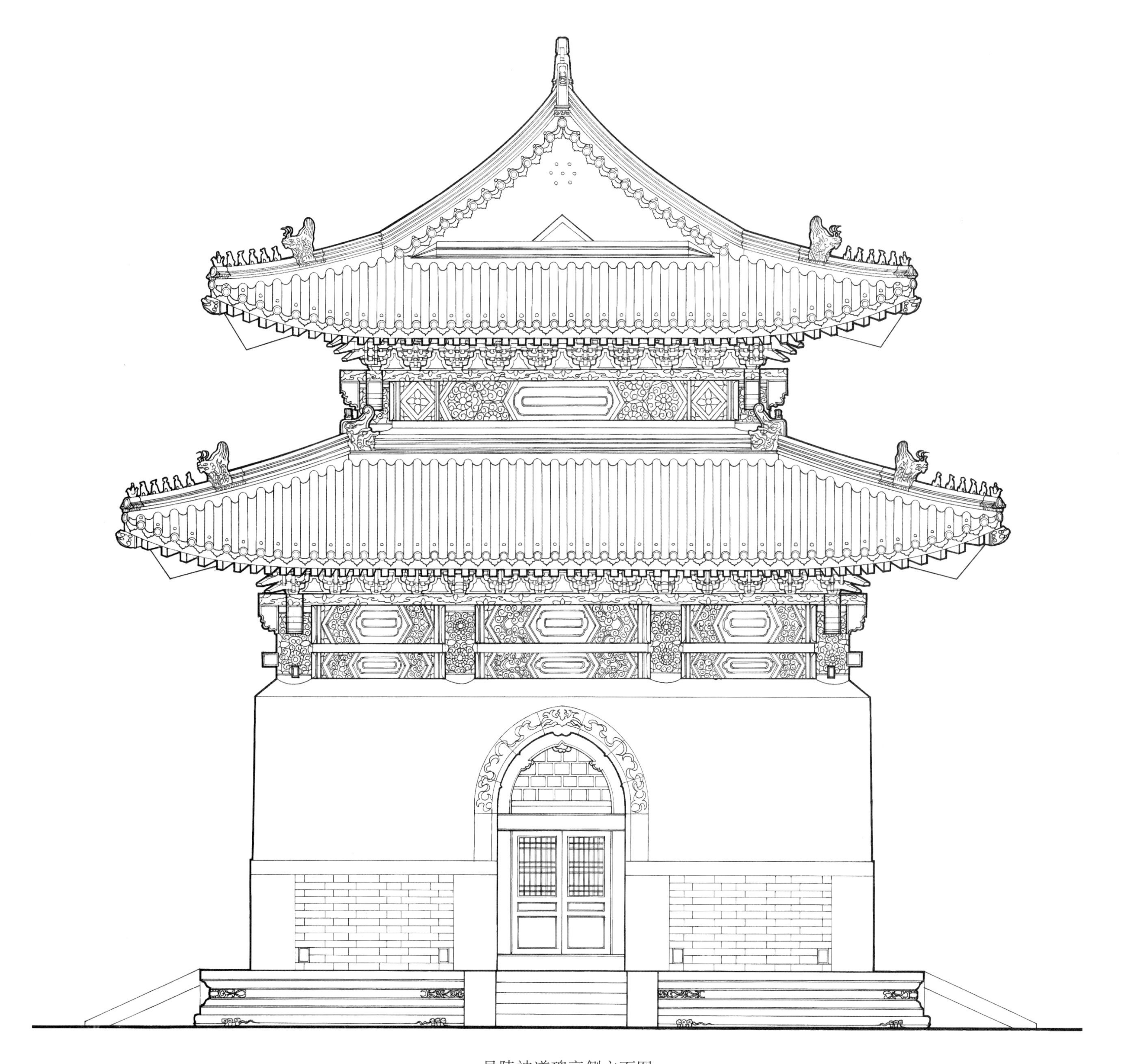

昌陵神道碑亭侧立面图

Side elevation of the Pavilion for the Stela on the Spirit Way,Tomb of Good Fortune

昌陵神道碑亭明间横剖面图

Cross section of the Central Chamber within the Pavilion for the Stela on the Spirit Way,Tomb of Good Fortune

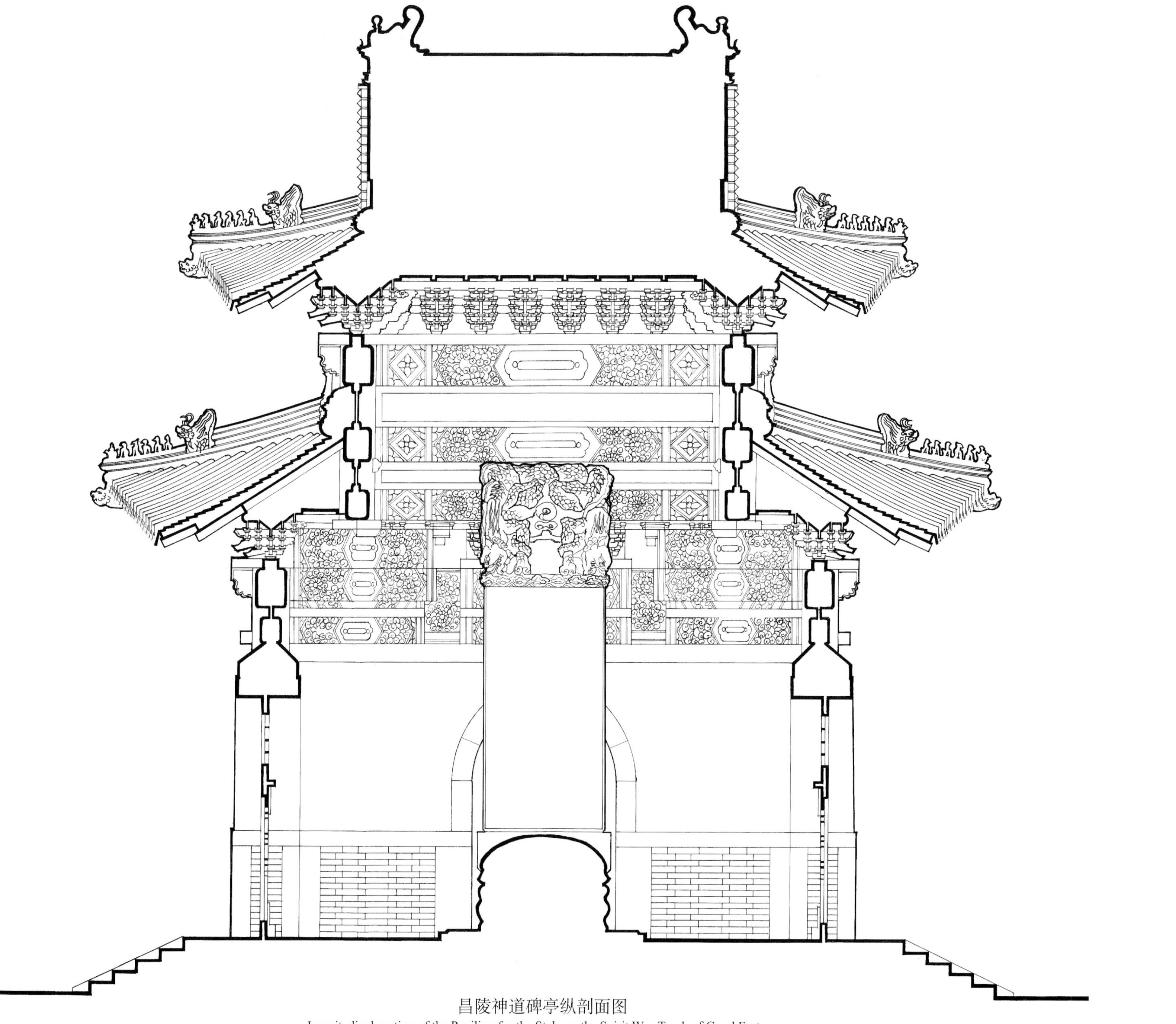

昌陵神道碑亭纵剖面图

Longitudinal section of the Pavilion for the Stela on the Spirit Way,Tomb of Good Fortune

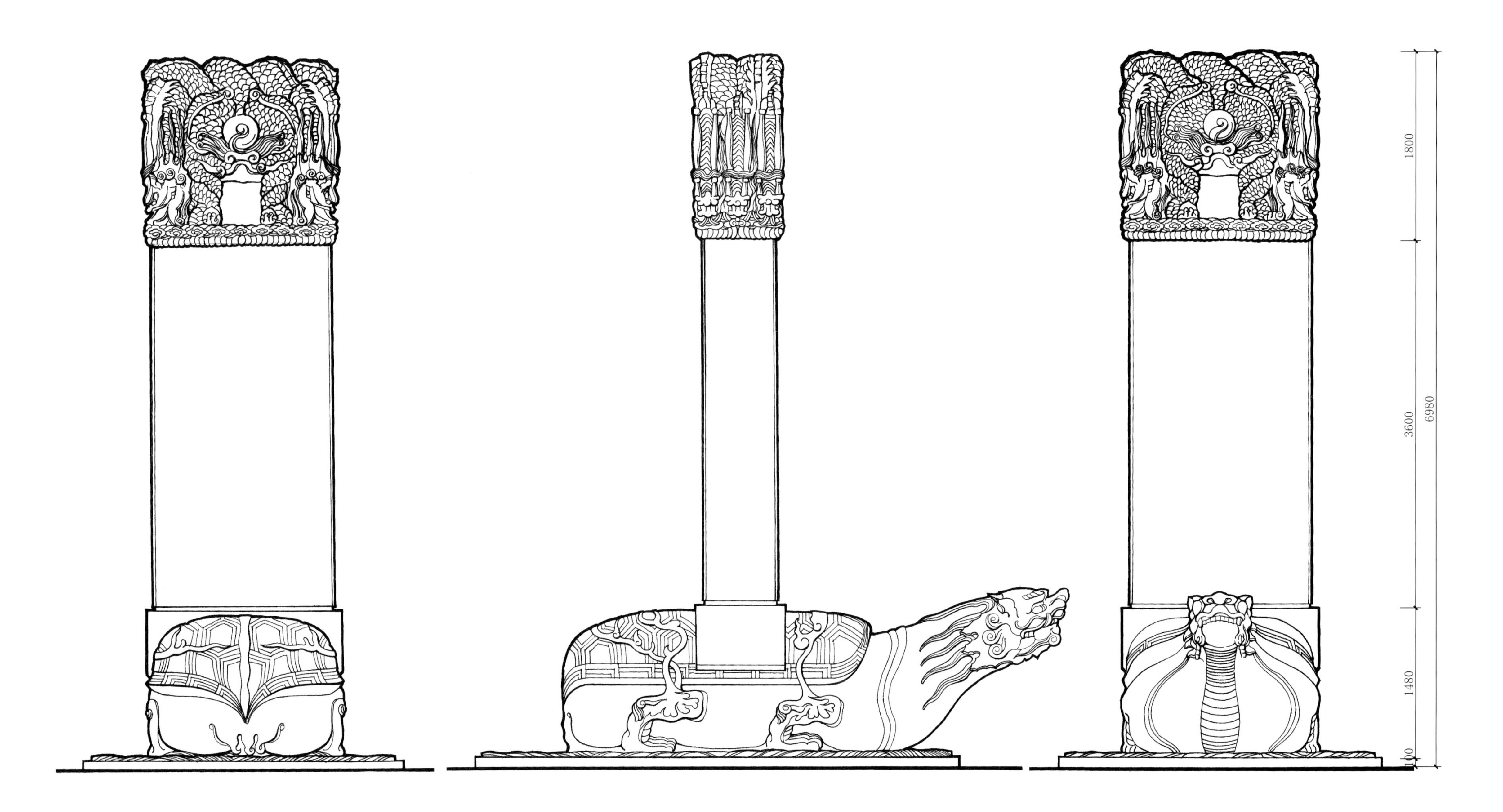

昌陵神道碑大样图（一）

Detailed drawing of the Tablet for the Stela on the Spirit Way, Tomb of Good Fortune (1)

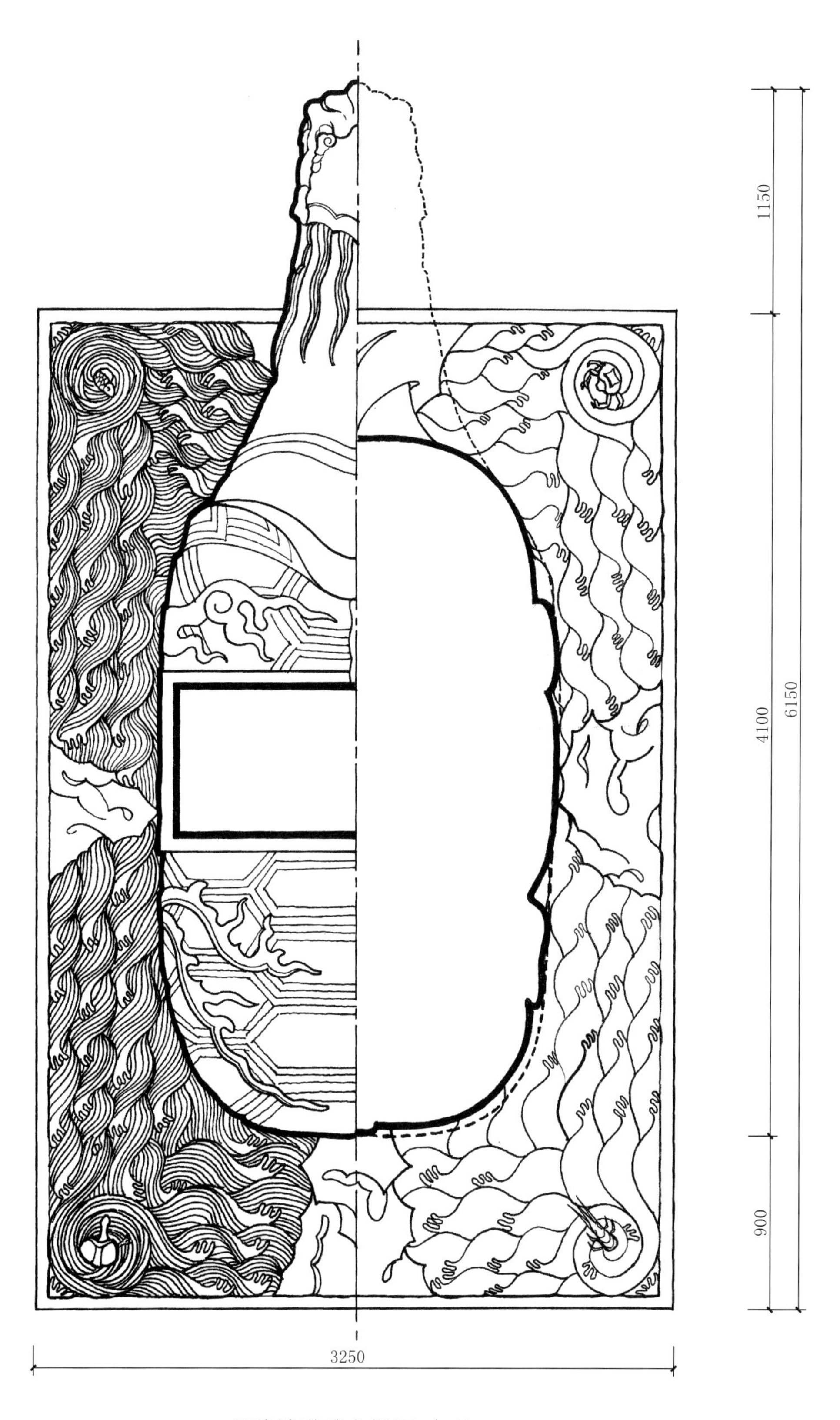

昌陵神道碑大样图（二）
Detailed drawing of the Tablet for the Stela on the Spirit Way, Tomb of Good Fortune (2)

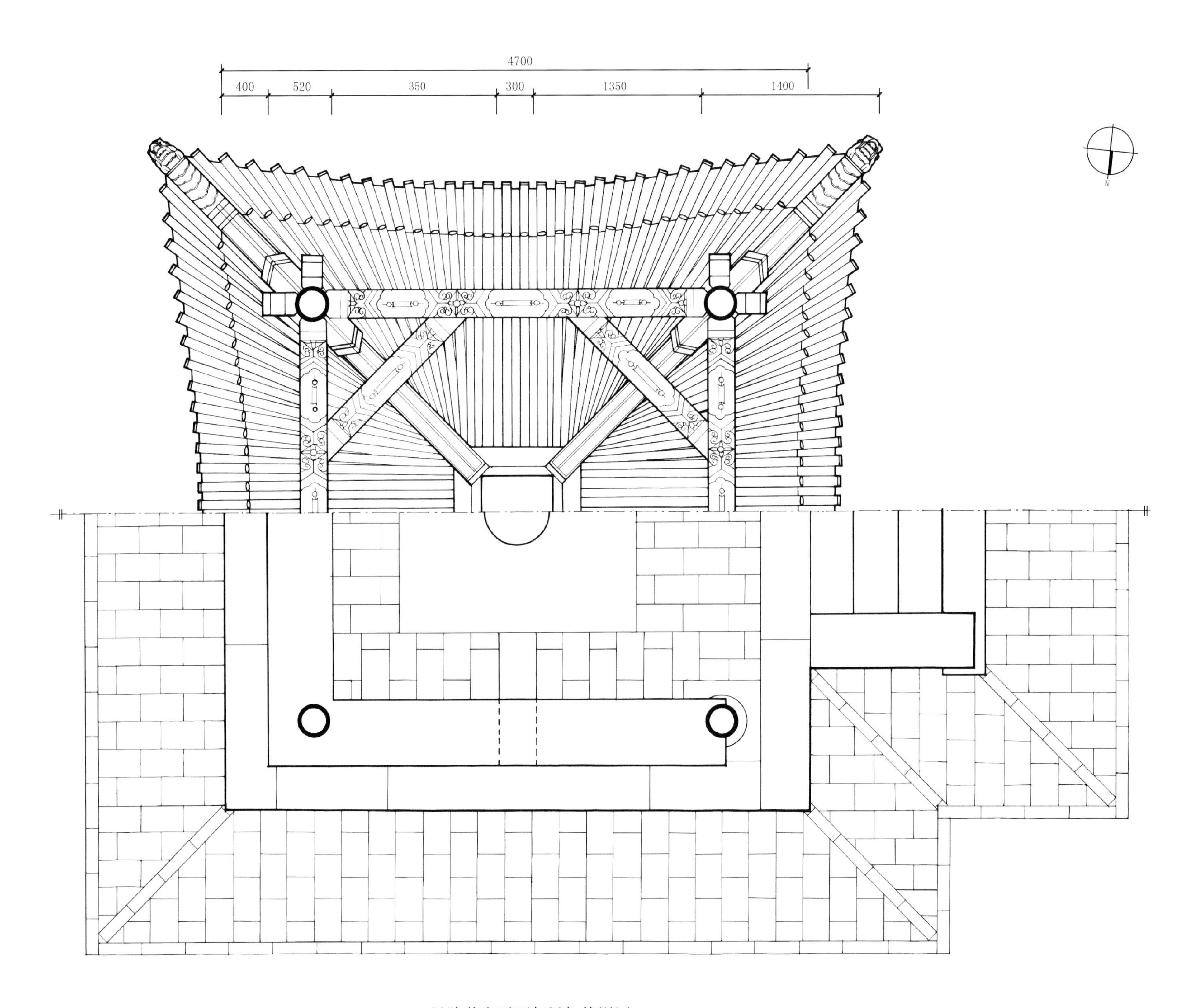

昌陵井亭平面与梁架仰视图
Plan and looking up at the Truss and Ceiling of the Pavilion for the Well,Tomb of Good Fortune

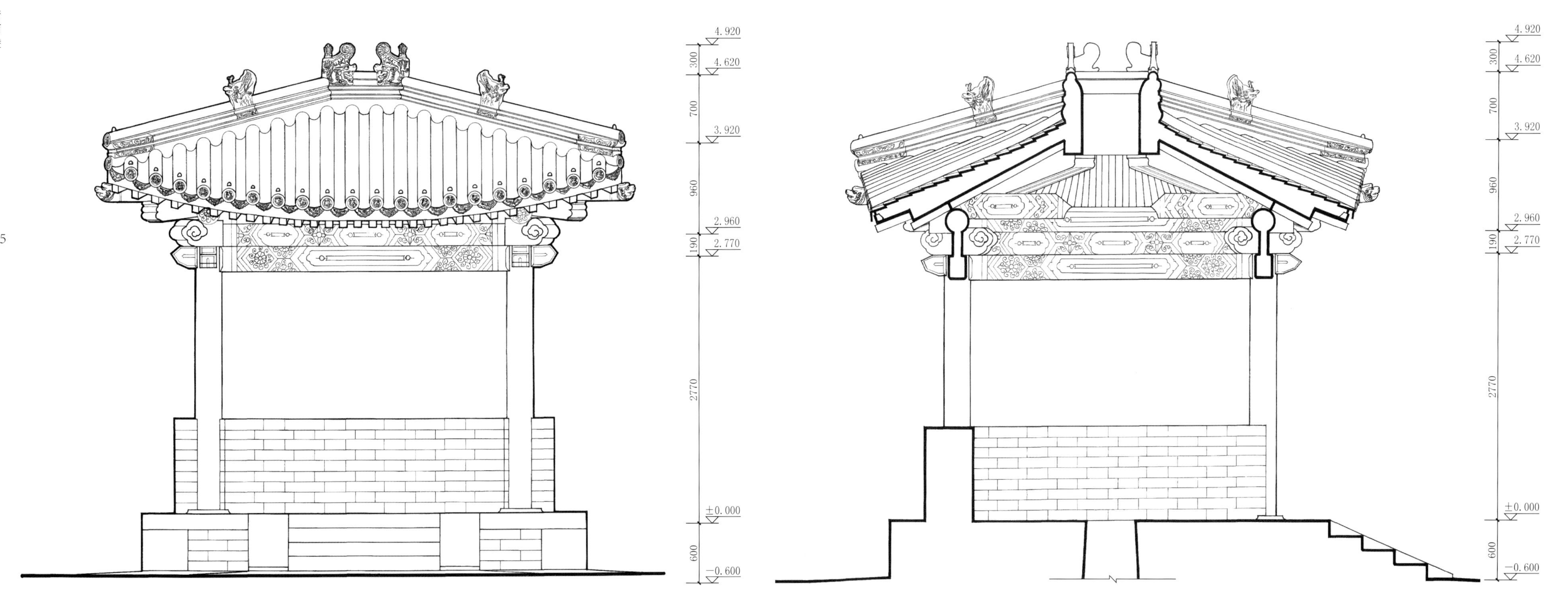

昌陵井亭正立面图
Front elevation of the Pavilion for the Well,Tomb of Good Fortune

昌陵井亭横剖面图
Cross section of the Pavilion for the Well,Tomb of Good Fortune

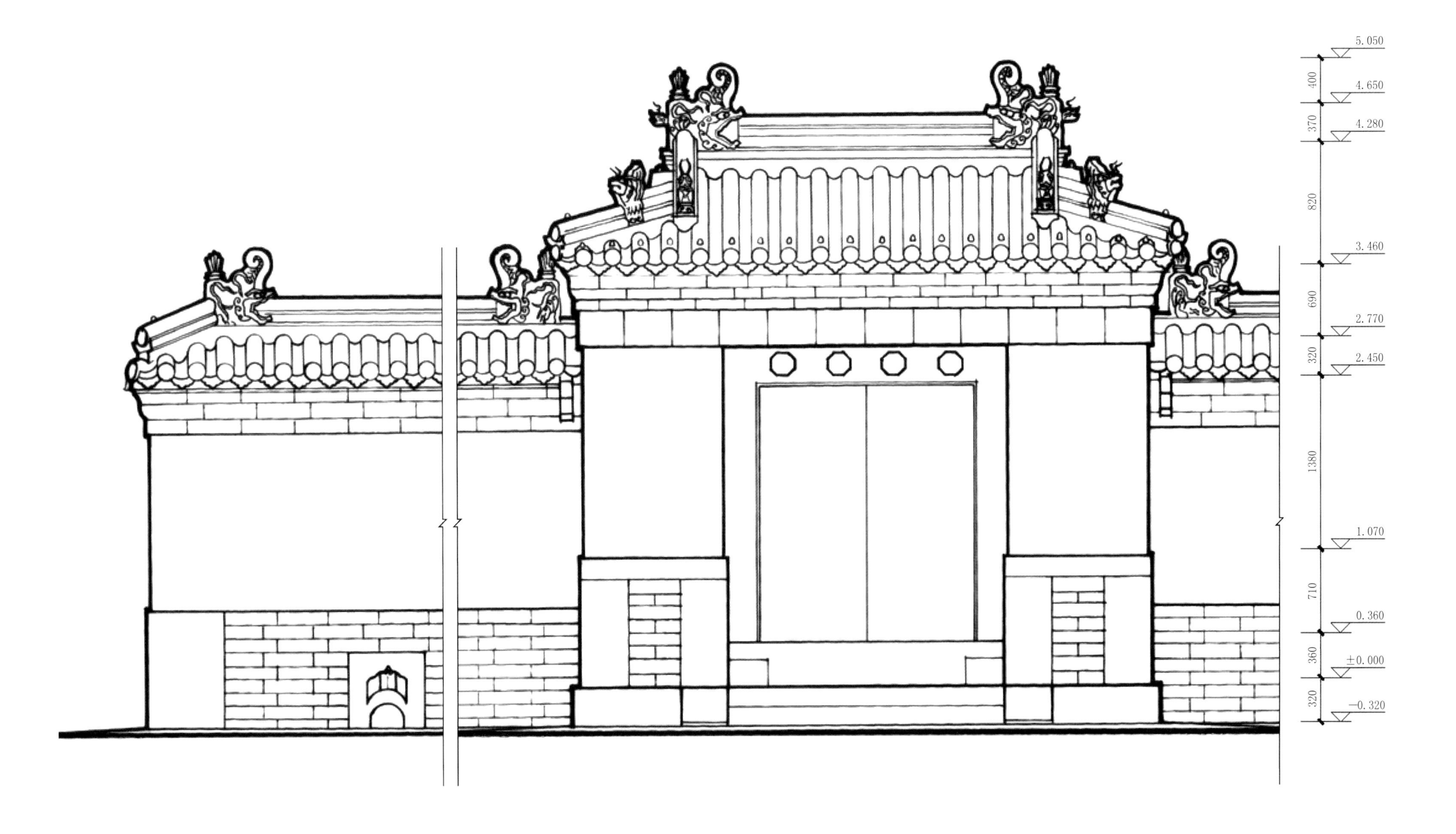

昌陵神厨库门正立面图

Front elevation of the Gate in the Culinary Courtyard for Sacrifices,Tomb of Good Fortune

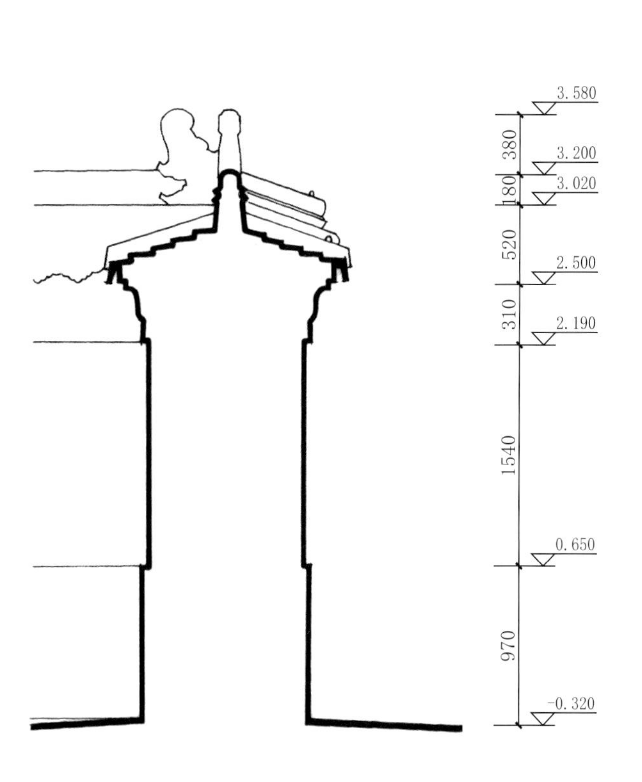

昌陵神厨库围墙横剖面图
Cross section of the wall around the Culinary Courtyard for Sacrifices,Tomb of Good Fortune

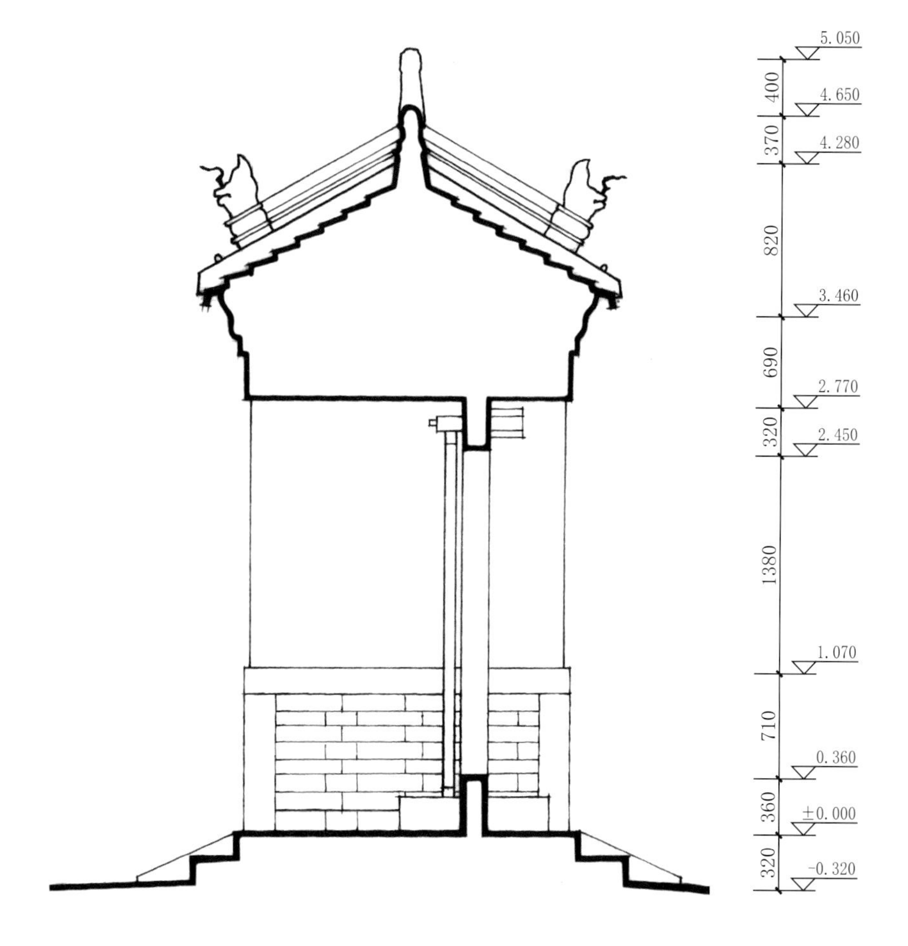

昌陵神厨库门横剖面图
Cross section of the Gate in the Culinary Courtyard for Sacrifices,Tomb of Good Fortune

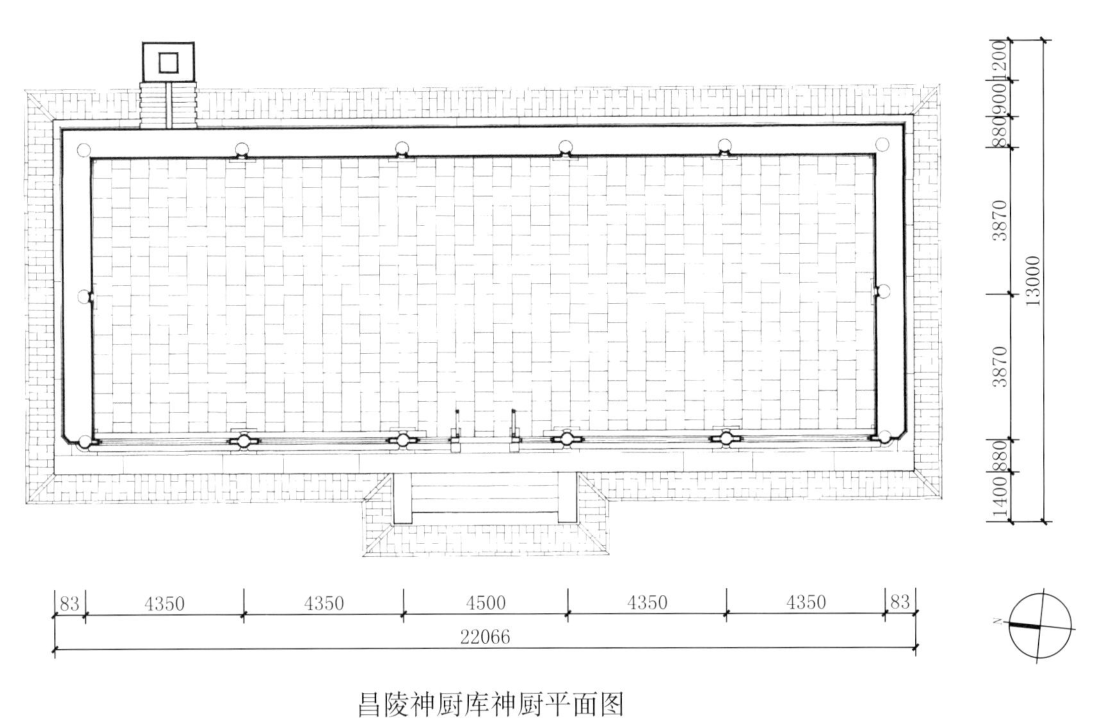

昌陵神厨库神厨平面图

Plan of the Kitchen in the Culinary Courtyard for Sacrifices,Tomb of Good Fortune

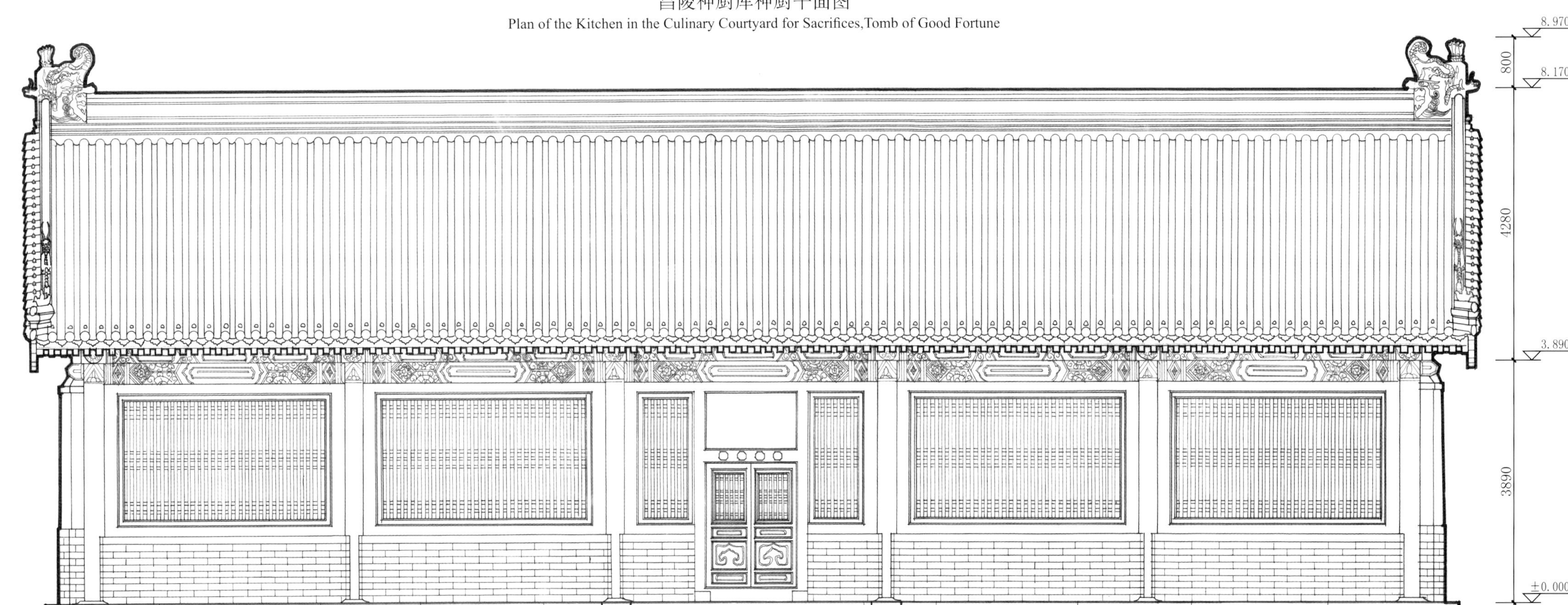

昌陵神厨库神厨正立面图

Front elevation of the Kitchen in the Culinary Courtyard for Sacrifices,Tomb of Good Fortune

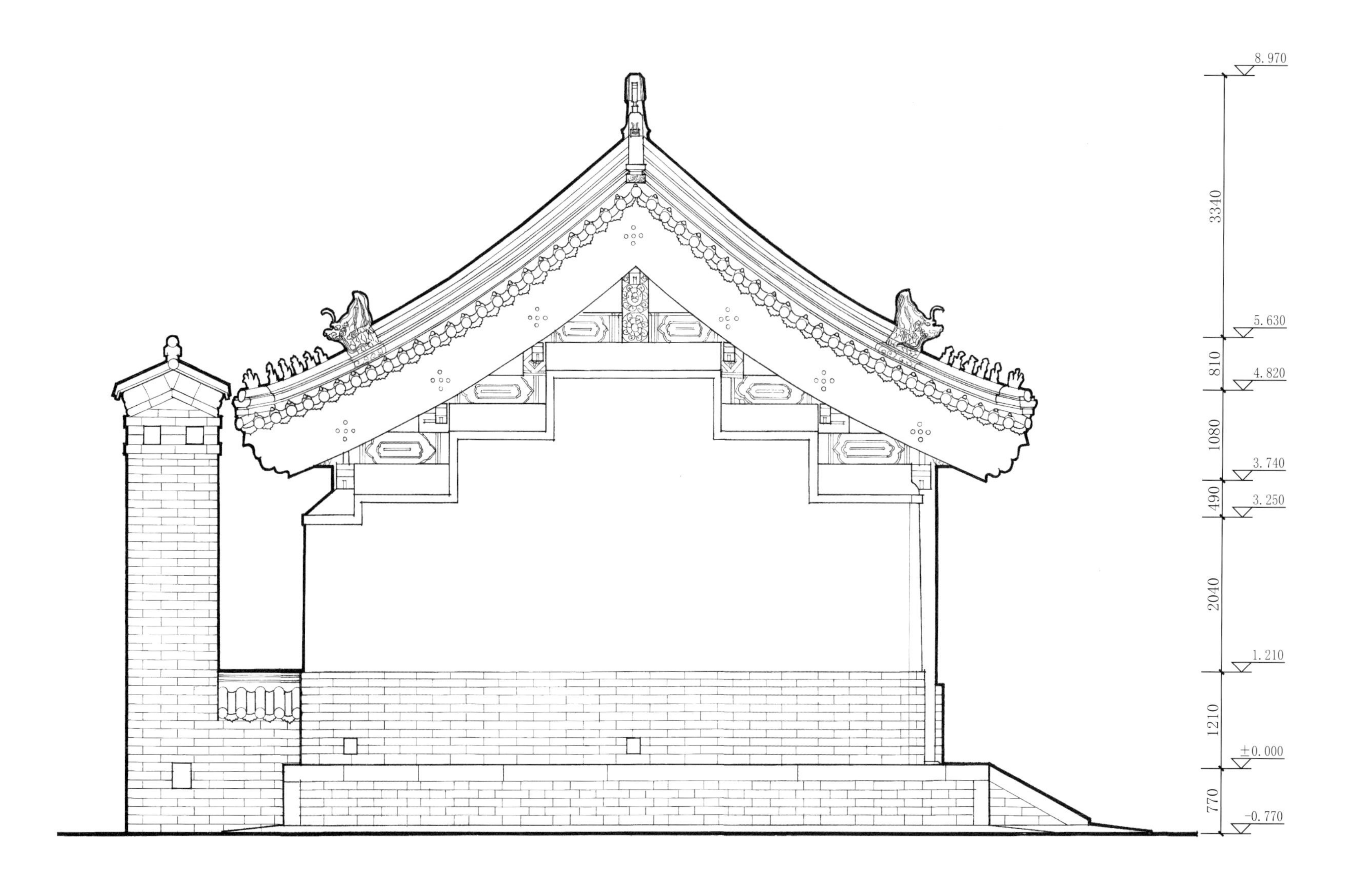

昌陵神厨库神厨侧立面图

Side elevation of the Kitchen in the Culinary Courtyard for Sacrifices,Tomb of Good Fortune

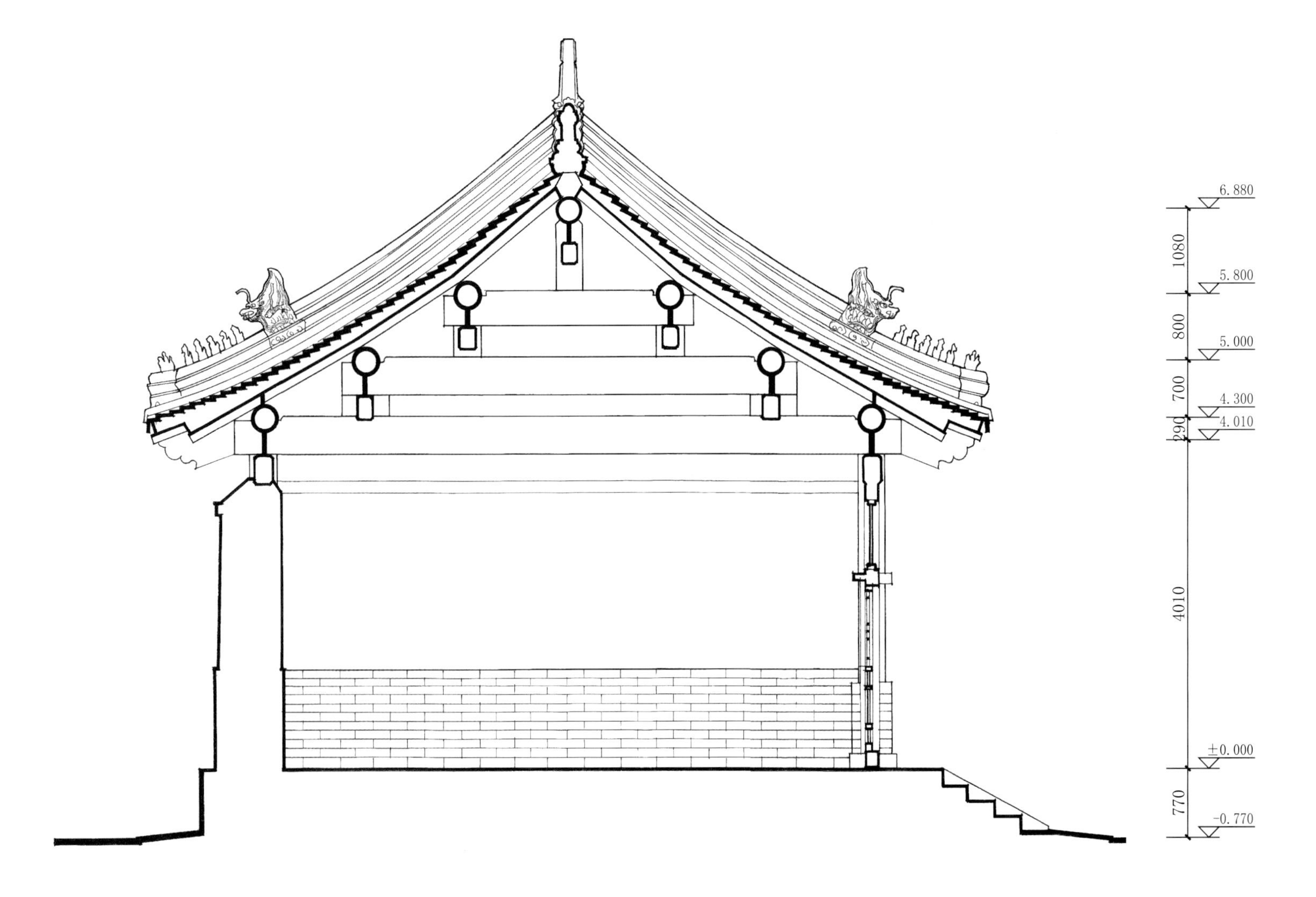

昌陵神厨库神厨明间横剖面图

Cross section of the Central Chamber within the Kitchen in the Culinary Courtyard for Sacrifices,Tomb of Good Fortune

昌陵神厨库神库平面图
Plan of the Repository in the Culinary Courtyard for Sacrifices,Tomb of Good Fortune

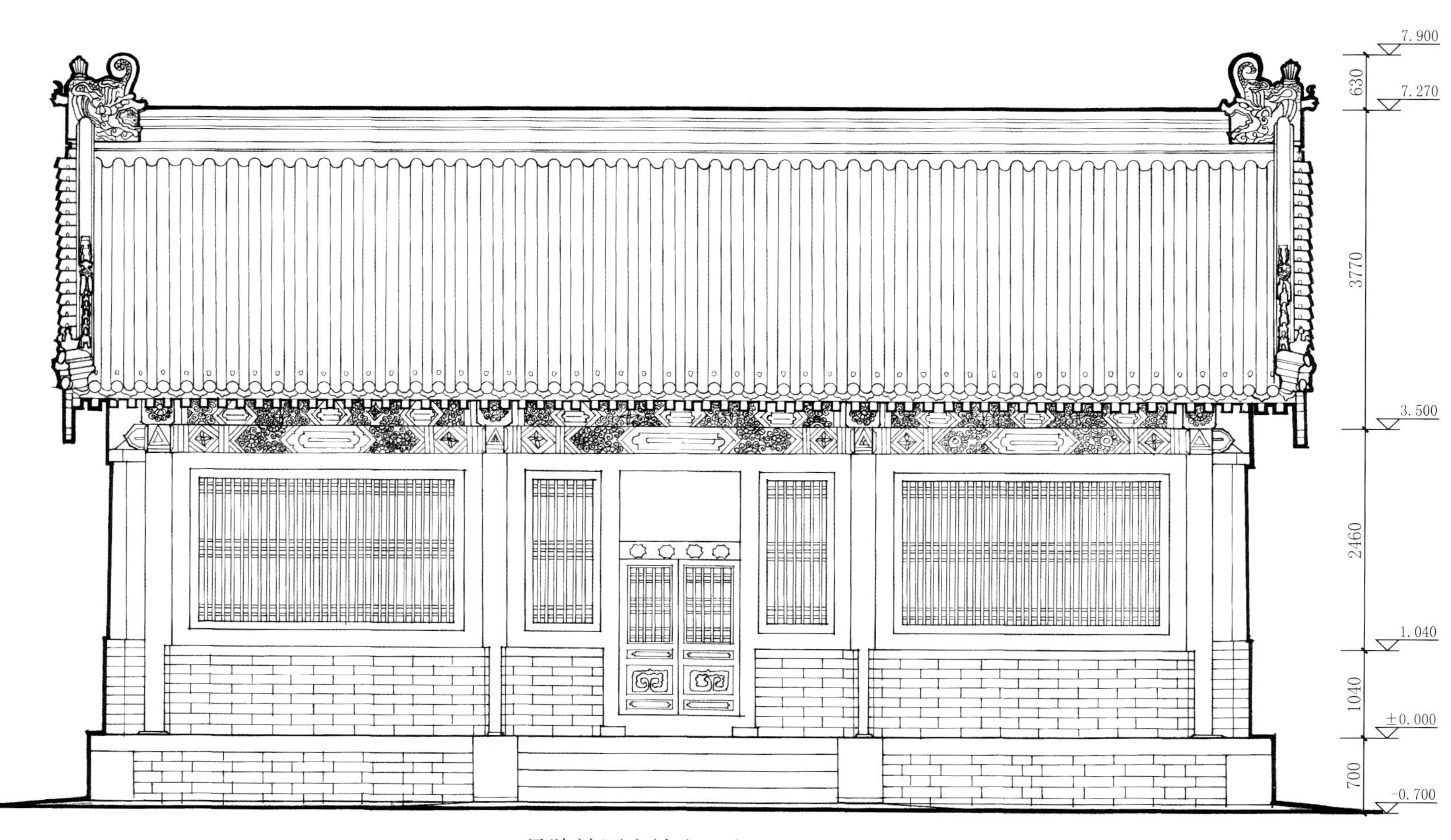

昌陵神厨库神库正立面图
Front elevation of the Repository in the Culinary Courtyard for Sacrifices,Tomb of Good Fortune

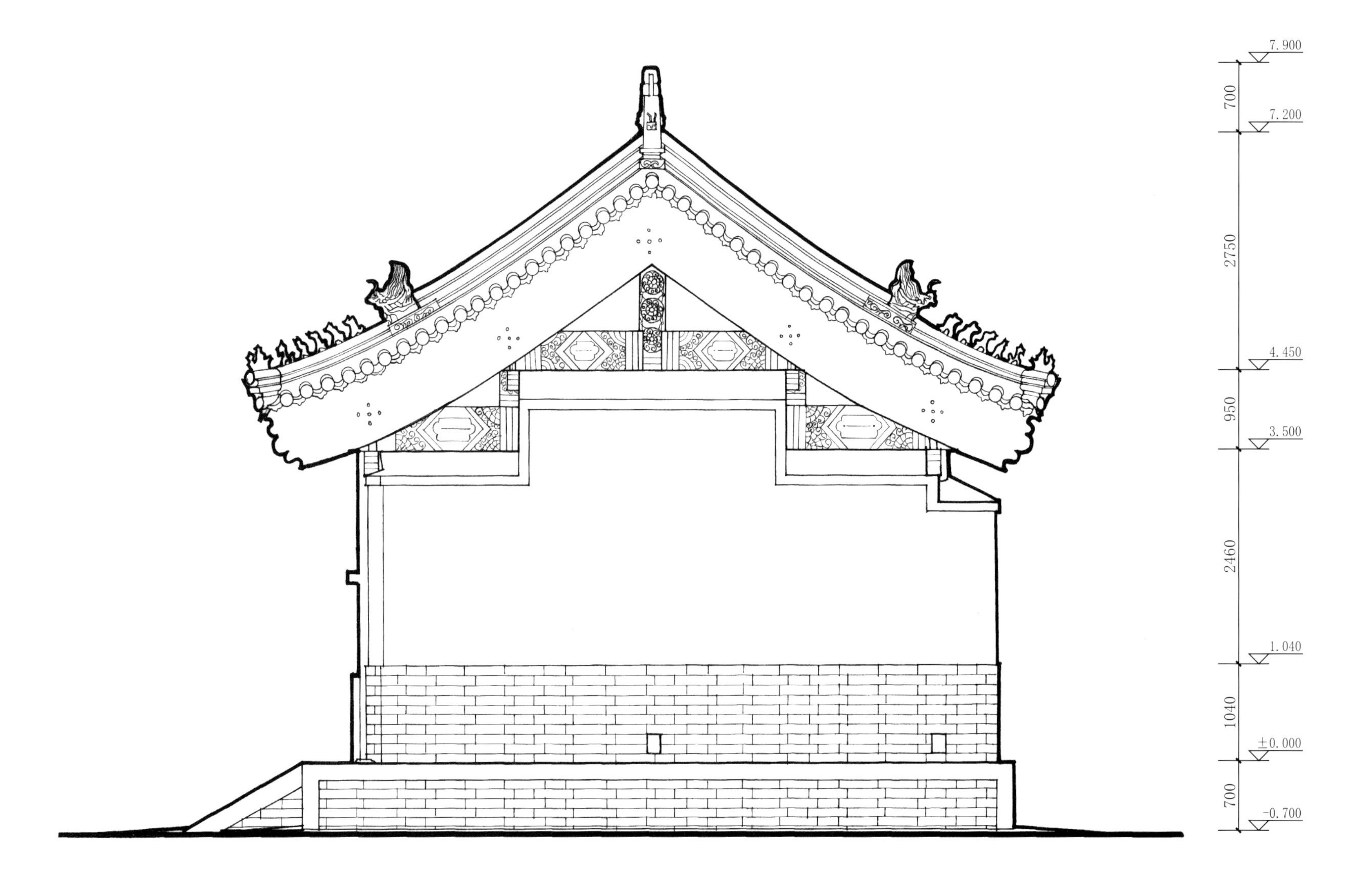

昌陵神厨库神库侧立面图

Side elevation of the Repository in the Culinary Courtyard for Sacrifices,Tomb of Good Fortune

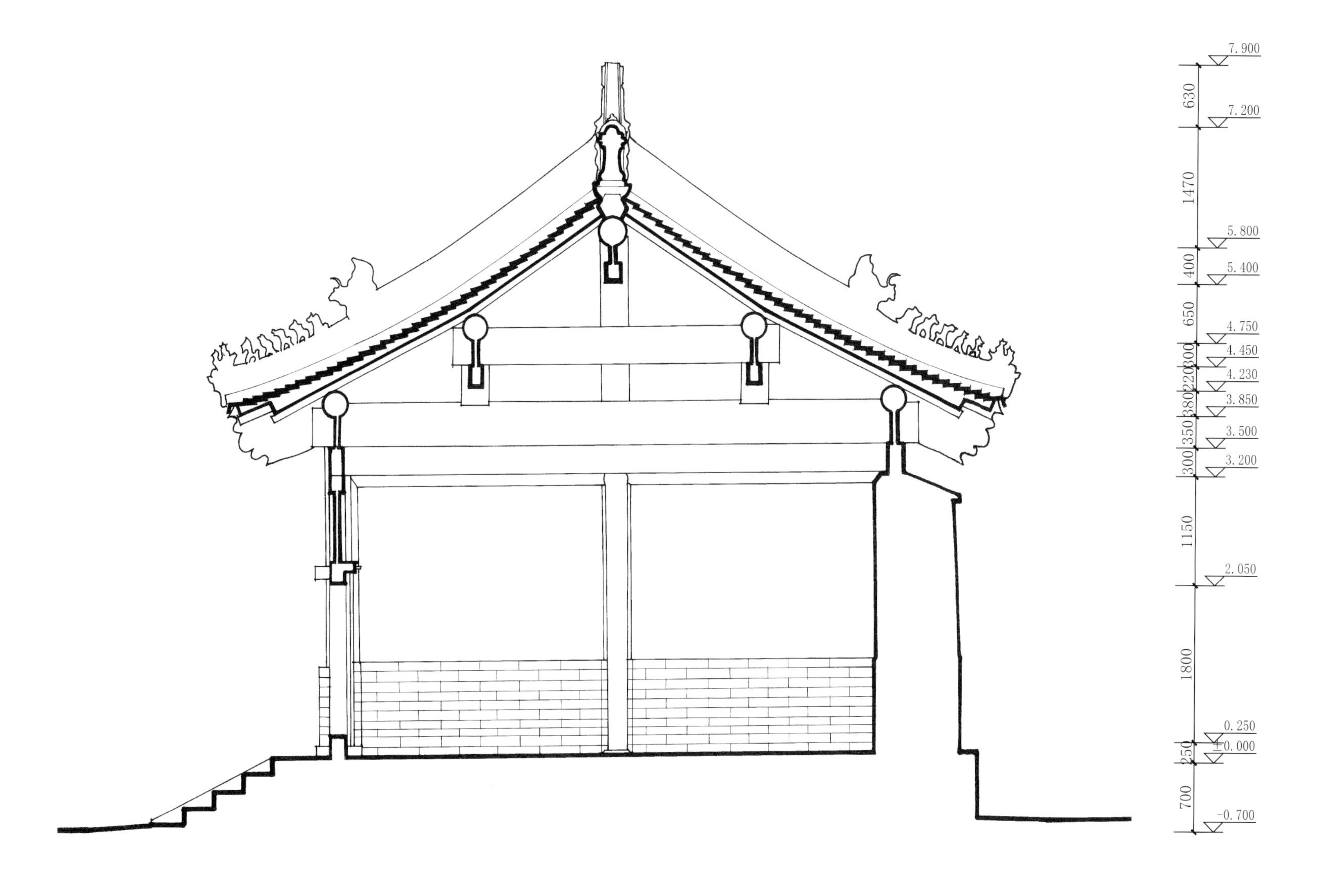

昌陵神厨库神库明间横剖面图

Cross section of the Central Chamber within the Repository in the Culinary Courtyard for Sacrifices,Tomb of Good Fortune

昌陵神厨库宰牲亭正立面图

Front elevation of the Ritual Abattoir in the Culinary Courtyard for Sacrifices,Tomb of Good Fortune

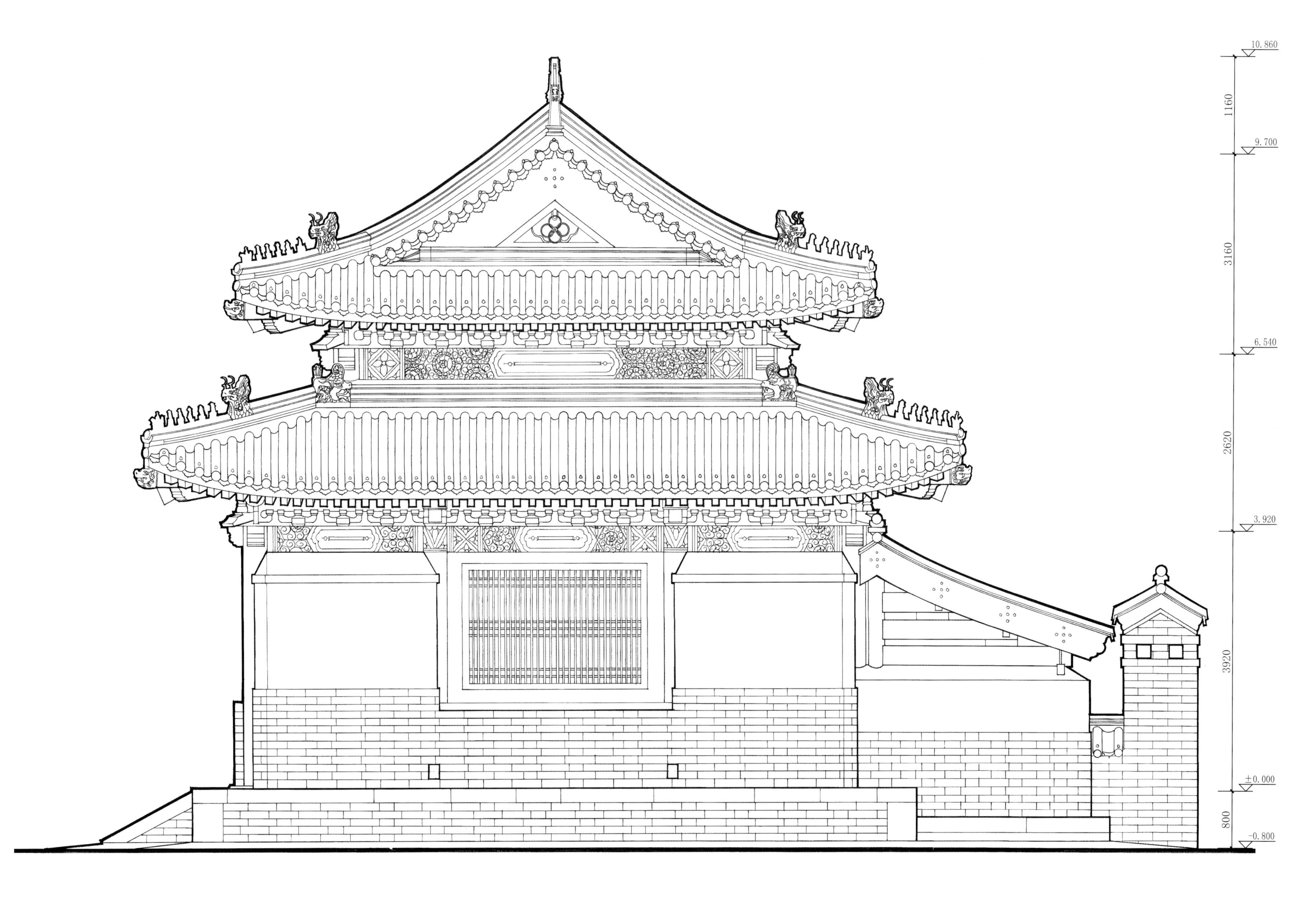

昌陵神厨库宰牲亭侧立面图
Side elevation of the Ritual Abattoir in the Culinary Courtyard for Sacrifices,Tomb of Good Fortune

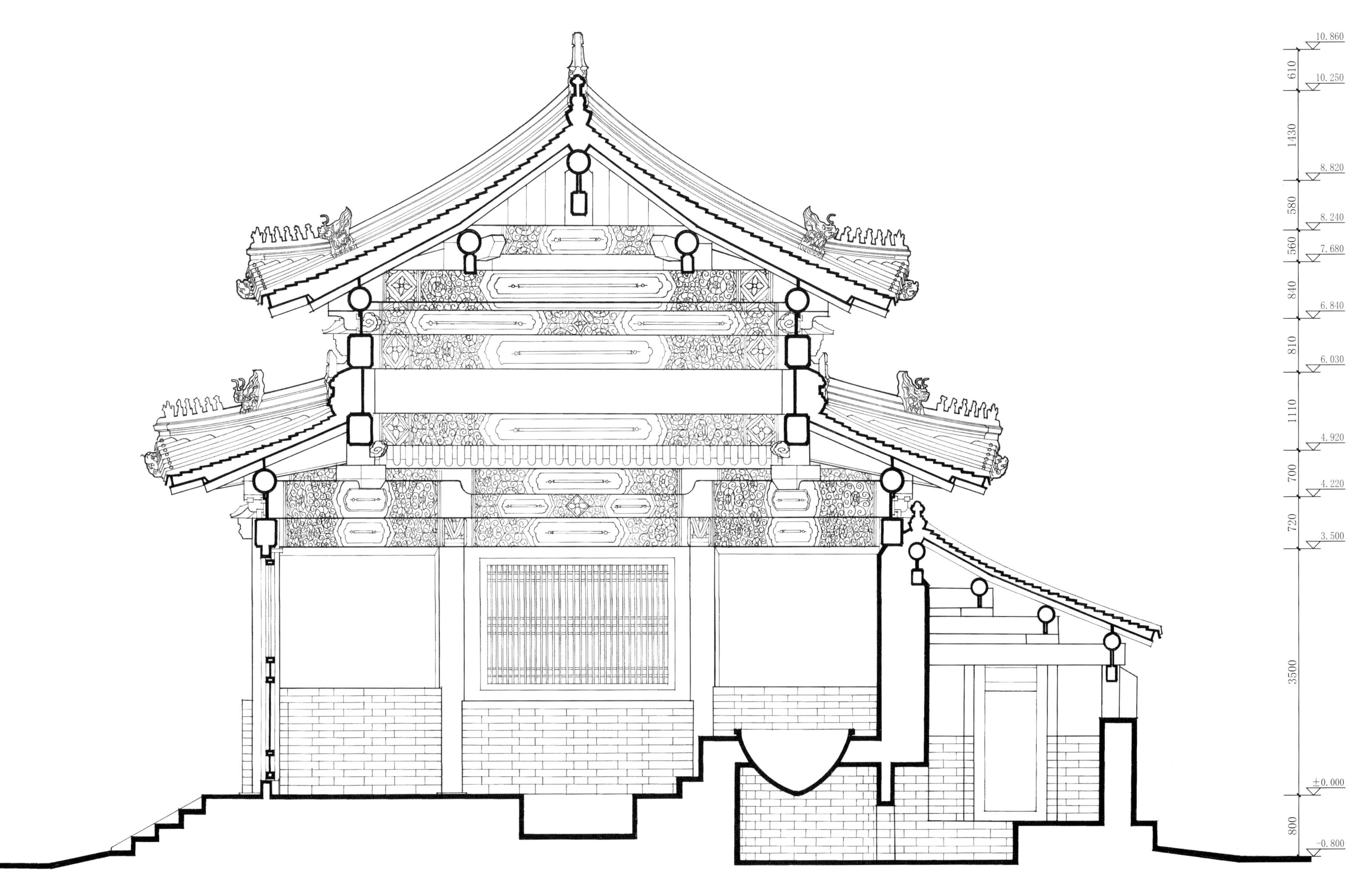

昌陵神厨库宰牲亭明间横剖面图

Cross section of the Central Chamber within the Ritual Abattoir in the Culinary Courtyard for Sacrifices,Tomb of Good Fortune

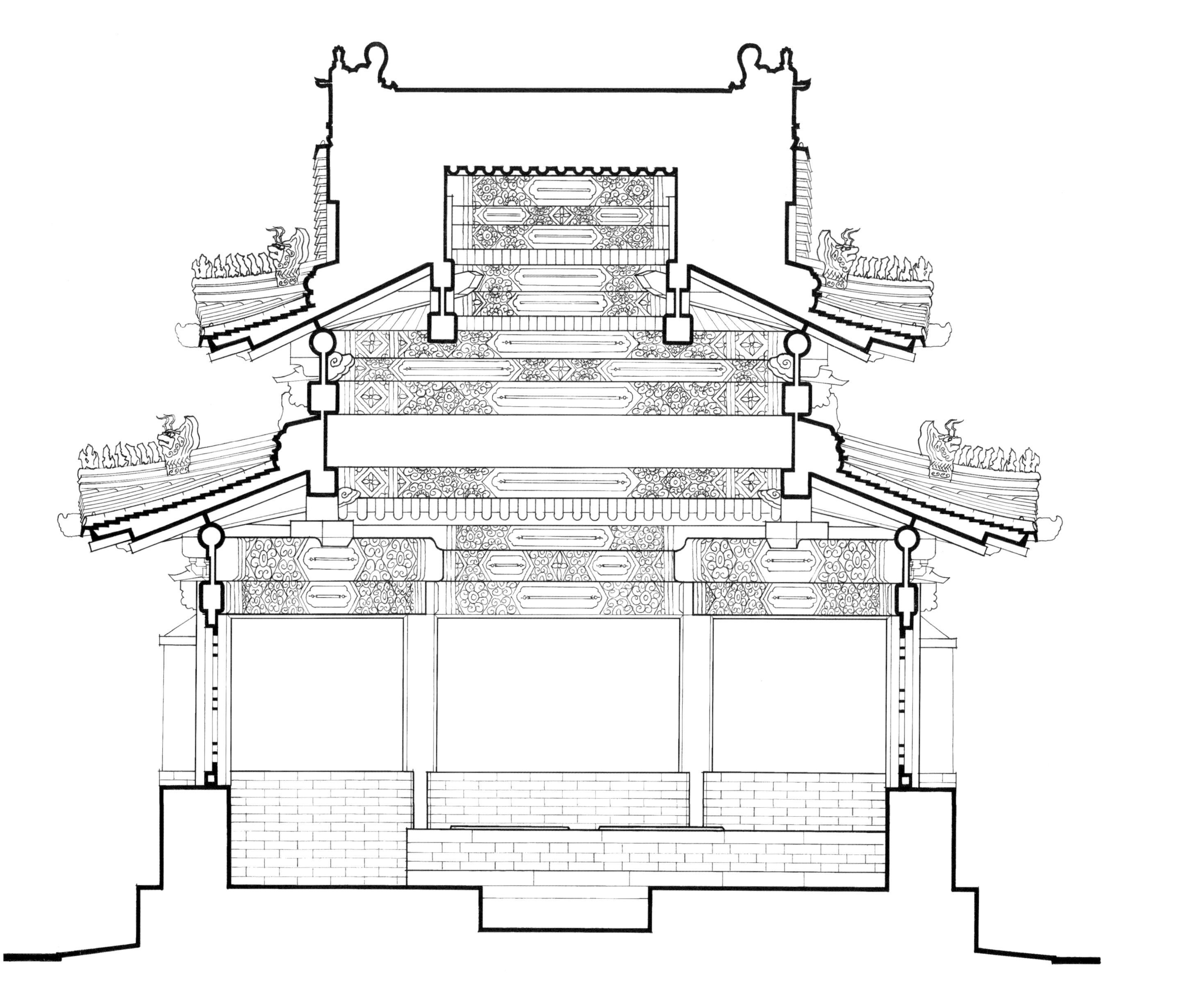

昌陵神厨库宰牲亭纵剖面图

Longitudinal section of the Ritual Abattoir in the Culinary Courtyard for Sacrifices,Tomb of Good Fortune

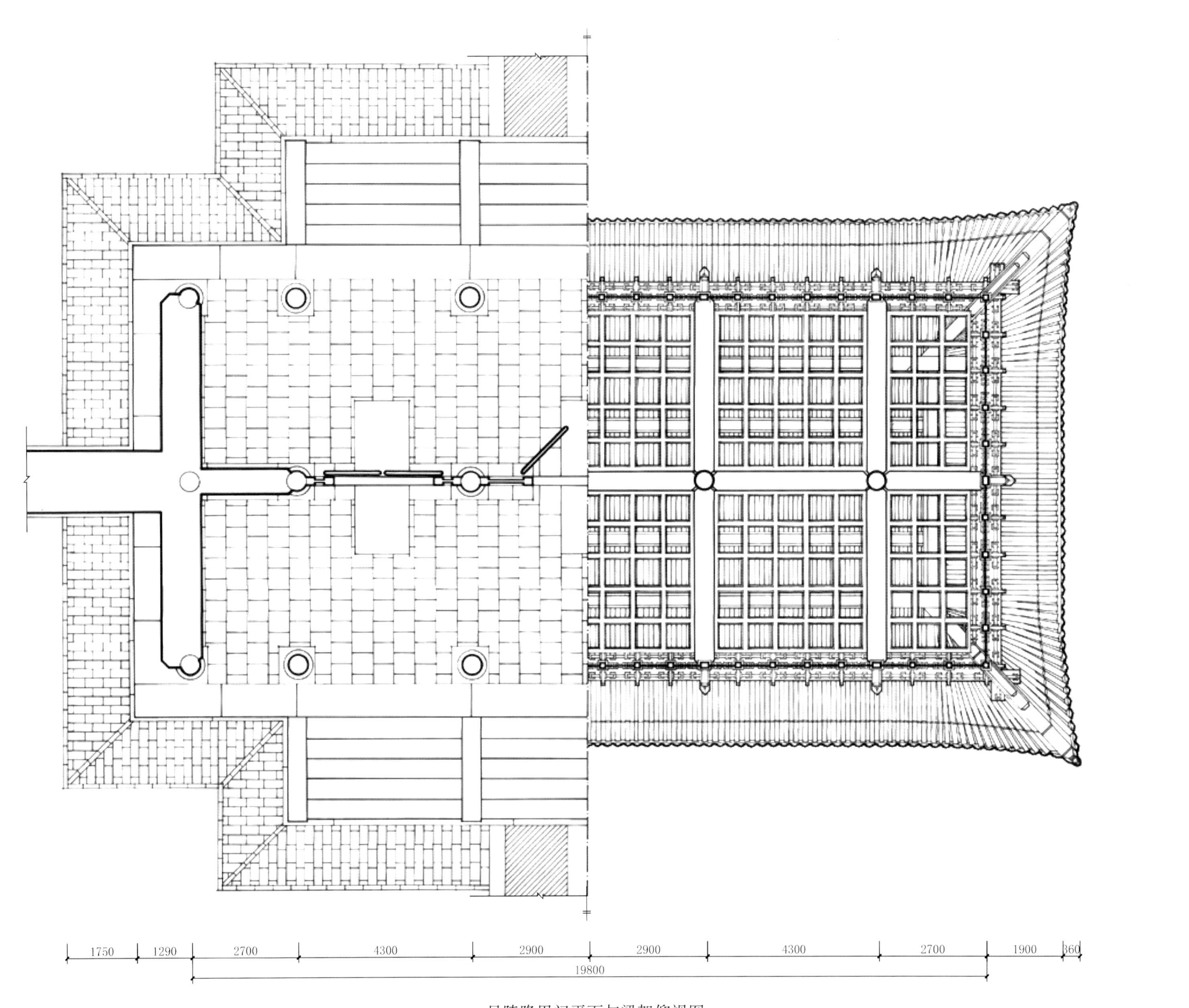

昌陵隆恩门平面与梁架仰视图

Plan and looking up at the Truss and Ceiling of the Gate of Monumental Grace,Tomb of Good Fortune

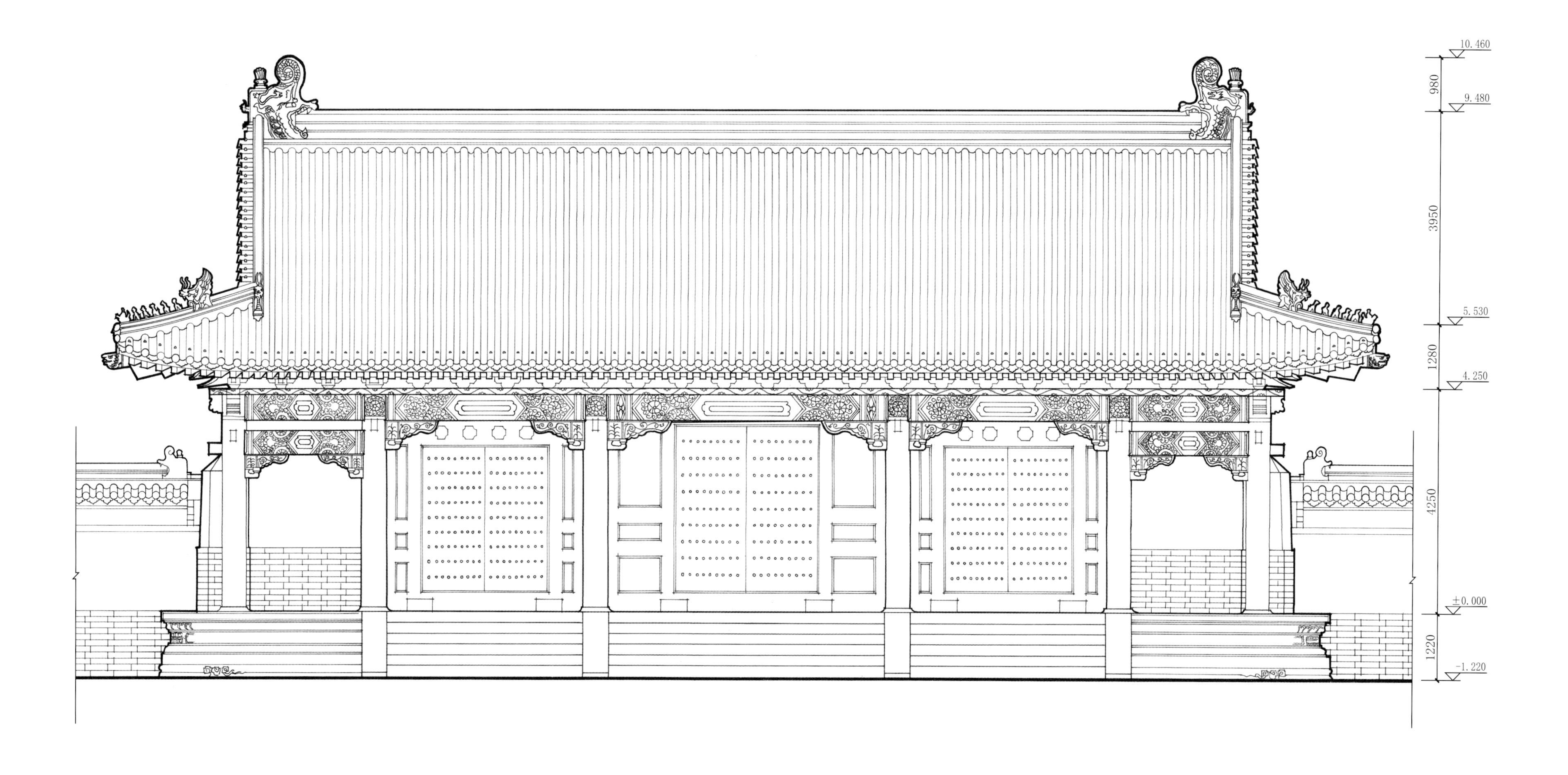

昌陵隆恩门正立面图

Front elevation of the Gate of Monumental Grace,Tomb of Good Fortune

昌陵隆恩门侧立面图

Side elevation of the Gate of Monumental Grace,Tomb of Good Fortune

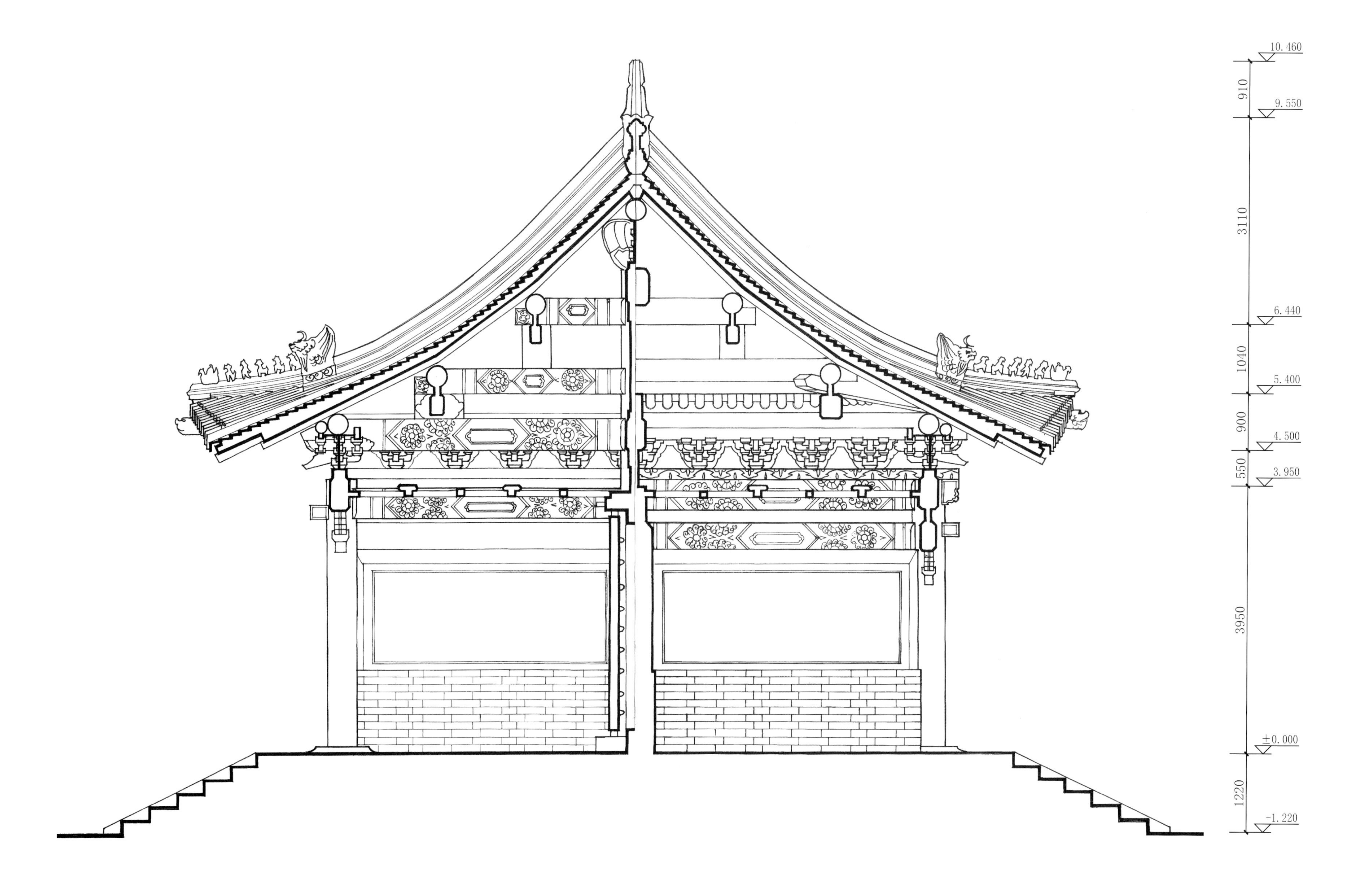

昌陵隆恩门明间稍间横剖面图

Cross section of the Central Chamber and Second-to-last Chamber within the Gate of Monumental Grace,Tomb of Good Fortune

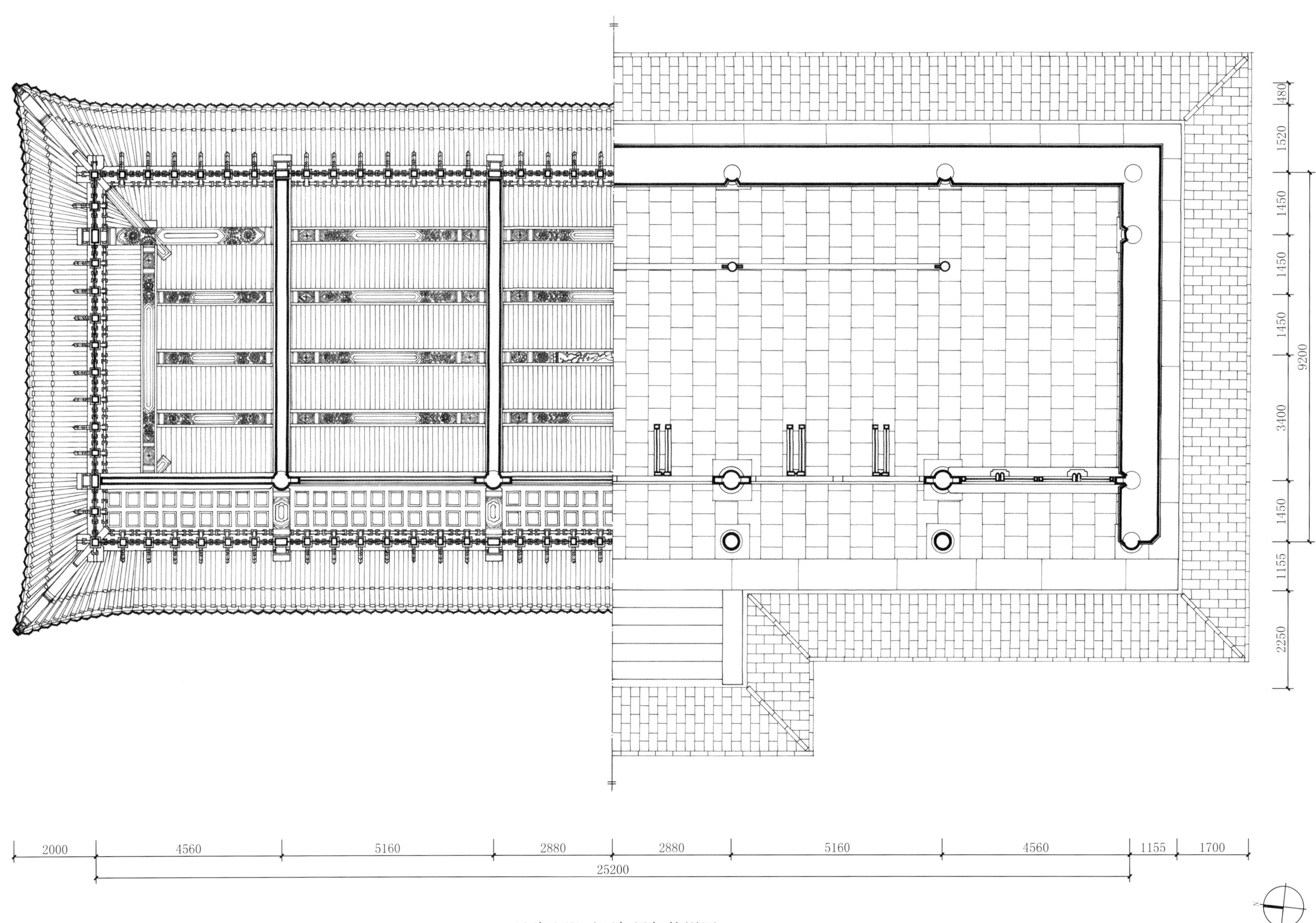

昌陵配殿平面与梁架仰视图

Plan and looking up at the Truss and Ceiling of the Side Hall,Tomb of Good Fortune

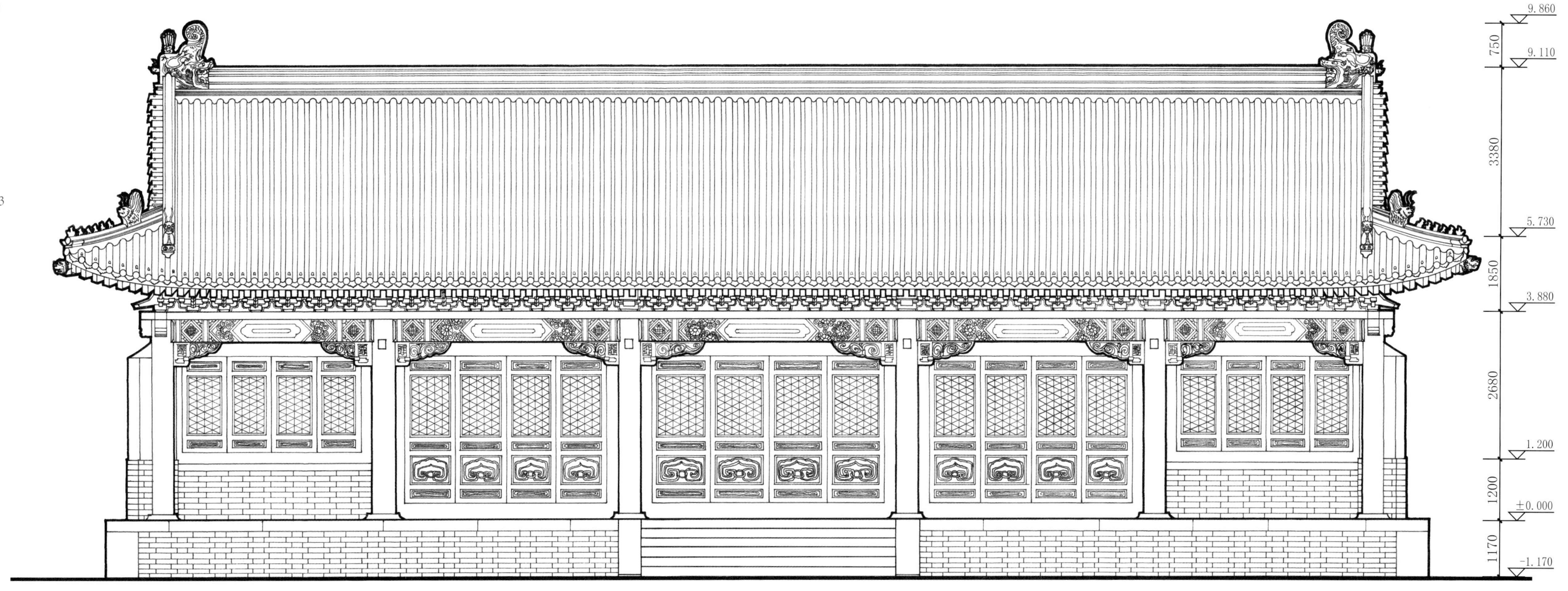

昌陵配殿正立面图

Front elevation of the Side Hall,Tomb of Good Fortune

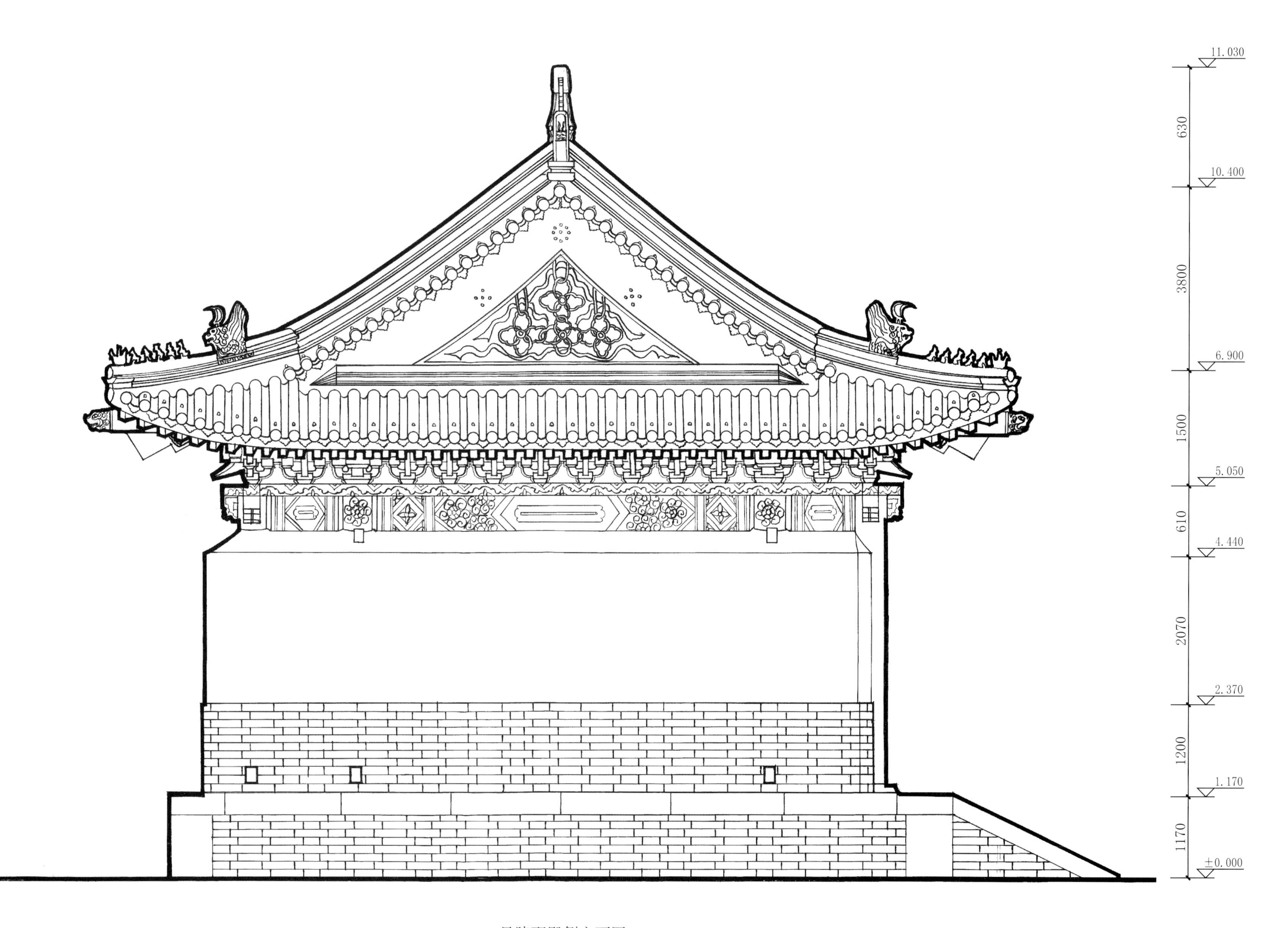

昌陵配殿侧立面图

Side elevation of the Side Hall,Tomb of Good Fortune

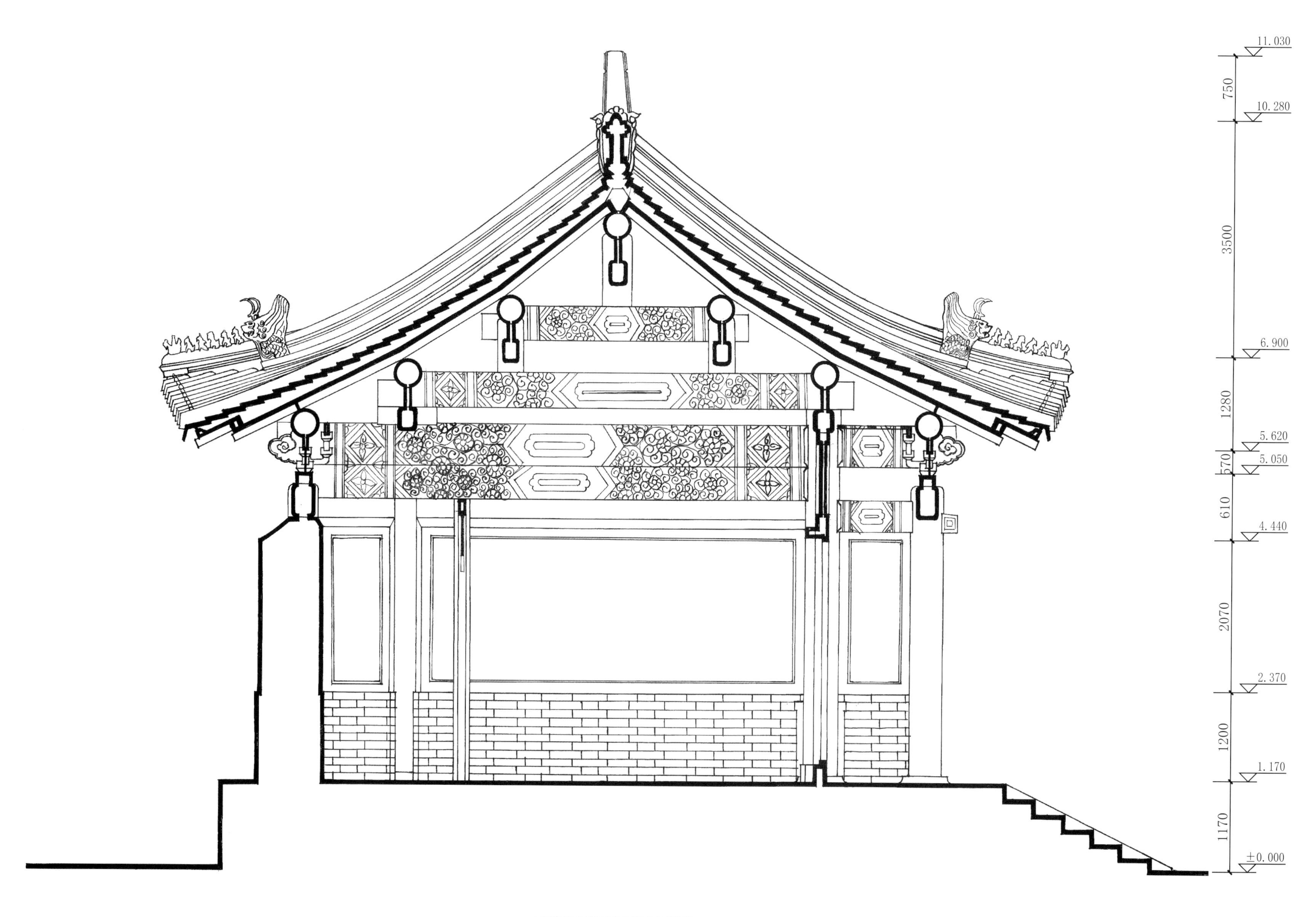

昌陵配殿明间横剖面图

Cross section of the Central Chamber within the Side Hall,Tomb of Good Fortune

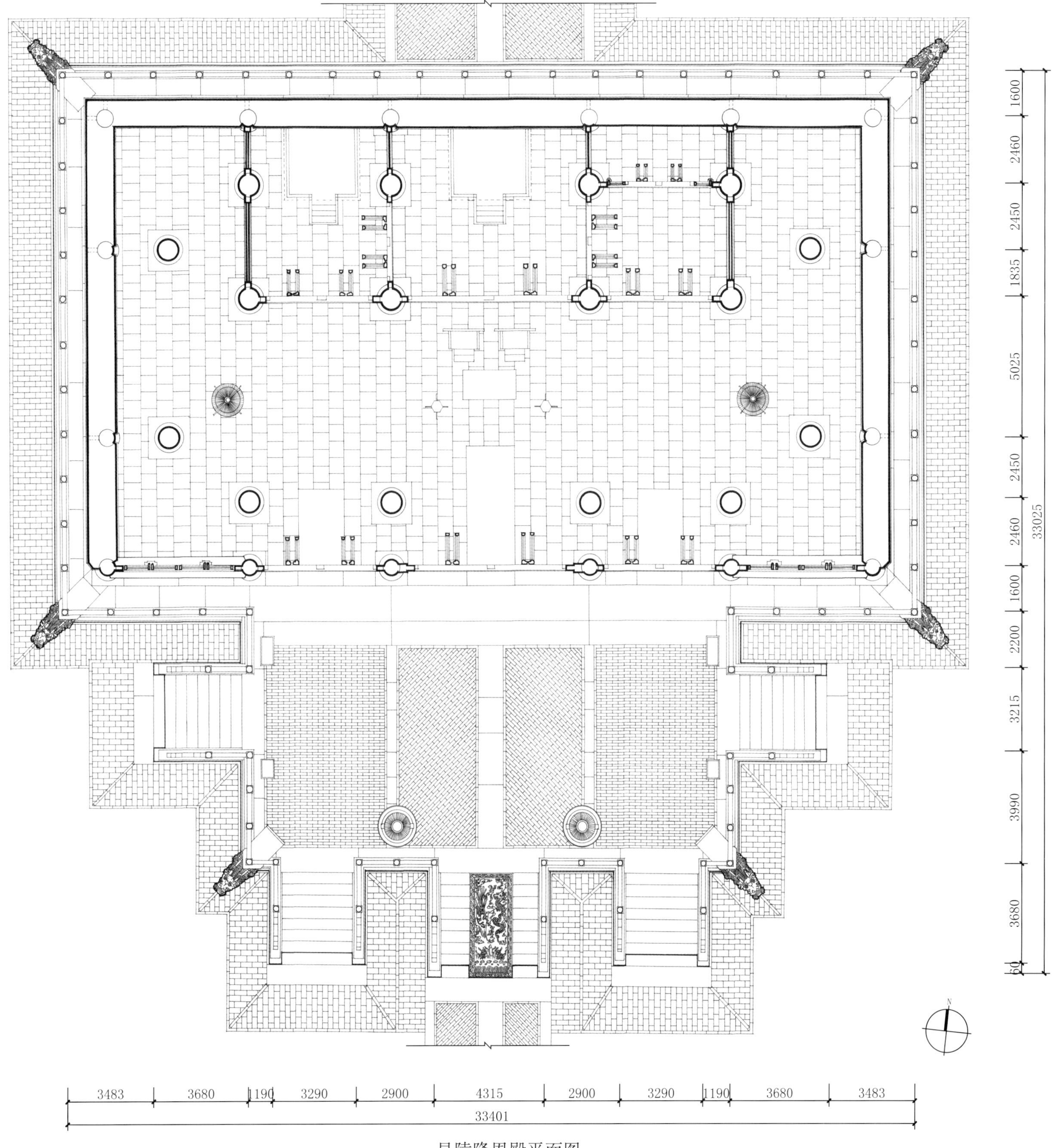

昌陵隆恩殿平面图

Plan of the Hall of Monumental Grace,Tomb of Good Fortune

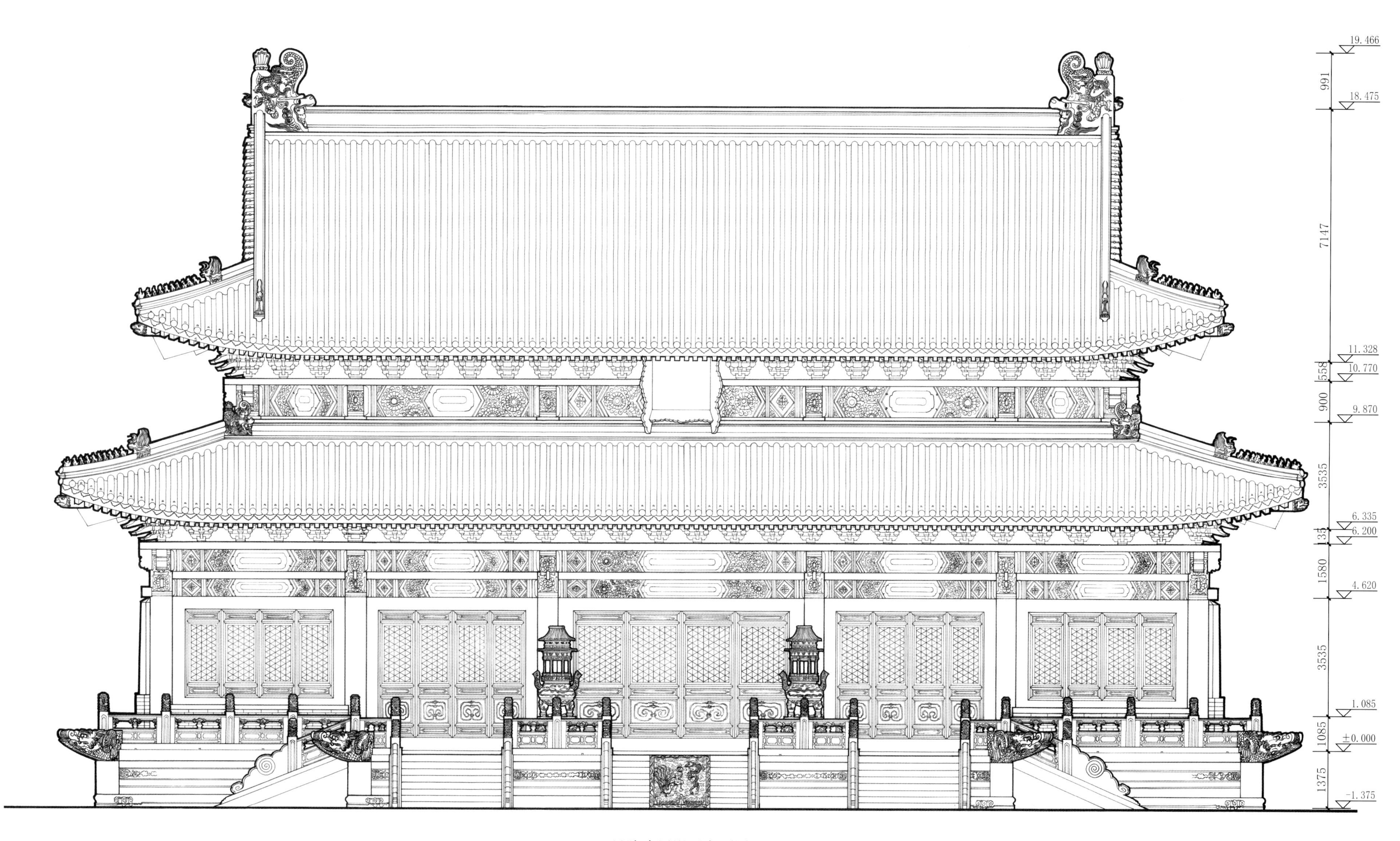

昌陵隆恩殿正立面图

Front elevation of the Hall of Monumental Grace,Tomb of Good Fortune

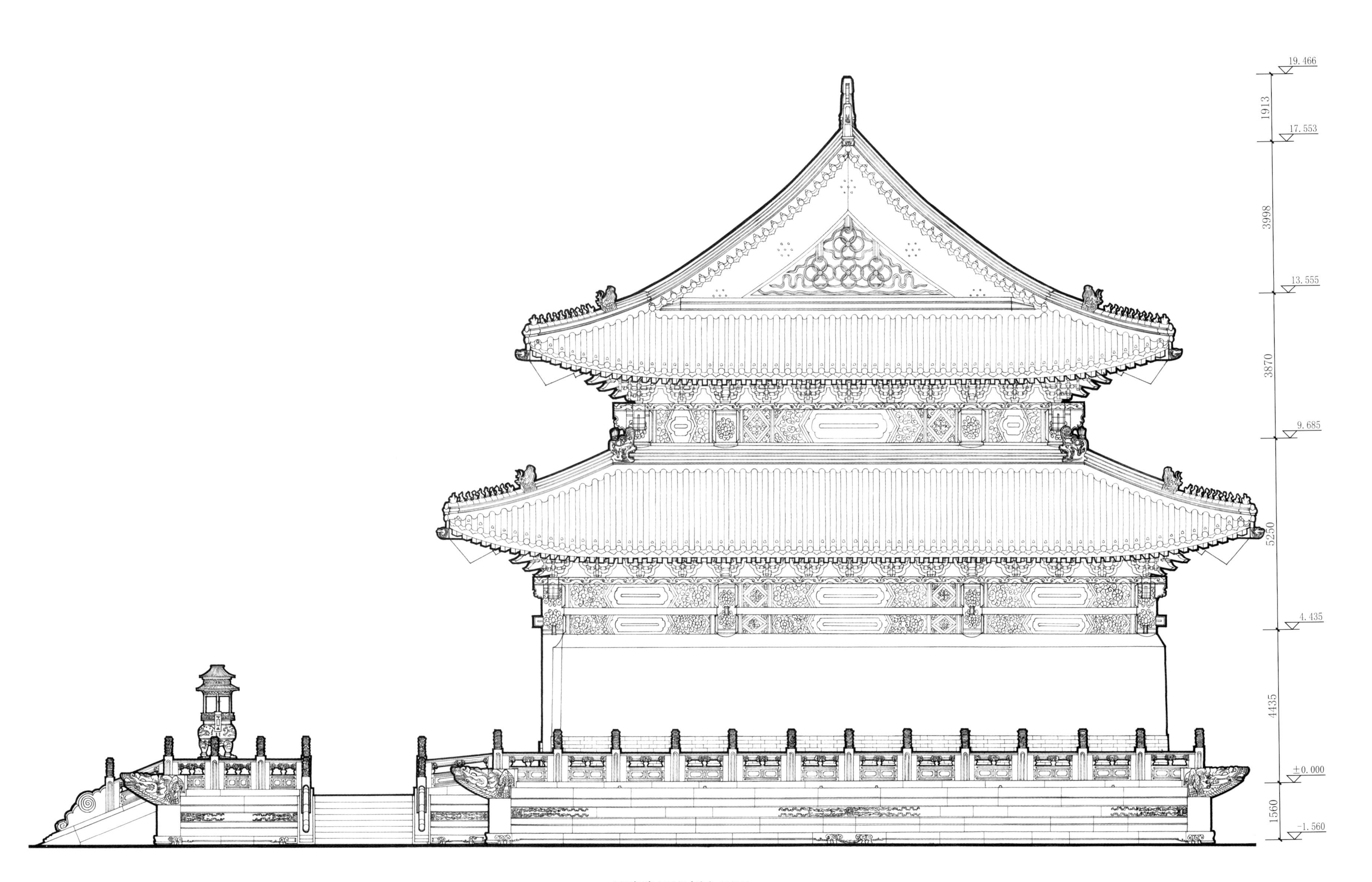

昌陵隆恩殿侧立面图

Side elevation of the Hall of Monumental Grace,Tomb of Good Fortune

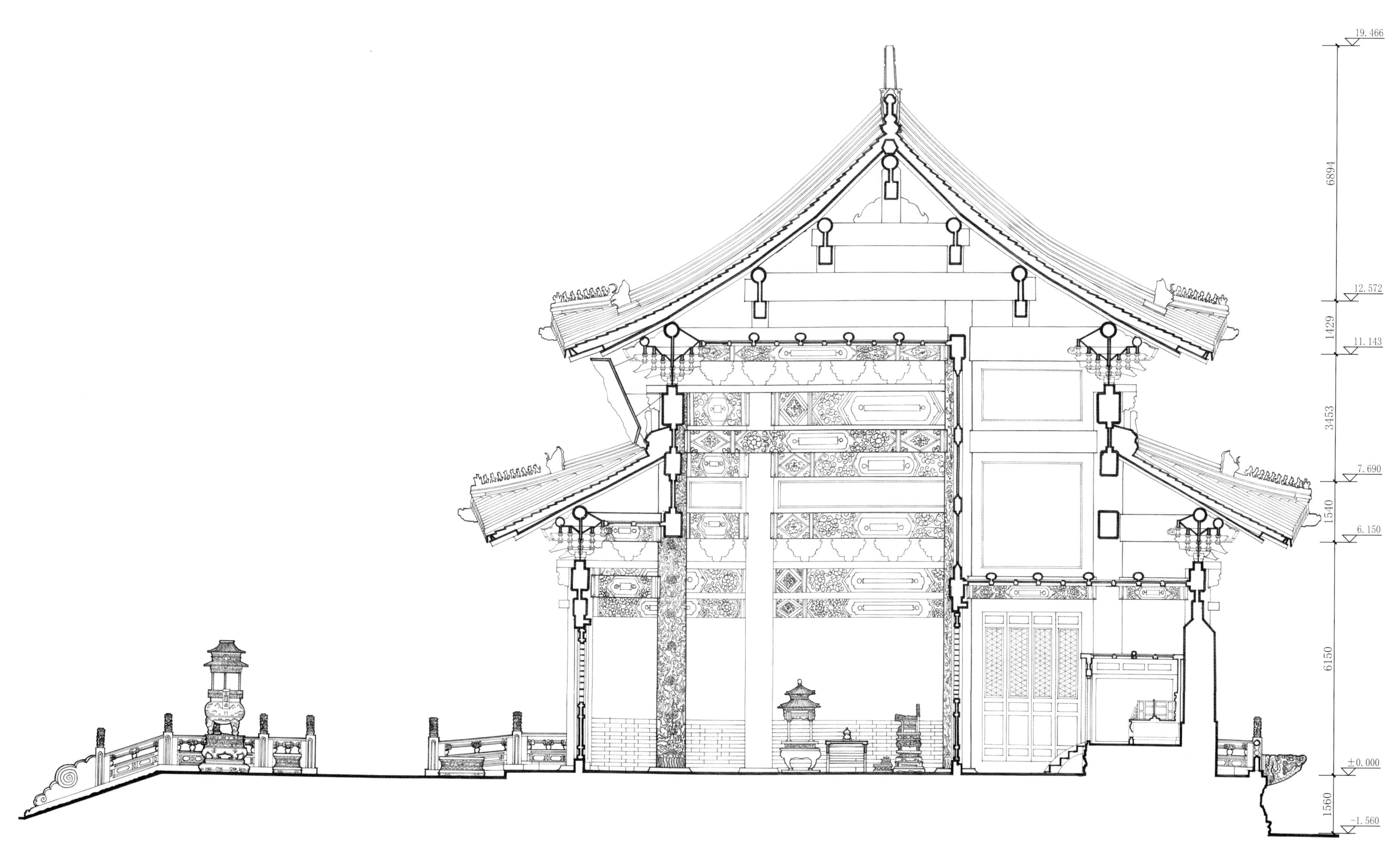

昌陵隆恩殿明间横剖面图

Cross section of the Central Chamber within the Hall of Monumental Grace,Tomb of Good Fortune

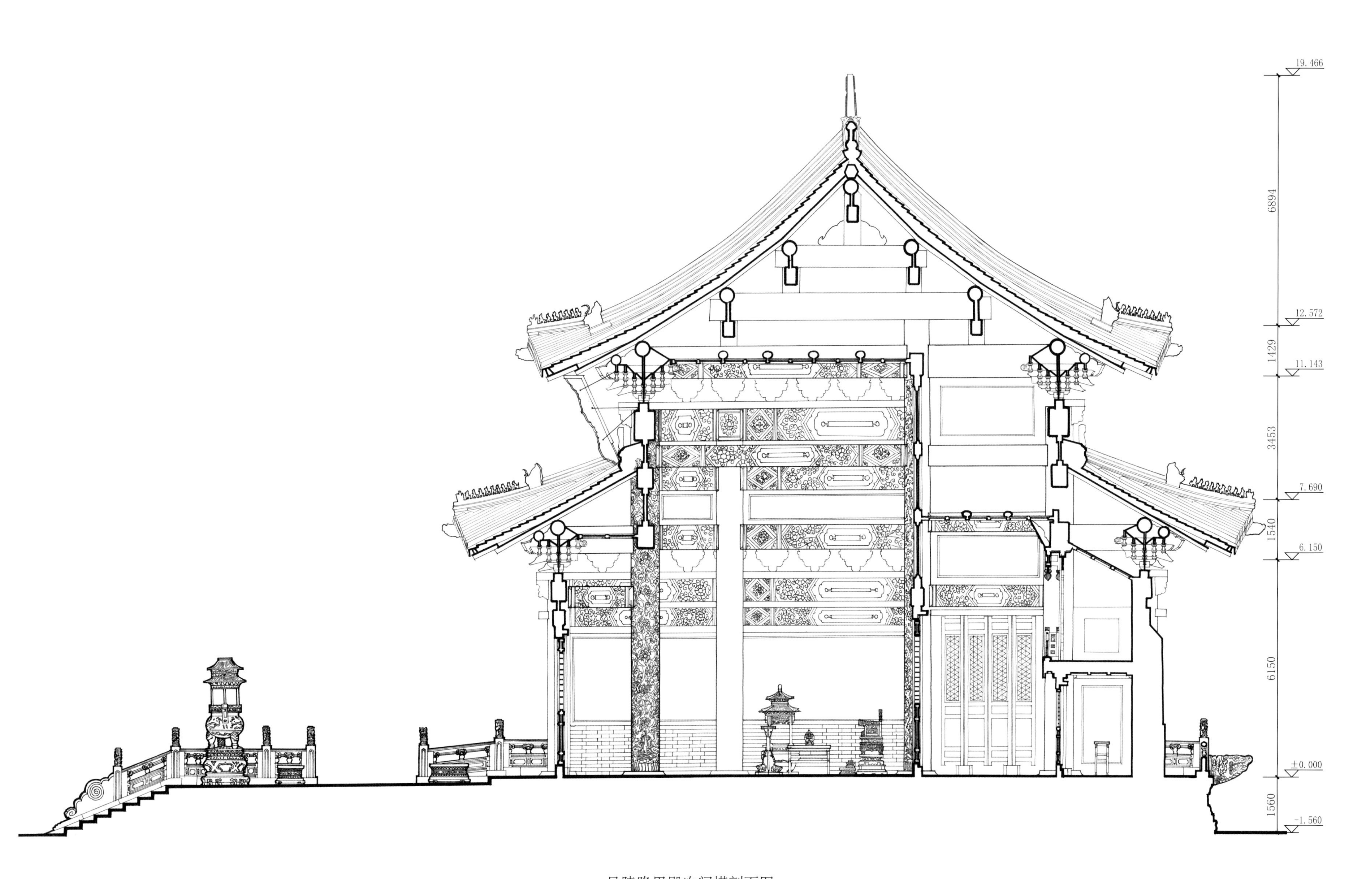

昌陵隆恩殿次间横剖面图

Cross section of the Side Chamber within the Hall of Monumental Grace,Tomb of Good Fortune

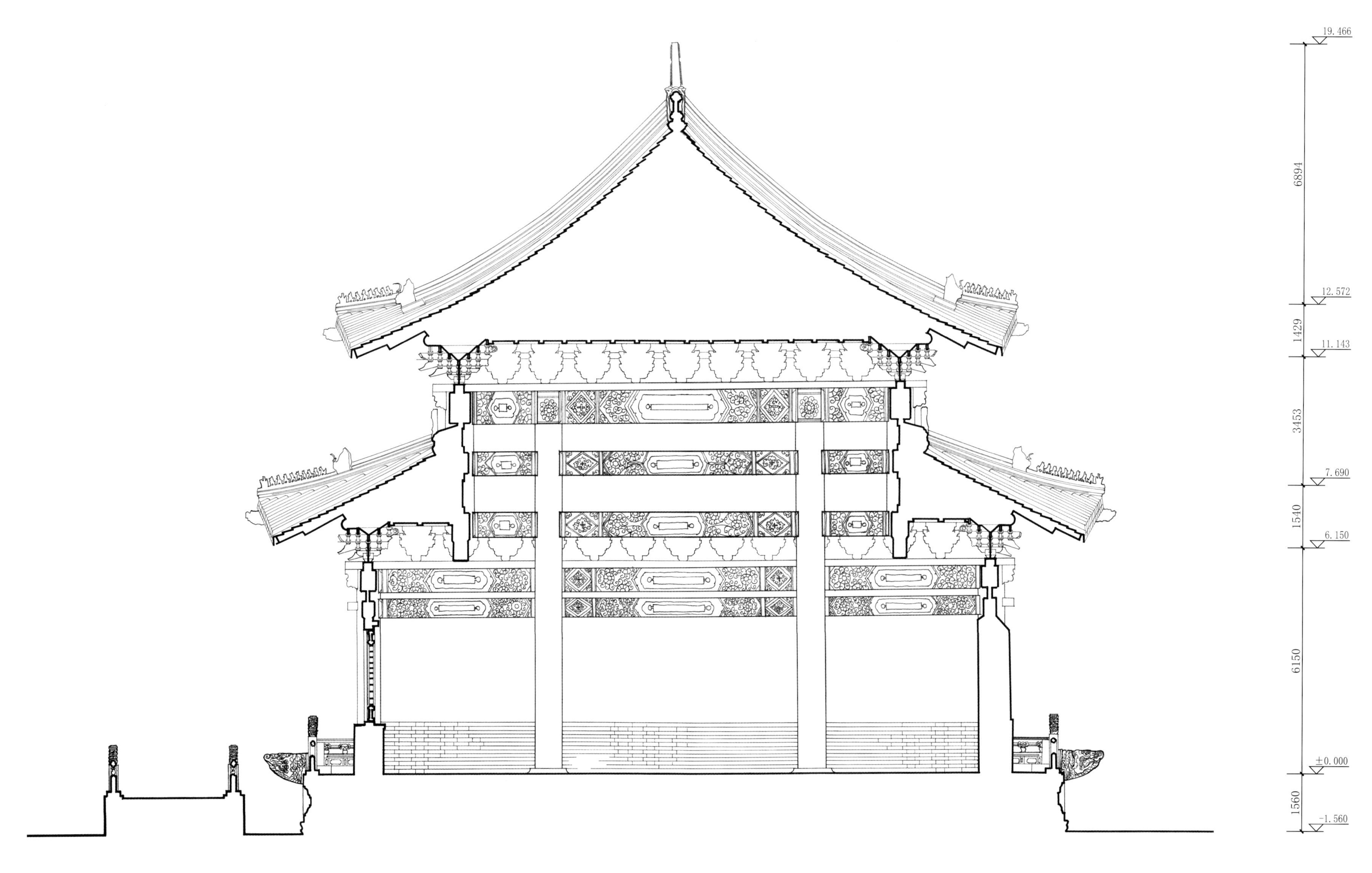

昌陵隆恩殿梢间横剖面图

Cross section of the Chamber at the end within the Hall of Monumental Grace,Tomb of Good Fortune

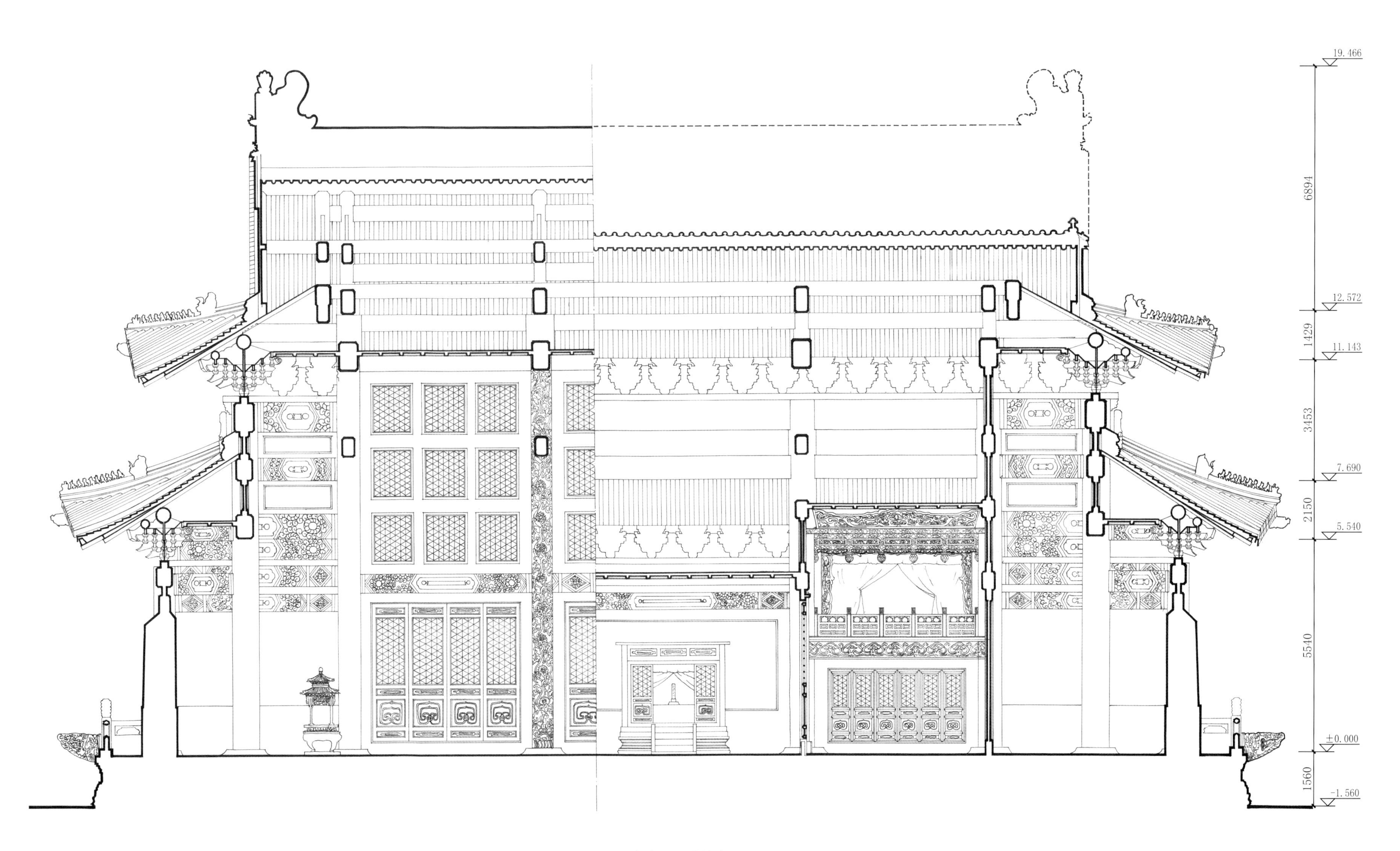

昌陵隆恩殿纵剖面图

Longitudinal section of the Hall of Monumental Grace,Tomb of Good Fortune

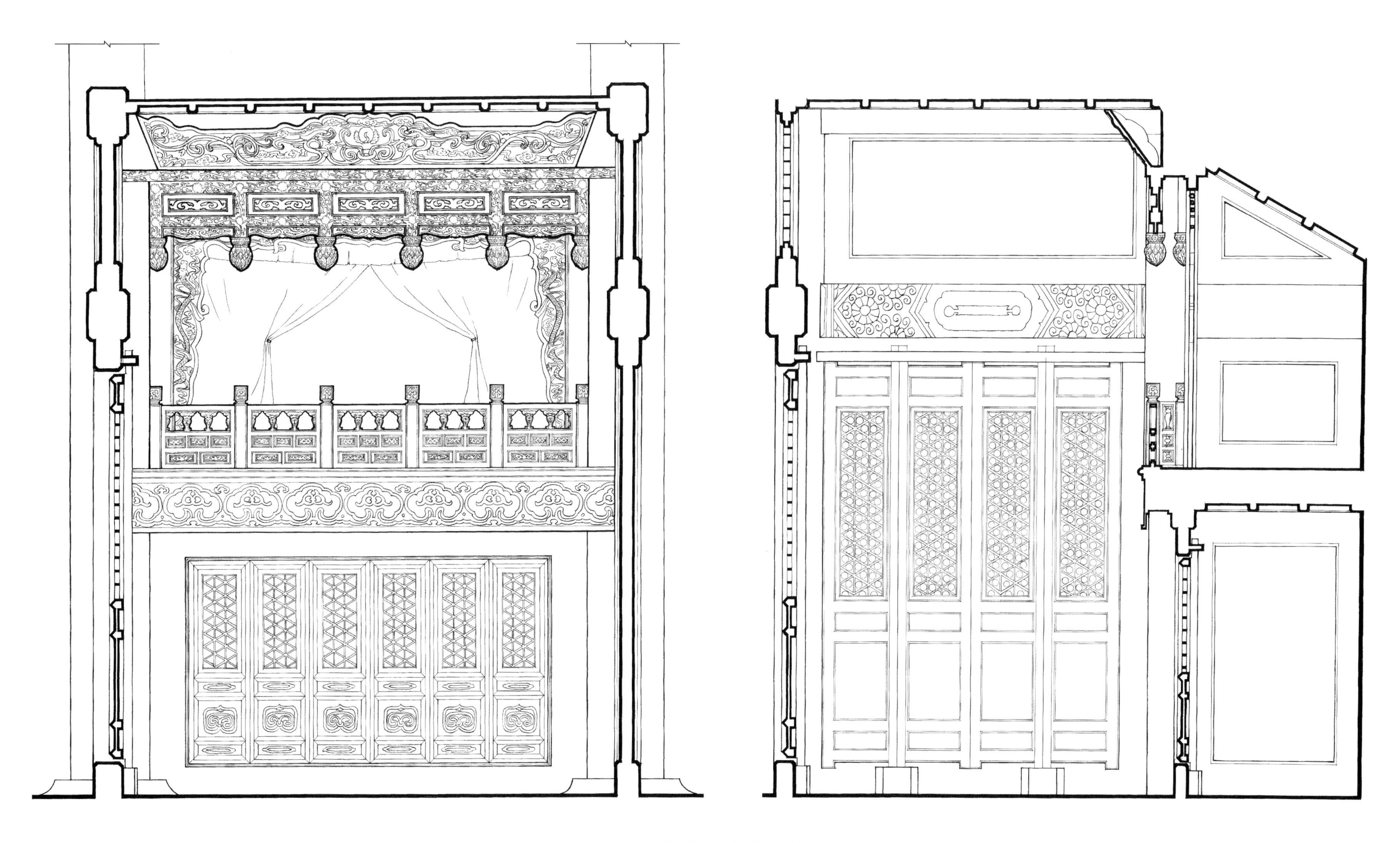

昌陵隆恩殿佛楼大样图

Detailed drawing of the Buddha Building of the Hall of Monumental Grace,Tomb of Good Fortune

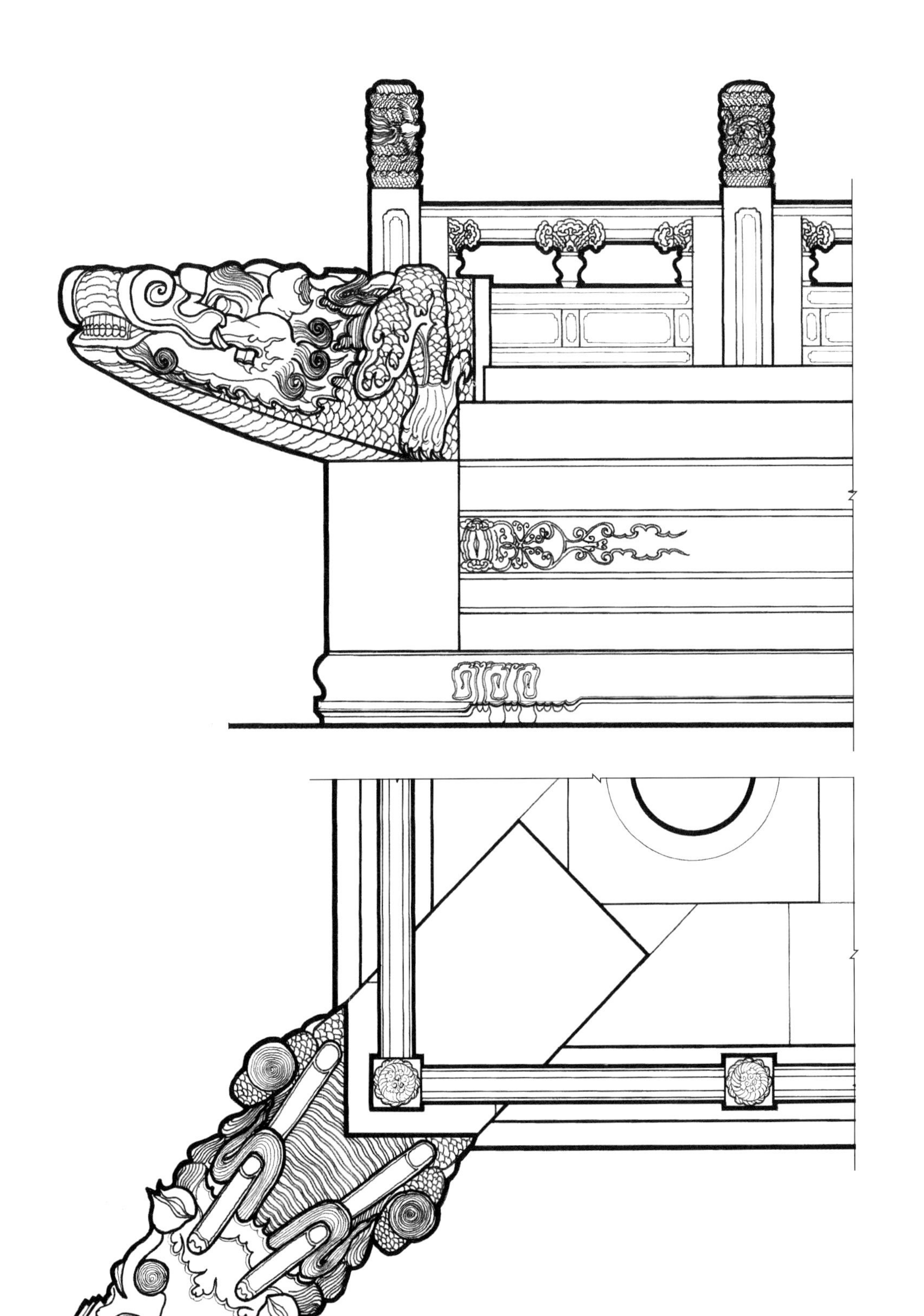

昌陵隆恩殿螭首大样图
Detailed drawing of the dragon gargoyle (chishou) outside the Hall of Monumental Grace,Tomb of Good Fortune

0 0.25 0.5 1m

昌陵隆恩殿香炉大样图
Detailed drawing of the Incense Burner of the Hall of Monumental Grace,Tomb of Good Fortune

0 0.1 0.2 0.4m

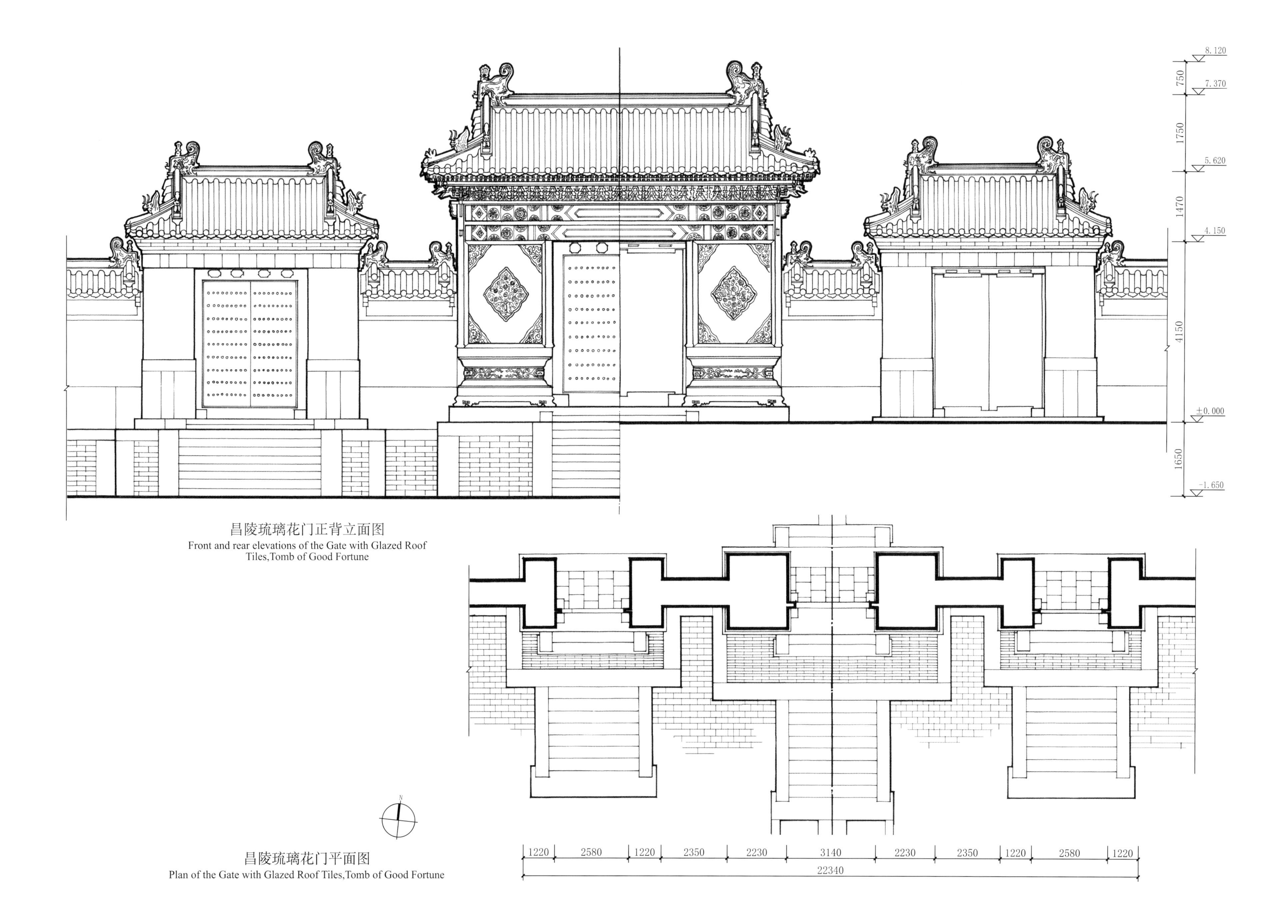

昌陵琉璃花门正背立面图
Front and rear elevations of the Gate with Glazed Roof Tiles,Tomb of Good Fortune

昌陵琉璃花门平面图
Plan of the Gate with Glazed Roof Tiles,Tomb of Good Fortune

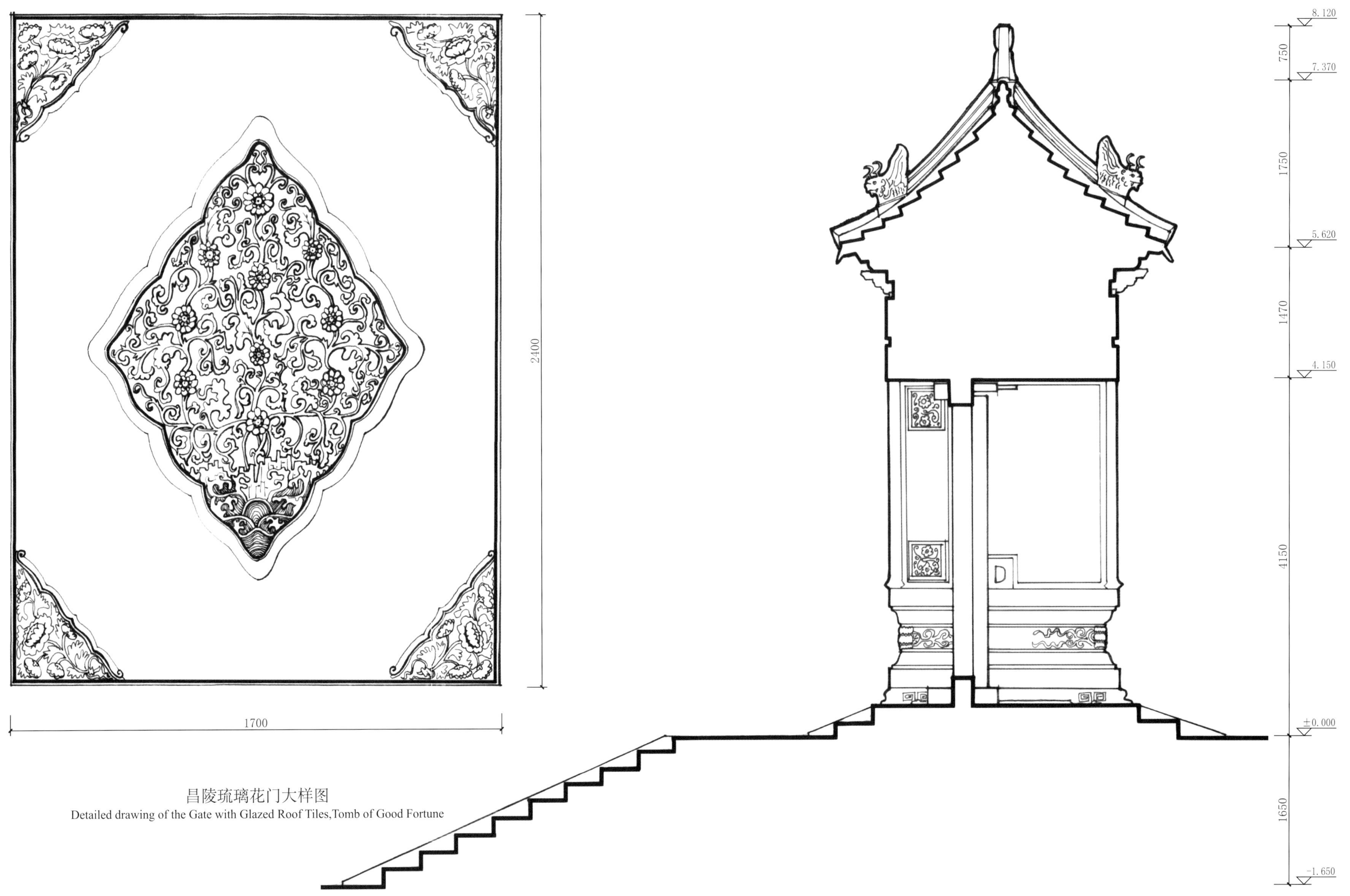

昌陵琉璃花门大样图
Detailed drawing of the Gate with Glazed Roof Tiles,Tomb of Good Fortune

昌陵琉璃花门横剖面图
Cross section of the Gate with Glazed Roof Tiles,Tomb of Good Fortune

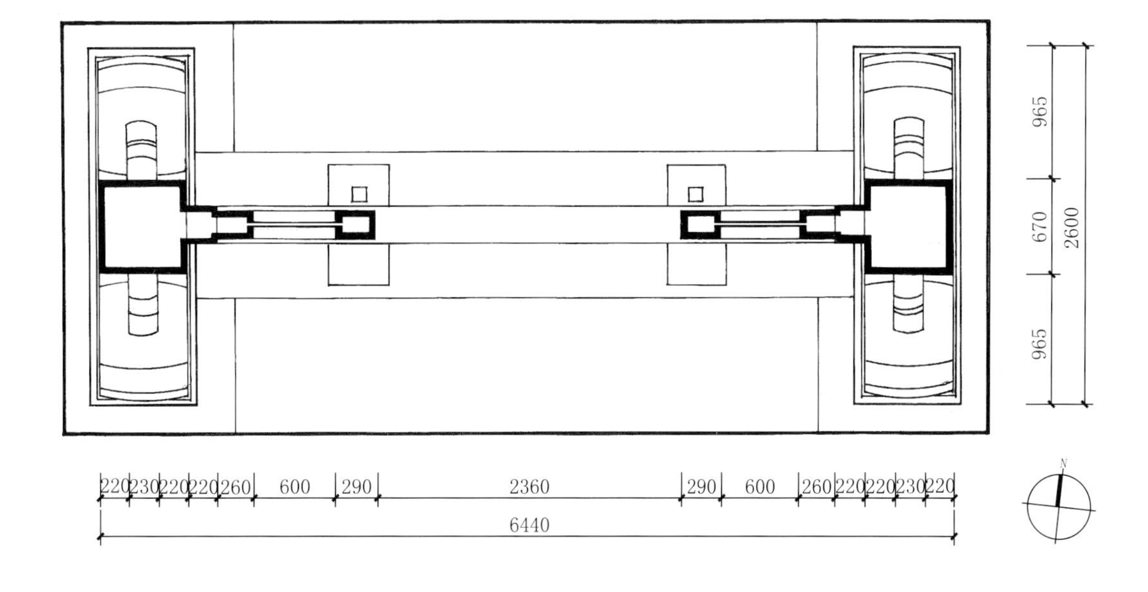

昌陵二柱门平面图
Plan of the Gate with Two Columns,Tomb of Good Fortune

昌陵二柱门大样图
Detailed drawing of the Gate with Two Columns,Tomb of Good Fortune

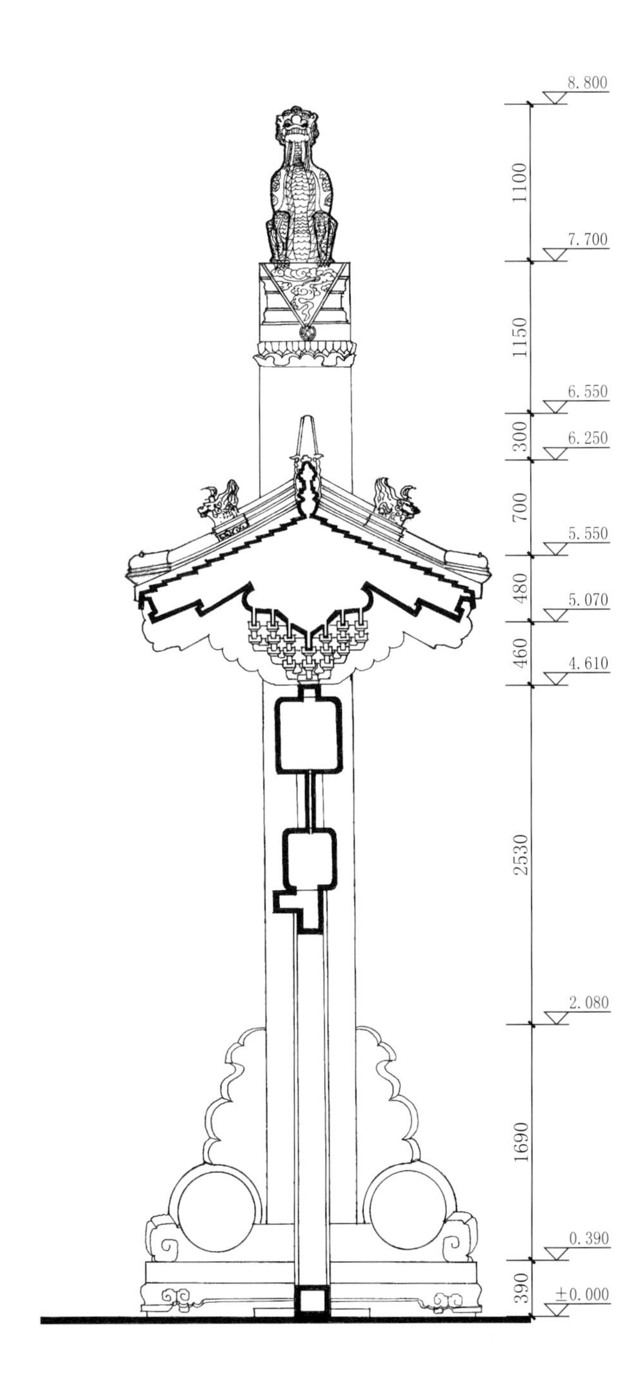

昌陵二柱门横剖面图
Cross section of the Gate with Two Columns,Tomb of Good Fortune

昌陵二柱门正立面图
Front elevation of the Gate with Two Columns,Tomb of Good Fortune

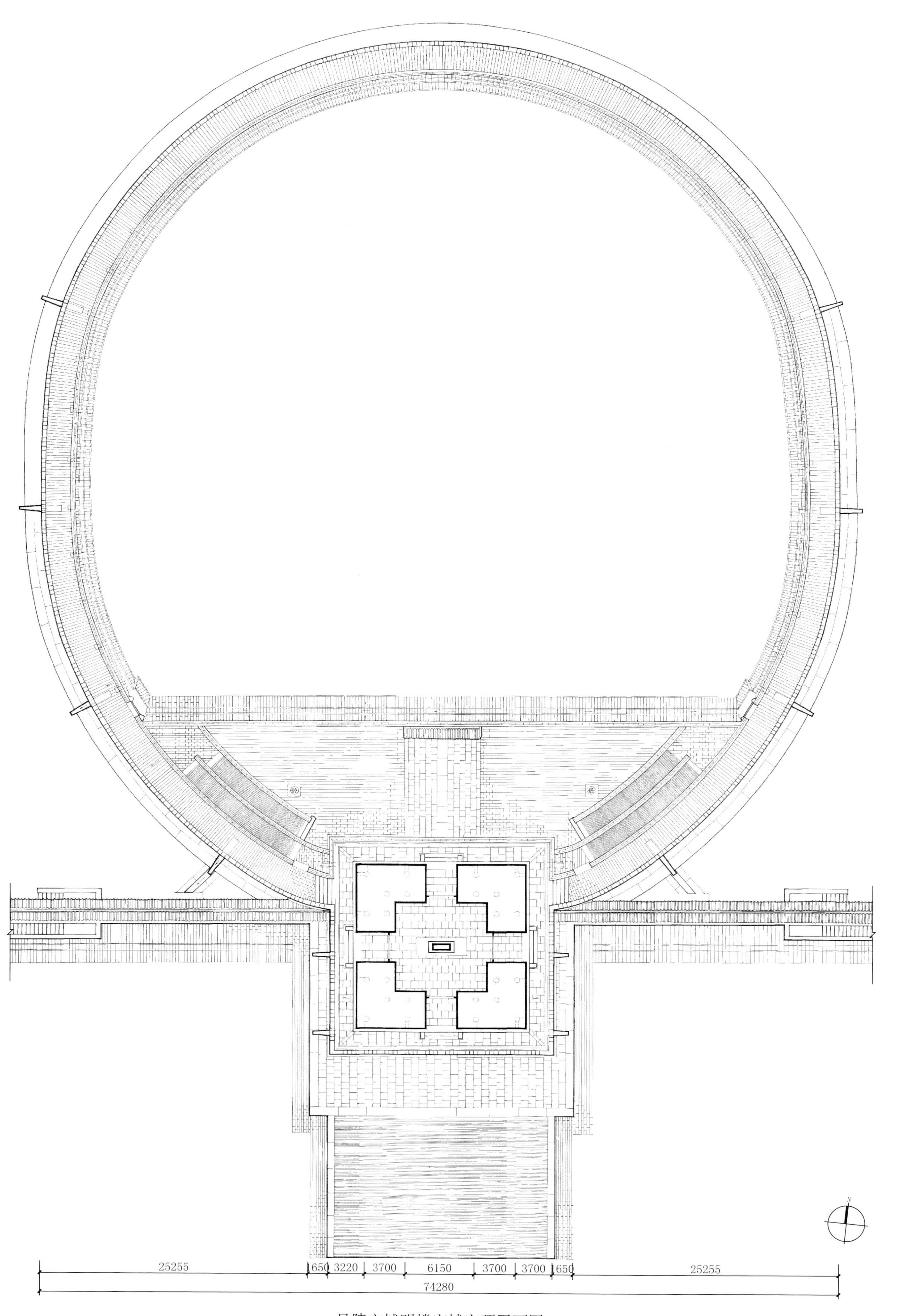

昌陵方城明楼宝城宝顶平面图

Plan of the Square Walled Terrace and the Memorial Tower as well as the Encircled Realm of Treasure and the Tumulus,Tomb of Good Fortune

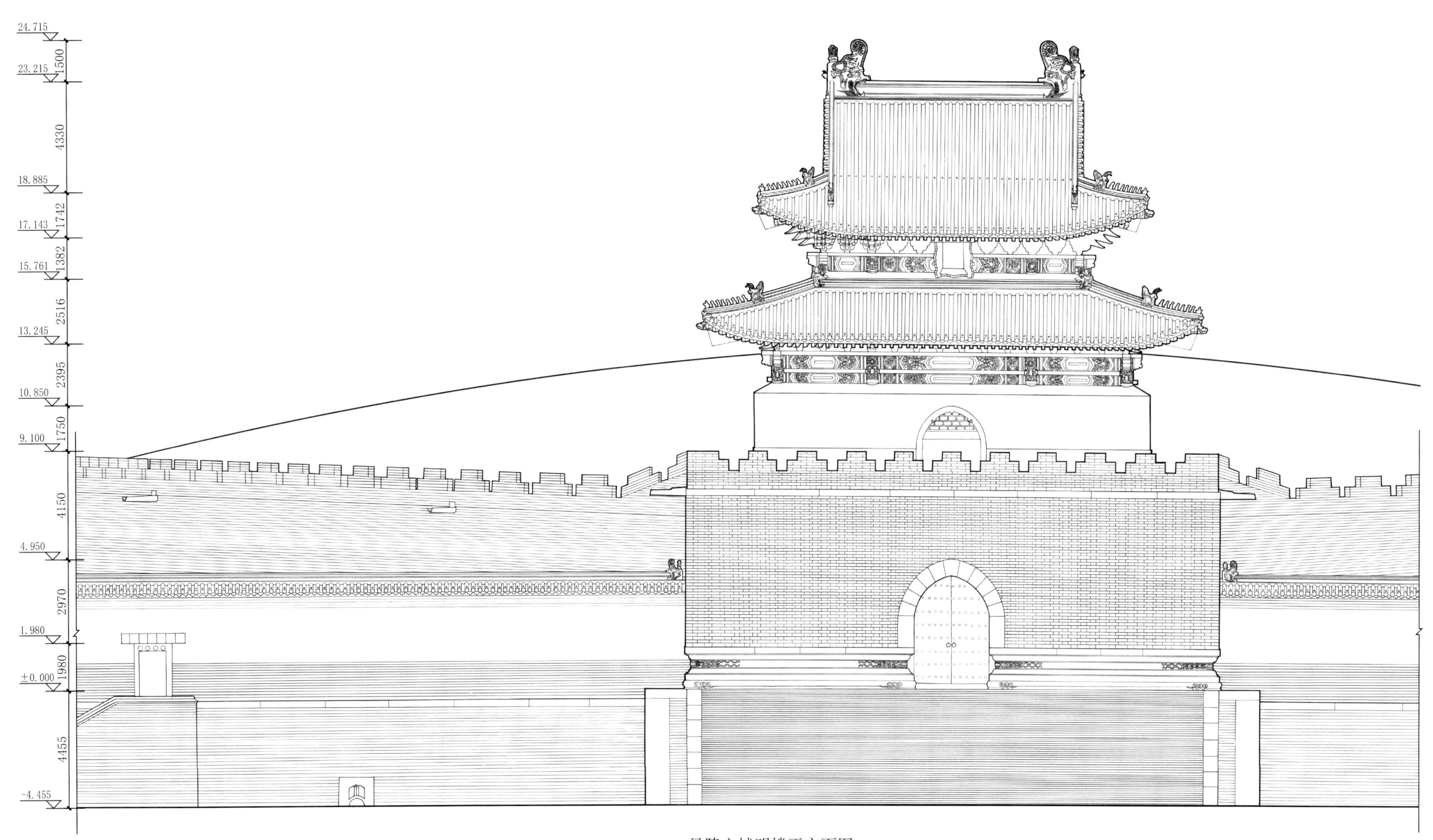

昌陵方城明楼正立面图

Front elevation of the Square Walled Terrace and the Memorial Tower,Tomb of Good Fortune

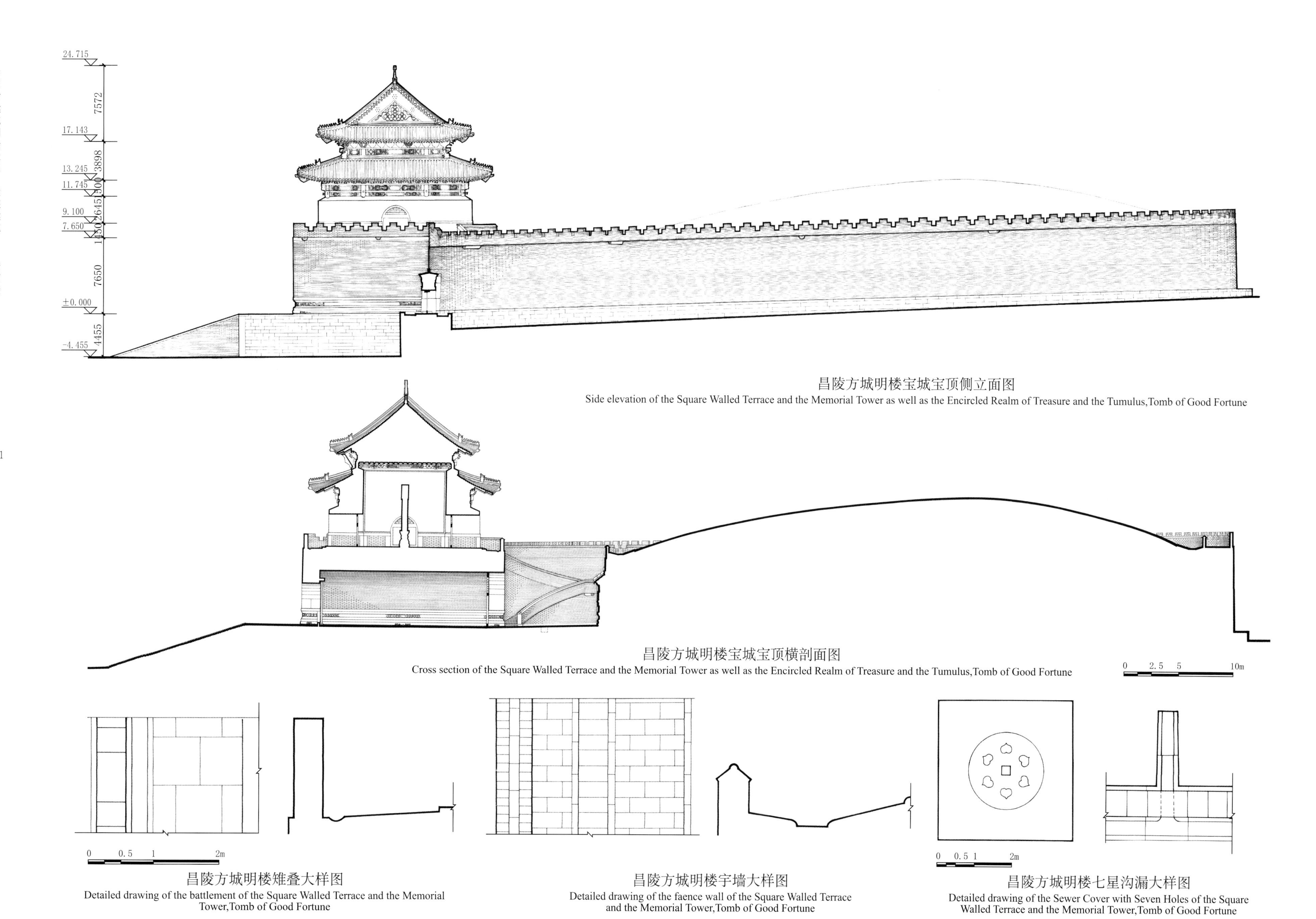

昌陵方城明楼宝城宝顶侧立面图

Side elevation of the Square Walled Terrace and the Memorial Tower as well as the Encircled Realm of Treasure and the Tumulus,Tomb of Good Fortune

昌陵方城明楼宝城宝顶横剖面图

Cross section of the Square Walled Terrace and the Memorial Tower as well as the Encircled Realm of Treasure and the Tumulus,Tomb of Good Fortune

昌陵方城明楼雉叠大样图

Detailed drawing of the battlement of the Square Walled Terrace and the Memorial Tower,Tomb of Good Fortune

昌陵方城明楼宇墙大样图

Detailed drawing of the faence wall of the Square Walled Terrace and the Memorial Tower,Tomb of Good Fortune

昌陵方城明楼七星沟漏大样图

Detailed drawing of the Sewer Cover with Seven Holes of the Square Walled Terrace and the Memorial Tower,Tomb of Good Fortune

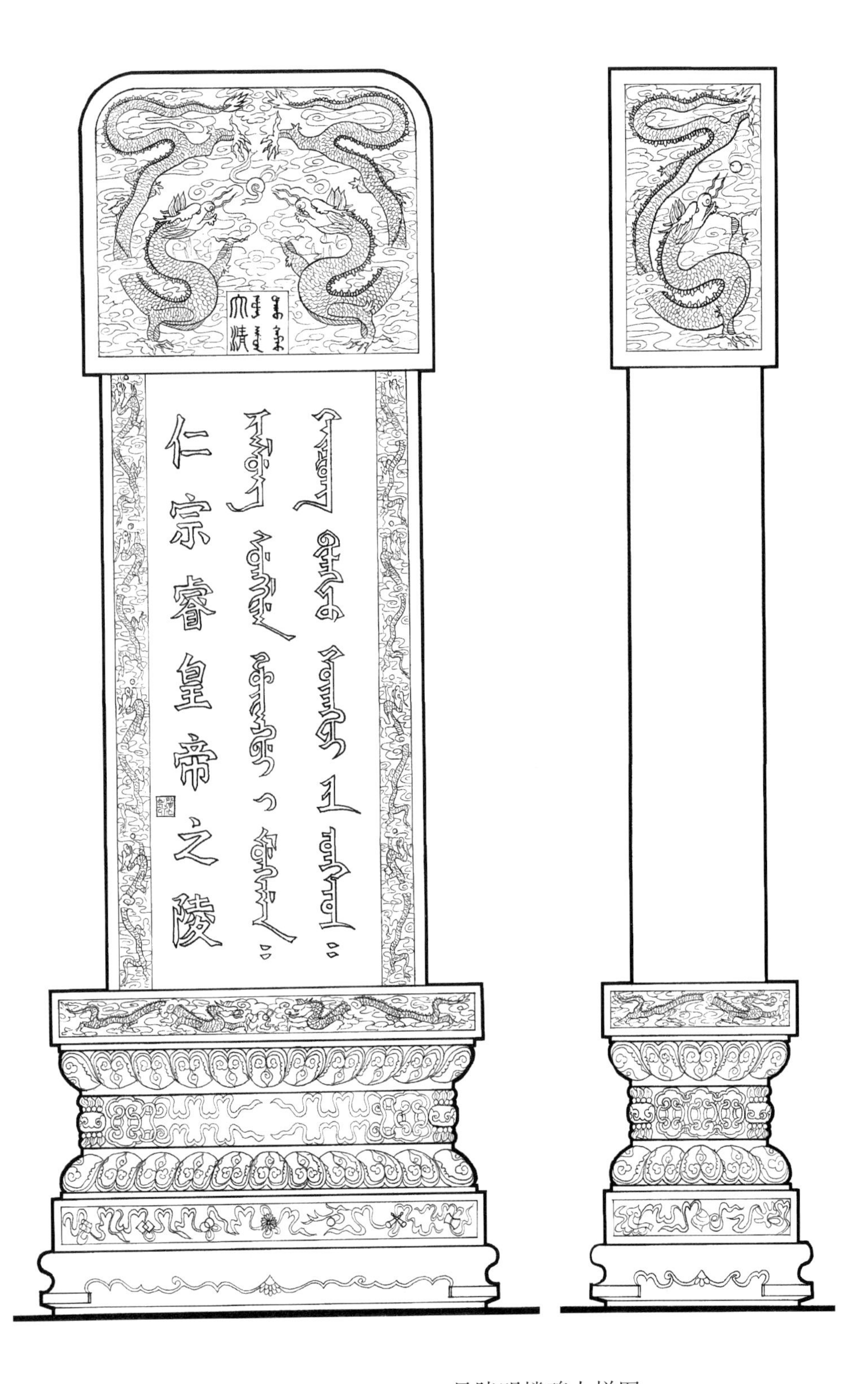

昌陵明楼碑大样图

Detailed drawing of the Tablet in the Memorial Tower,Tomb of Good Fortune

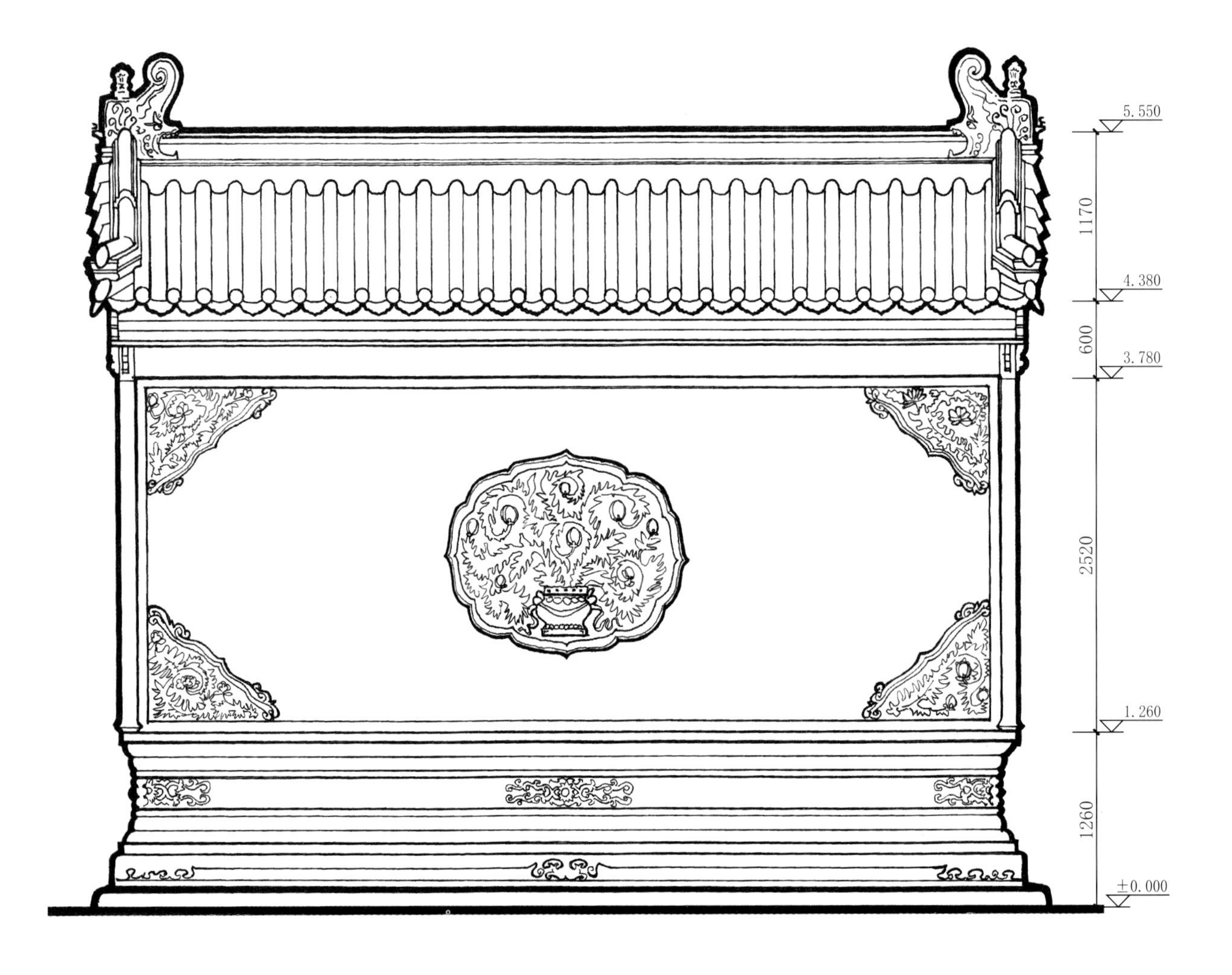

昌陵方城明楼琉璃影壁正立面图

Front elevation of the Screen Wall of Glazed Tiles in the Square Walled Terrace and the Memorial Tower,Tomb of Good Fortune

昌西陵
West Tomb of Good Fortune (Chang Xiling)

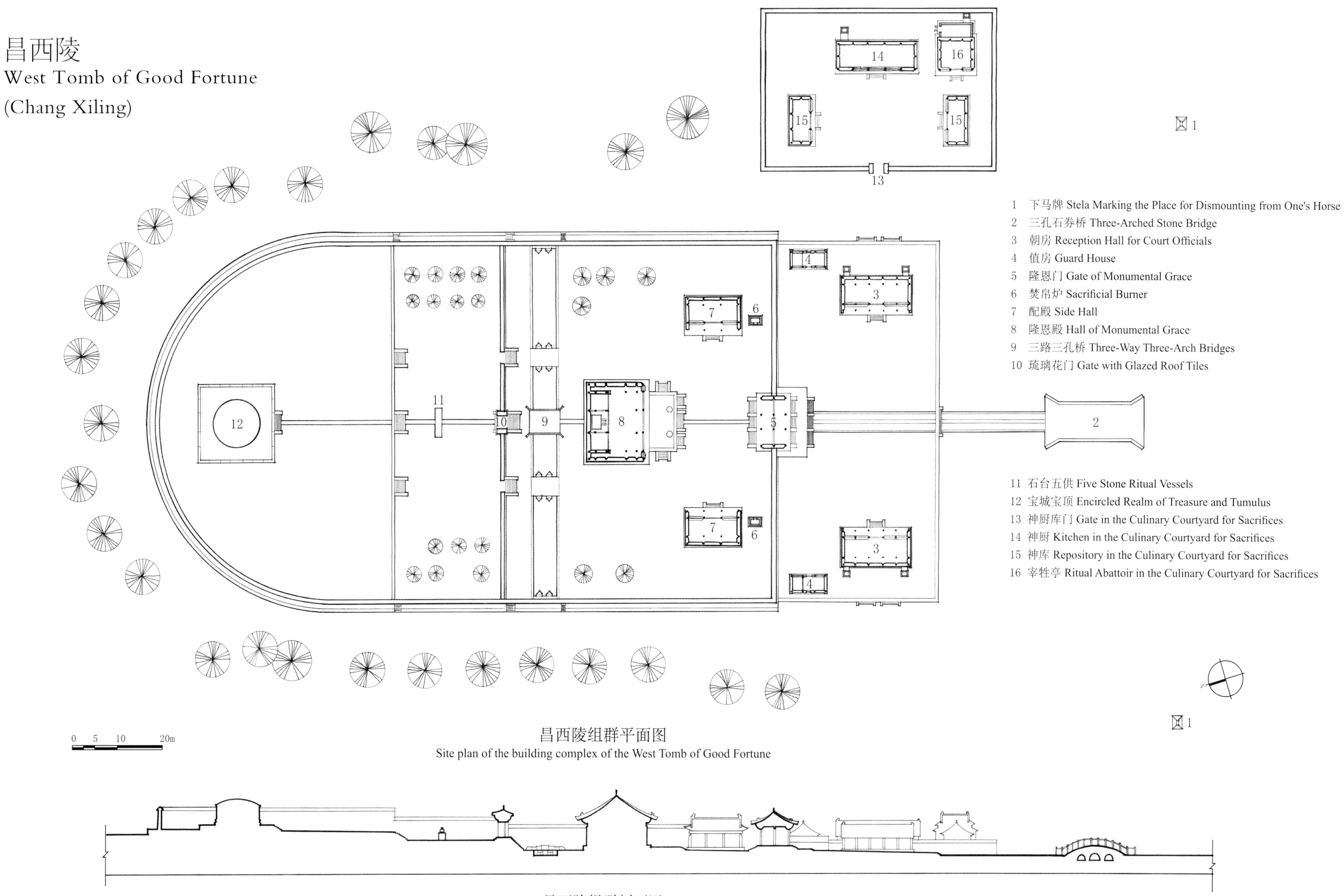

昌西陵组群平面图
Site plan of the building complex of the West Tomb of Good Fortune

昌西陵组群剖面图
Site section of the building complex of the West Tomb of Good Fortune

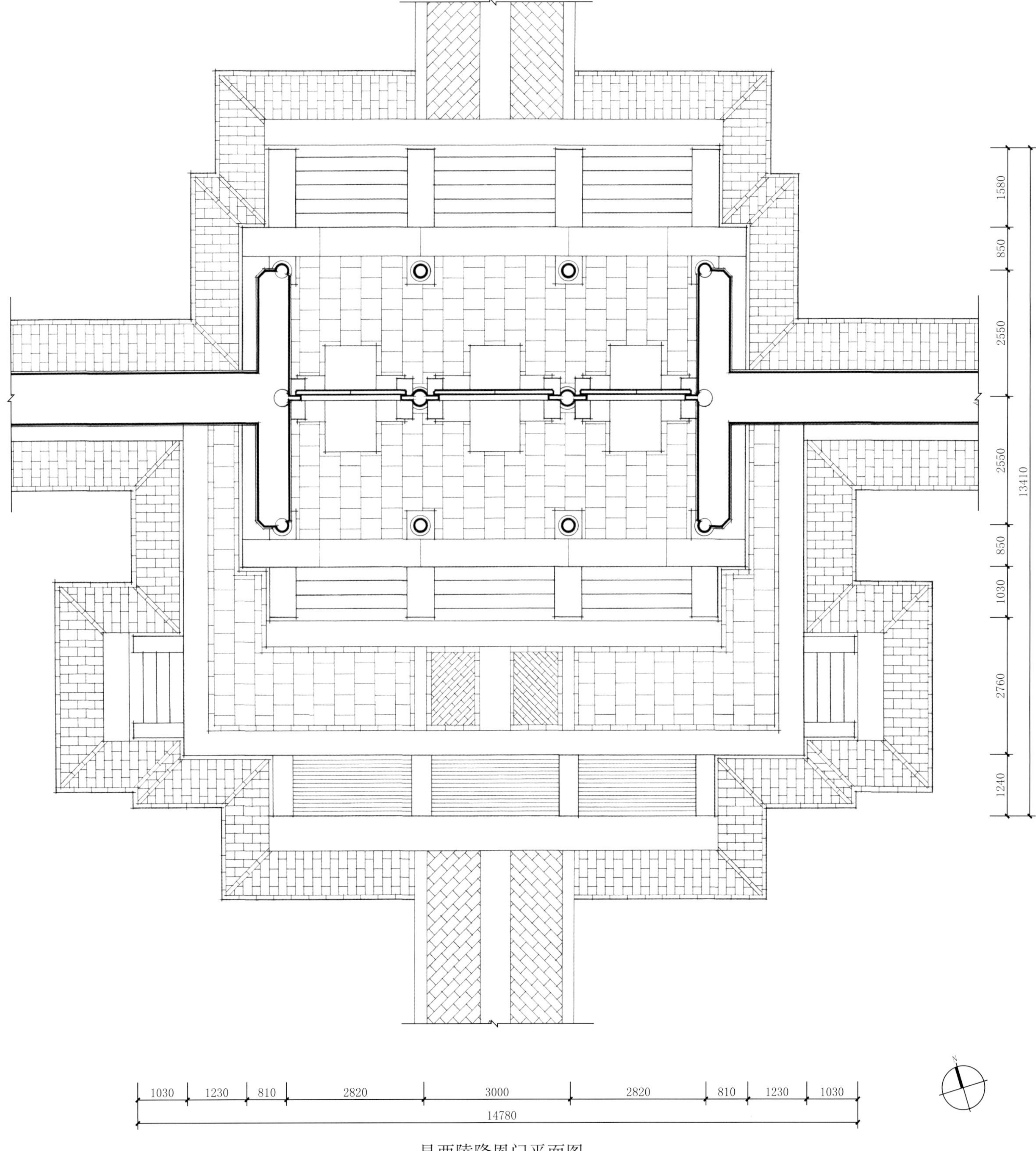

昌西陵隆恩门平面图

Plan of the Gate of Monumental Grace, West Tomb of Good Fortune

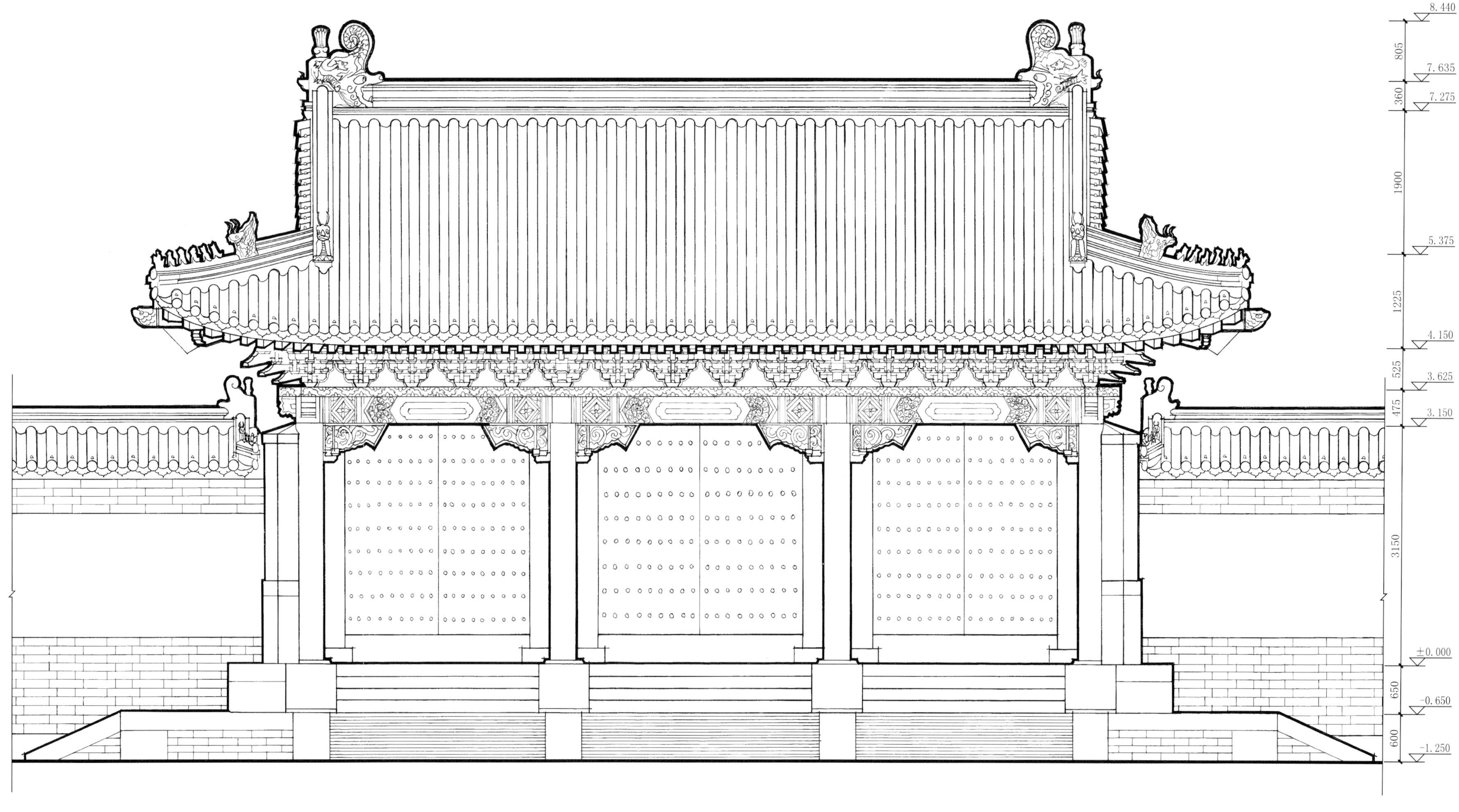

昌西陵隆恩门正立面图

Front elevation of the Gate of Monumental Grace, West Tomb of Good Fortune

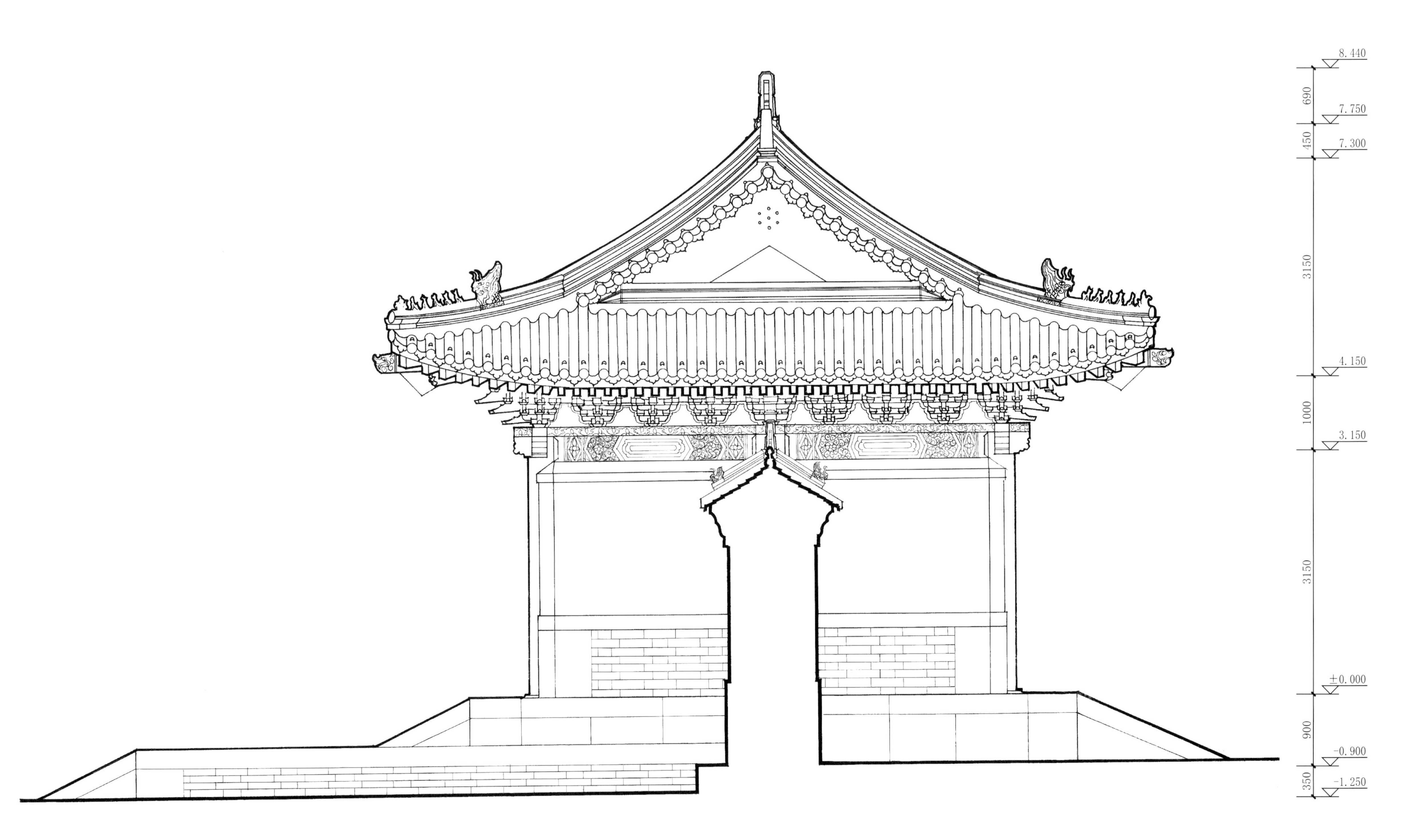

昌西陵隆恩门侧立面图

Side elevation of the Gate of Monumental Grace,West Tomb of Good Fortune

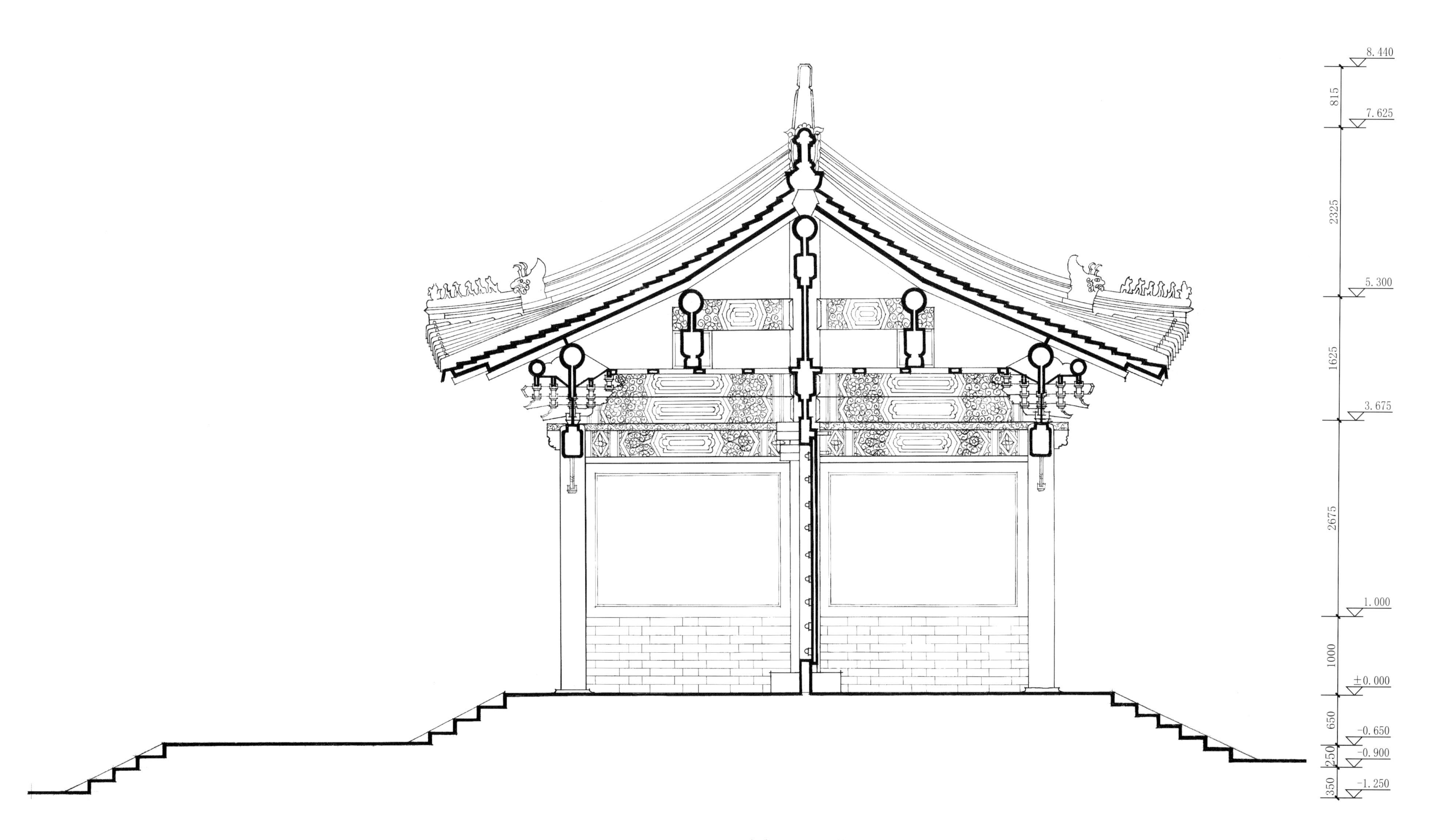

昌西陵隆恩门明间横剖面图

Cross section of the Central Chamber within the Gate of Monumental Grace,West Tomb of Good Fortune

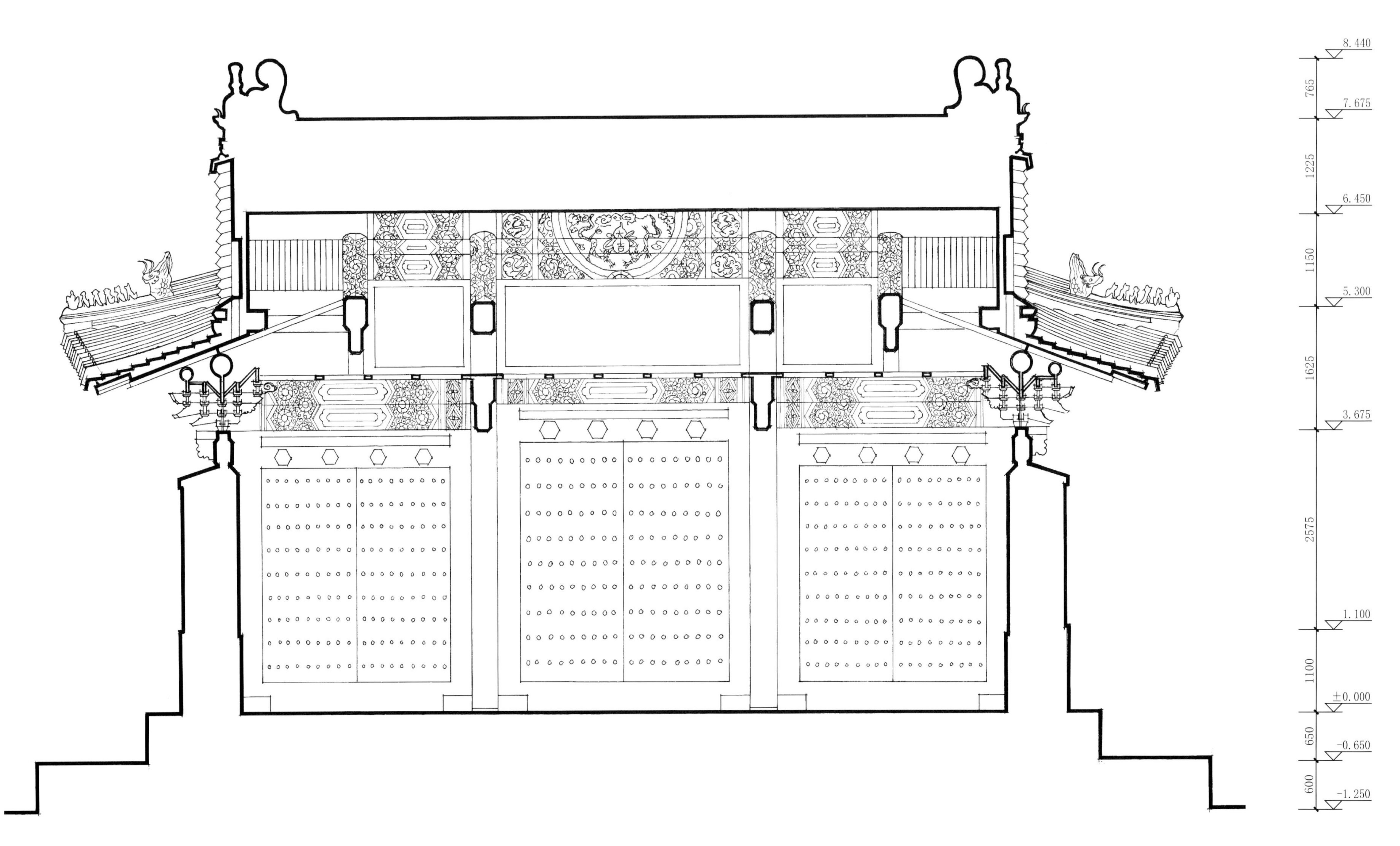

昌西陵隆恩门纵剖面图

Longitudinal section of the Gate of Monumental Grace, West Tomb of Good Fortune

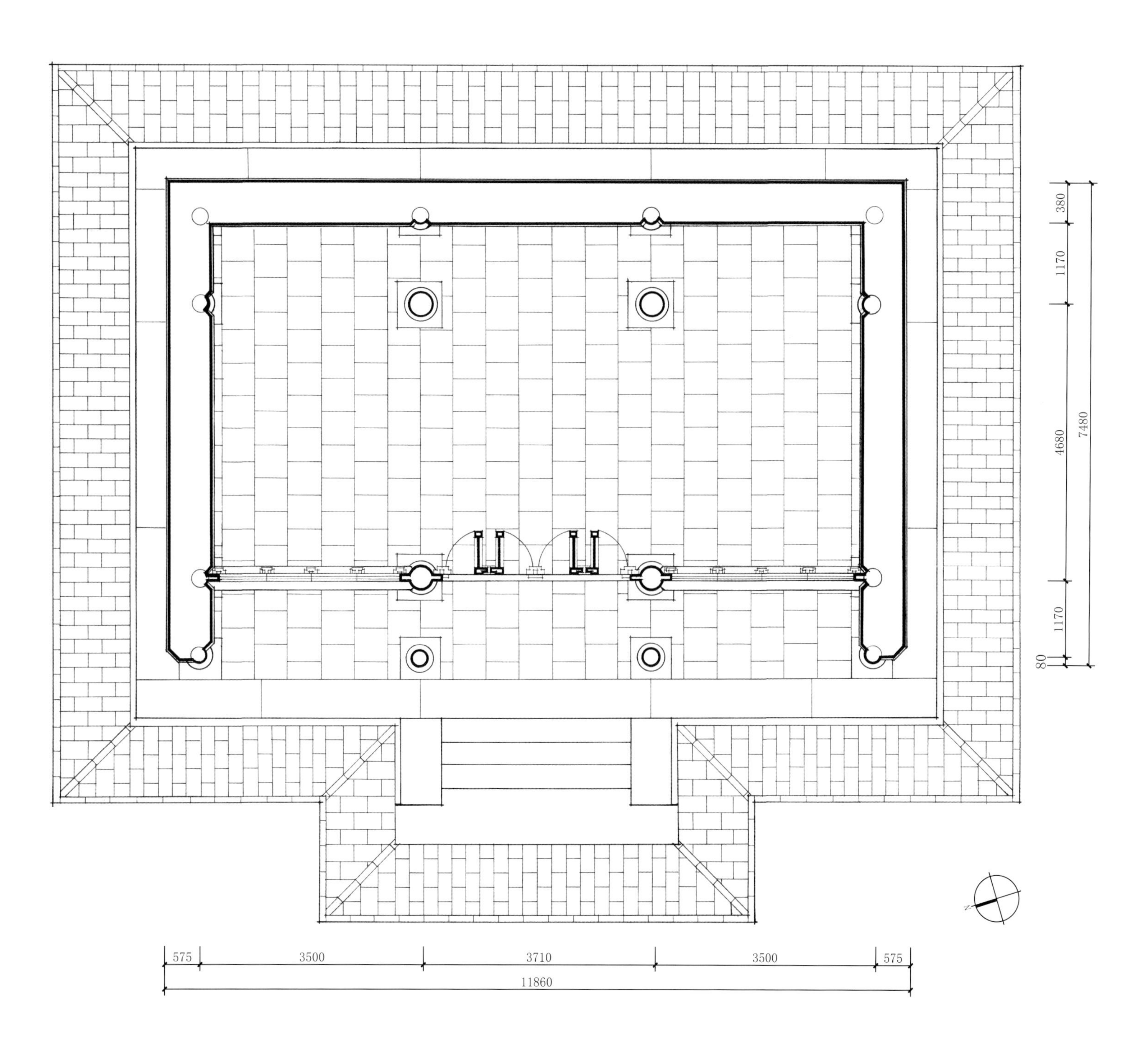

昌西陵配殿平面图
Plan of the Side Hall,West Tomb of Good Fortune

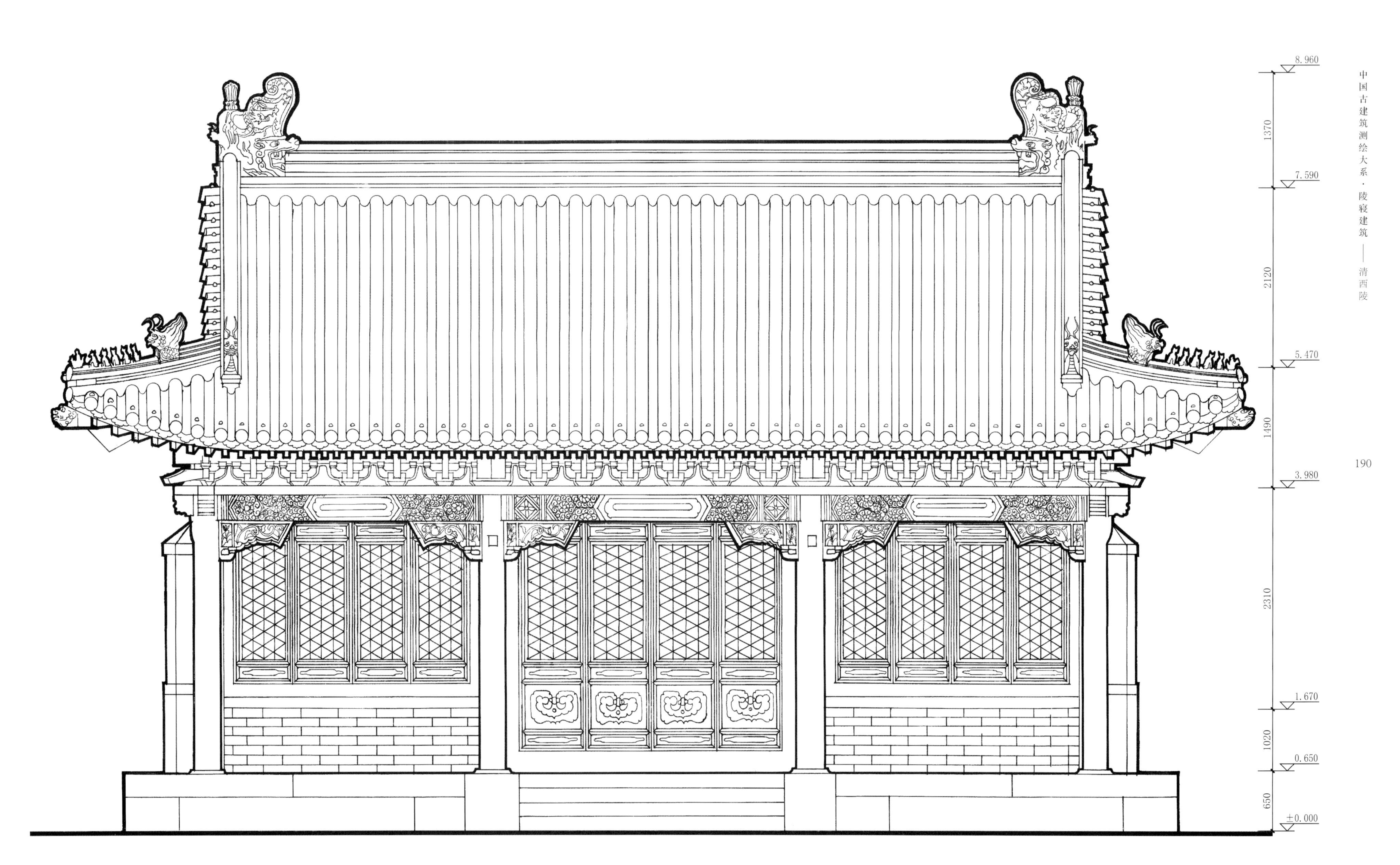

昌西陵配殿正立面图

Front elevation of the Side Hall,West Tomb of Good Fortune

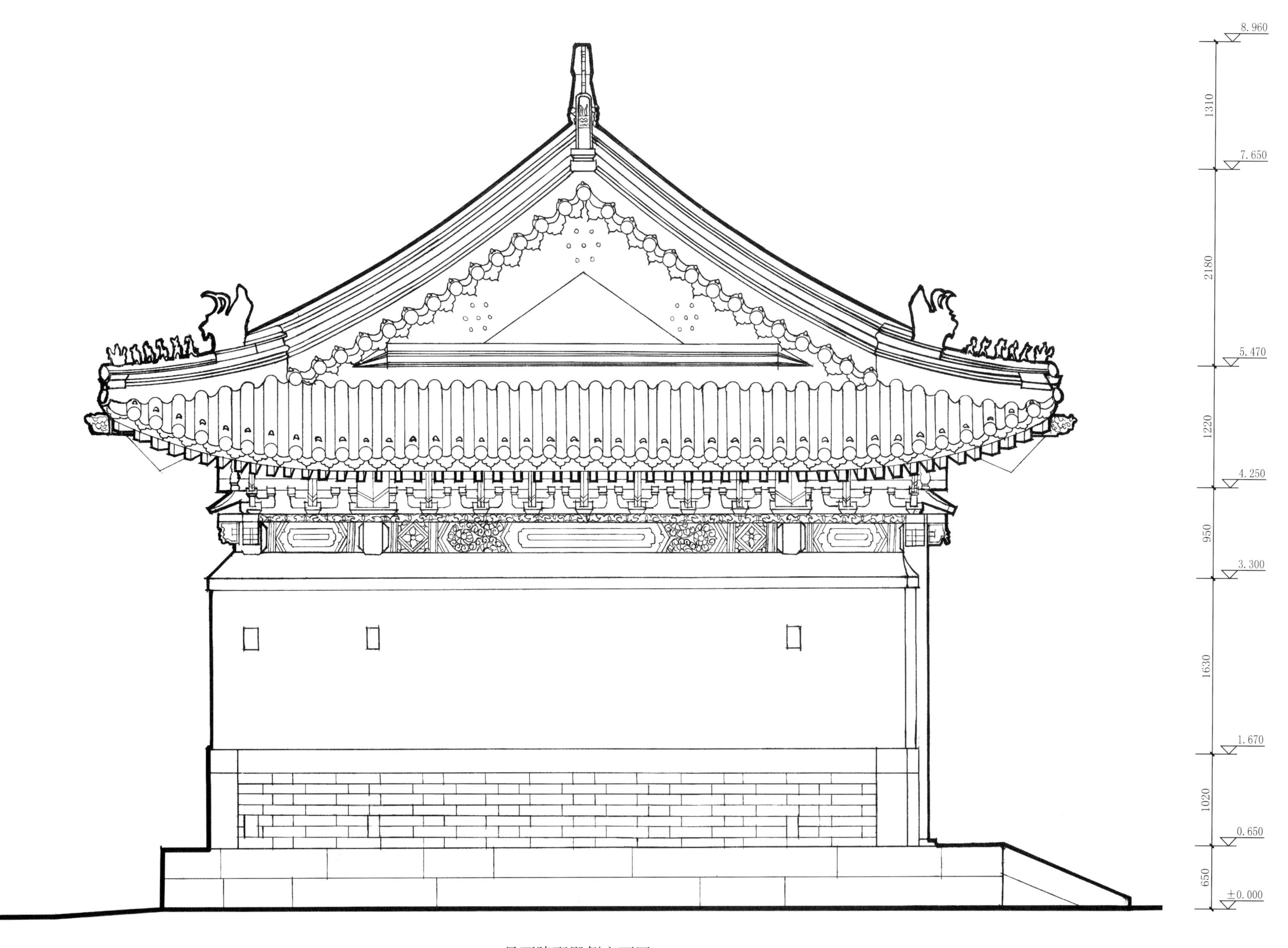

昌西陵配殿侧立面图
Side elevation of the Side Hall,West Tomb of Good Fortune

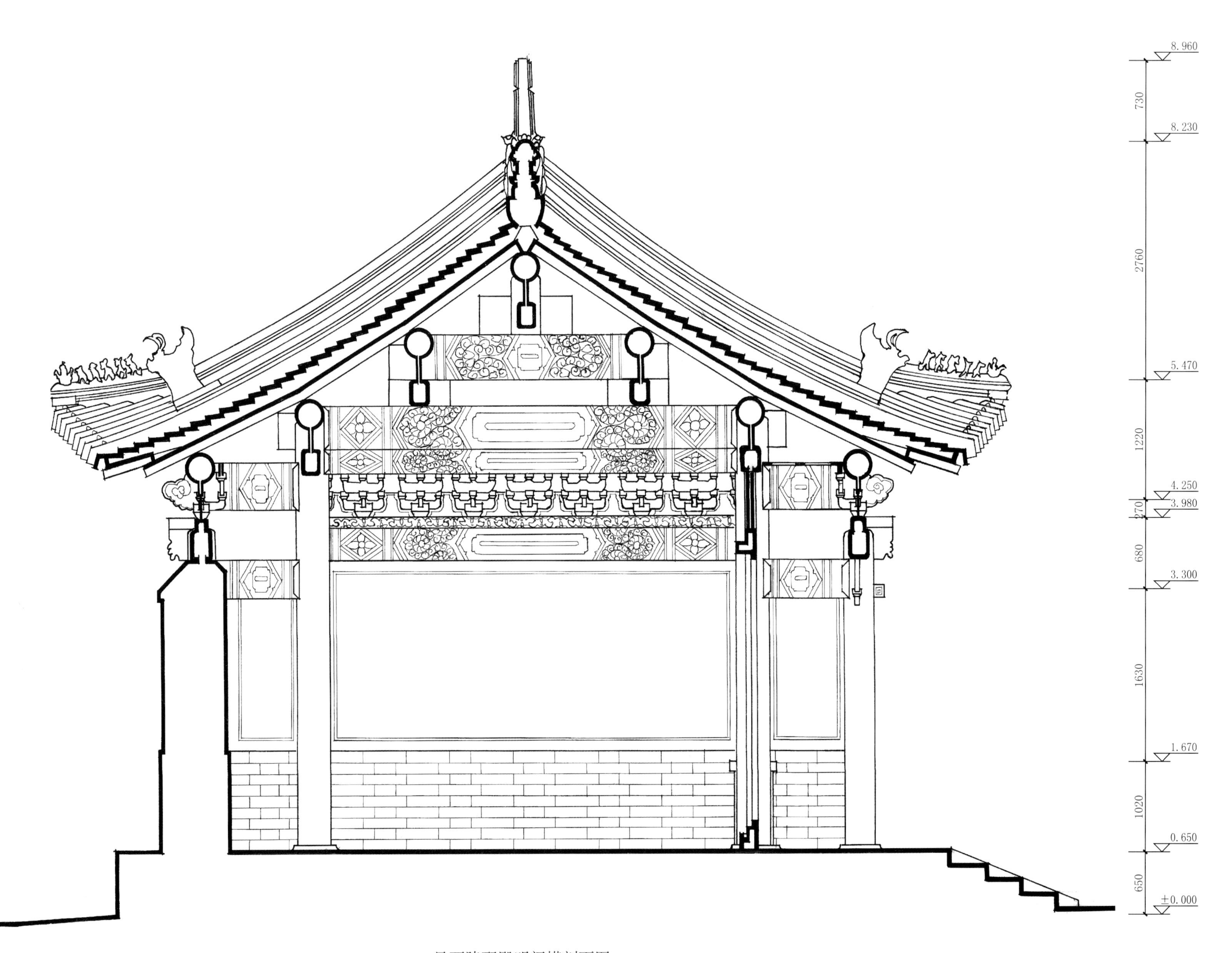

昌西陵配殿明间横剖面图
Cross section of the Central Chamber within the Side Hall,West Tomb of Good Fortune

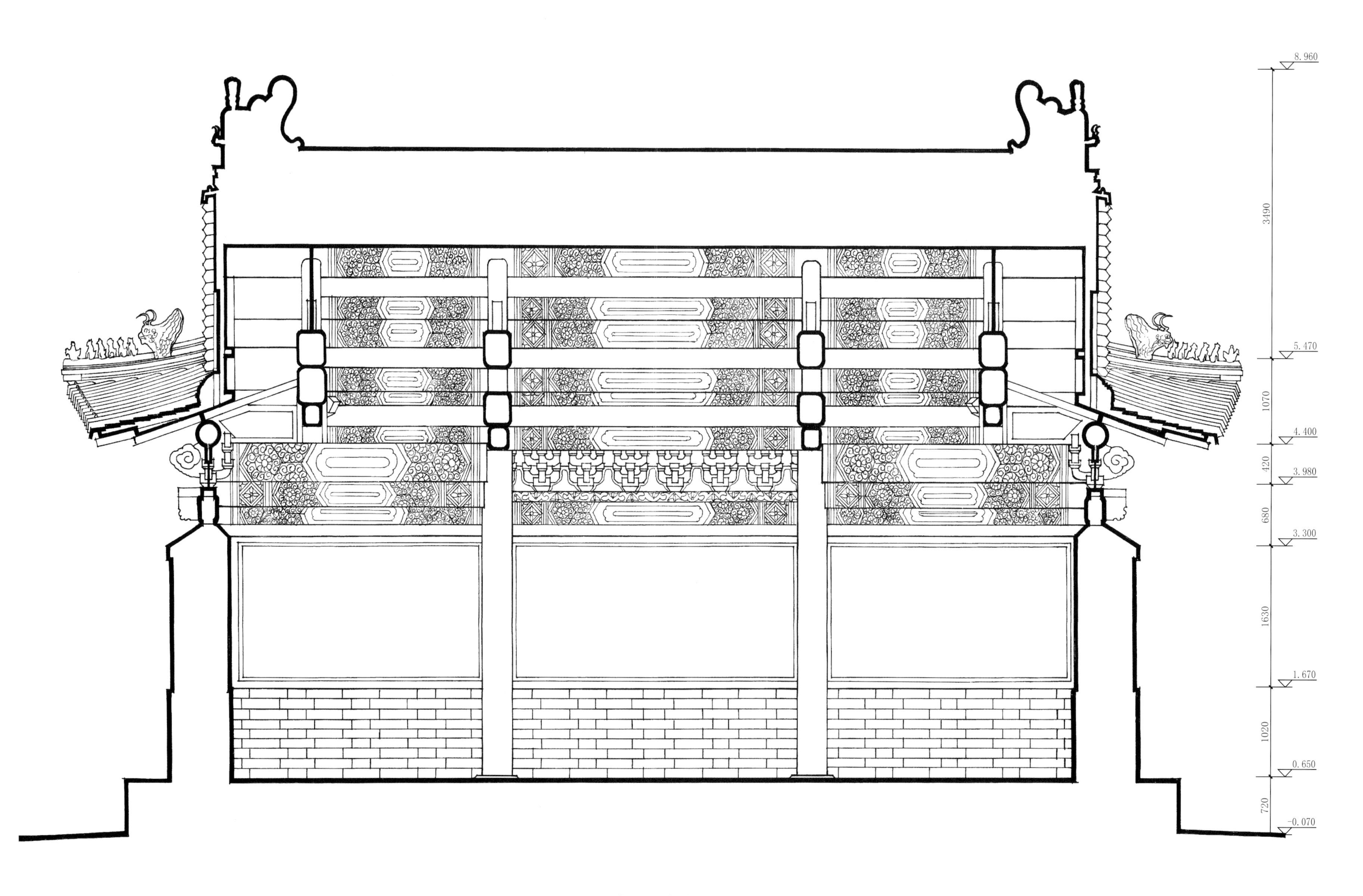

昌西陵配殿纵剖面图

Longitudinal section of the Side Hall,West Tomb of Good Fortune

昌西陵隆恩殿平面图

Plan of the Hall of Monumental Grace, West Tomb of Good Fortune

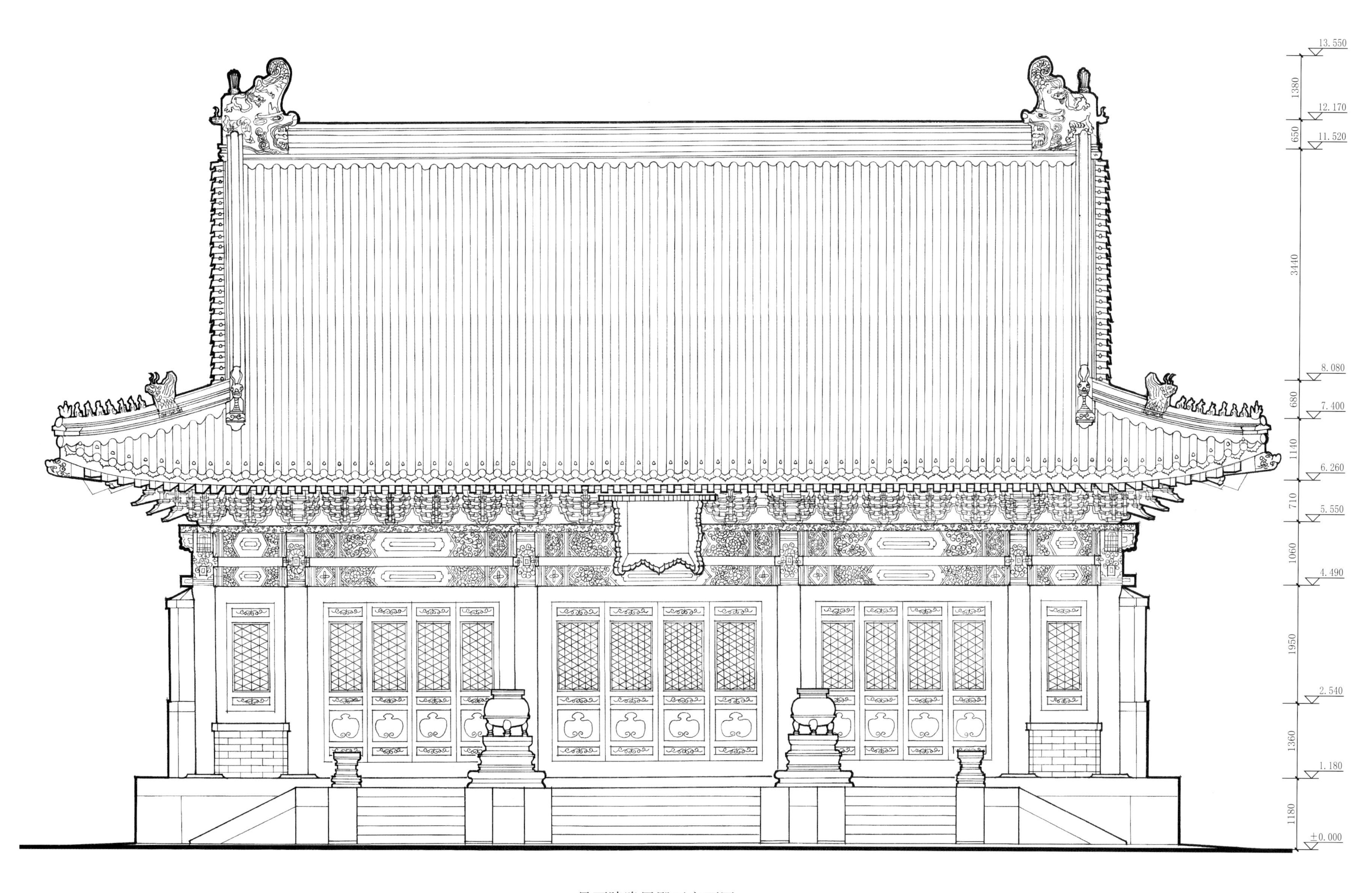

昌西陵隆恩殿正立面图

Front elevation of the Hall of Monumental Grace,West Tomb of Good Fortune

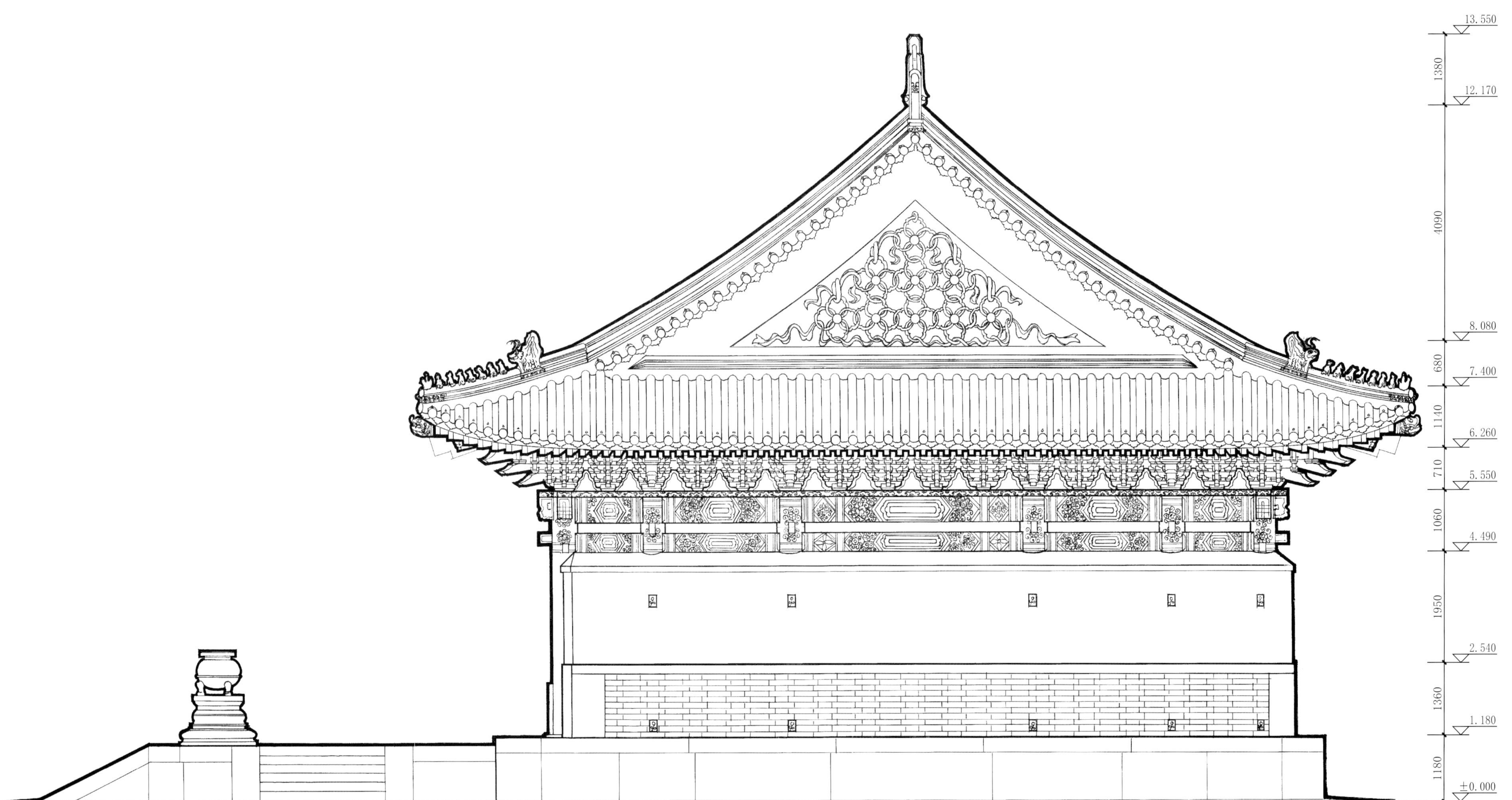

昌西陵隆恩殿侧立面图

Side elevation of the Hall of Monumental Grace,West Tomb of Good Fortune

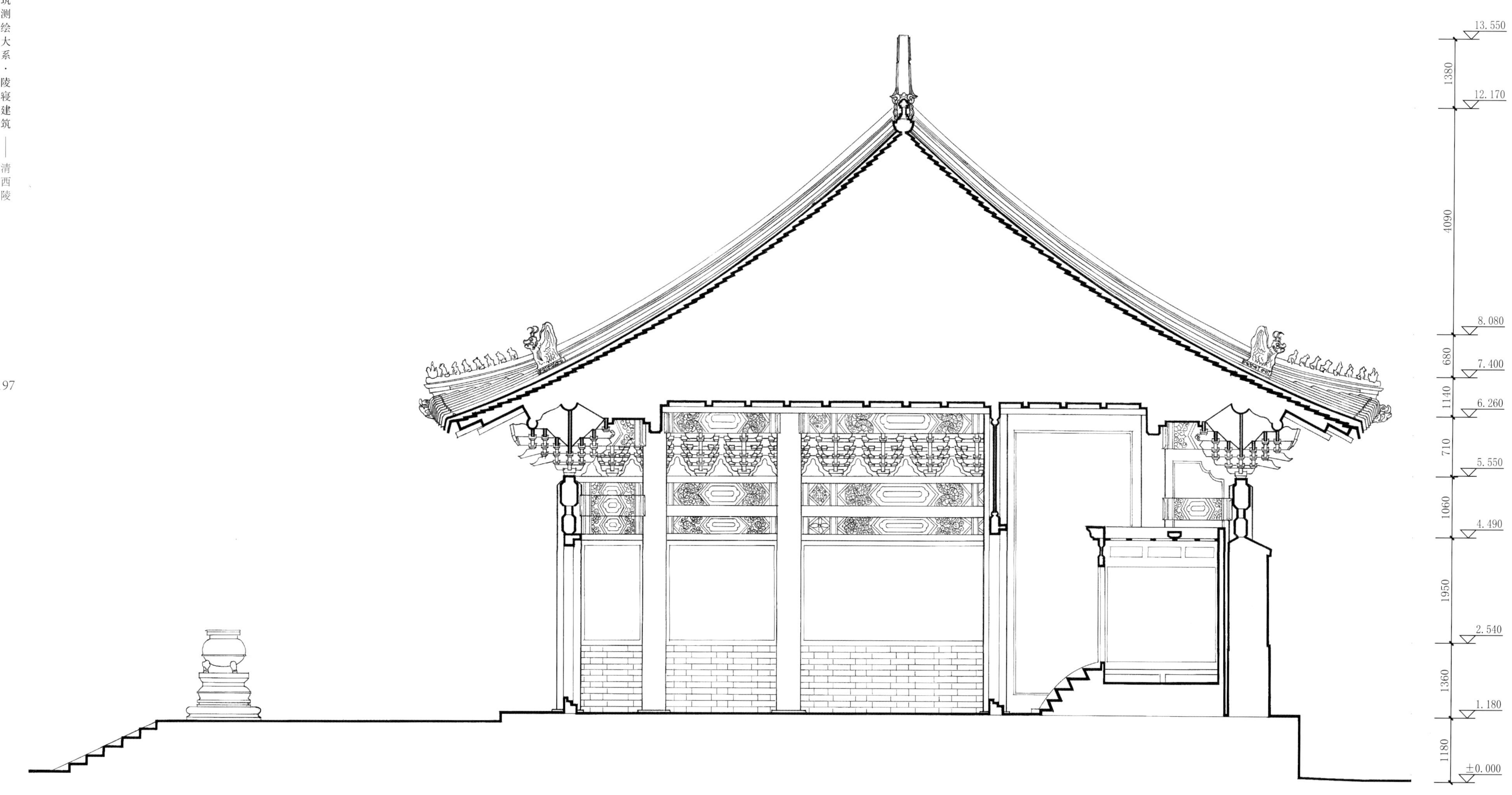

昌西陵隆恩殿明间横剖面图

Cross section of the Central Chamber within the Hall of Monumental Grace,West Tomb of Good Fortune

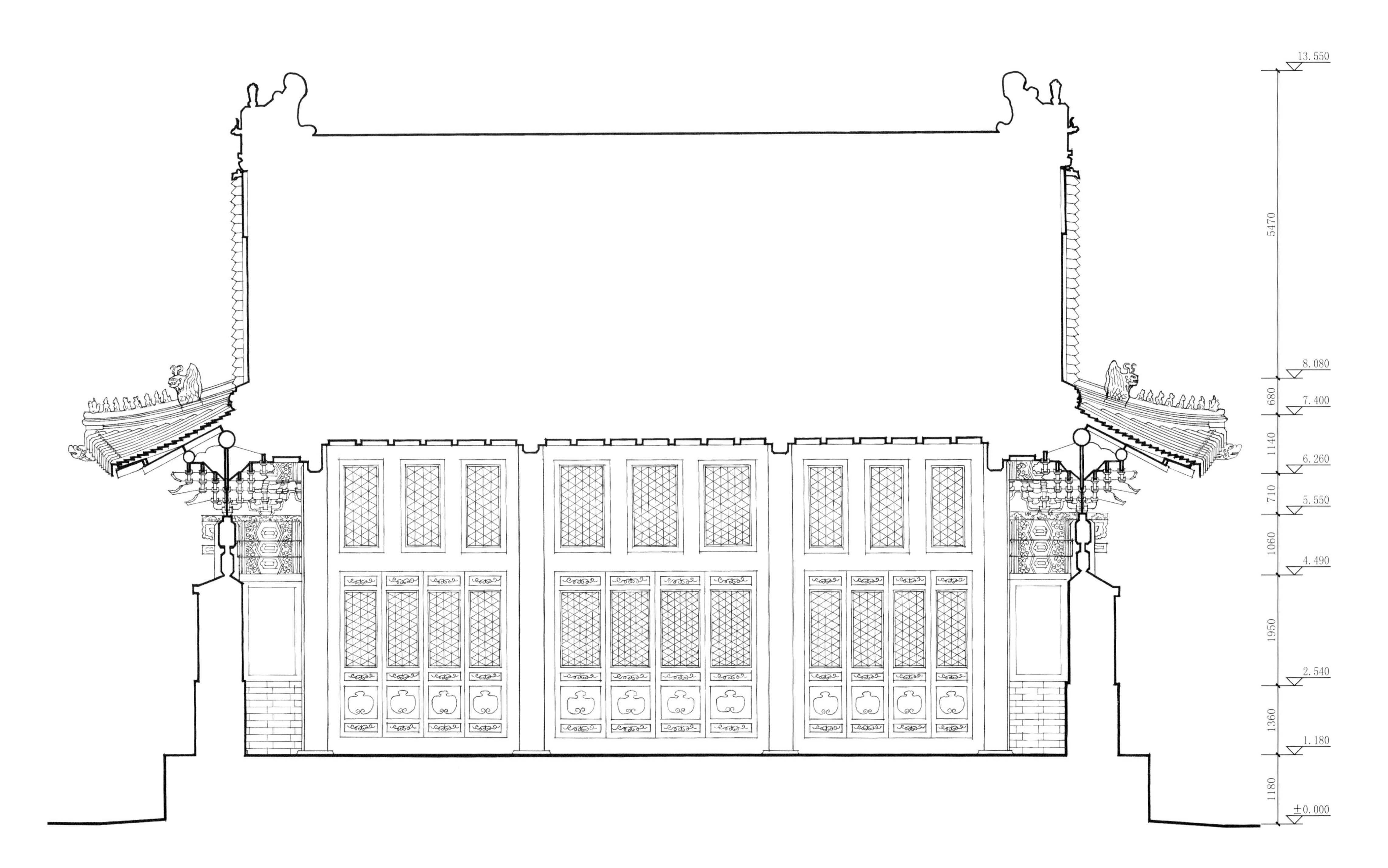

昌西陵隆恩殿纵剖面图

Longitudinal section of the Hall of Monumental Grace,West Tomb of Good Fortune

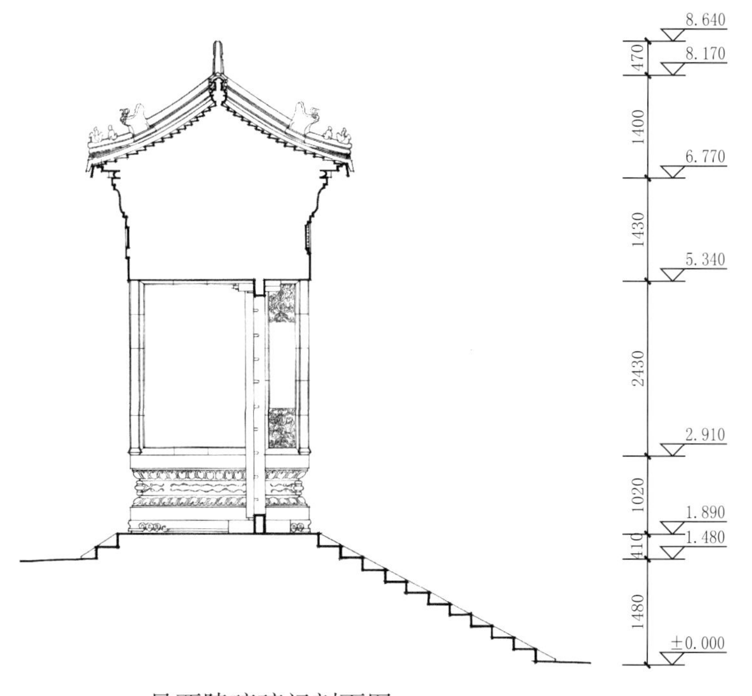

昌西陵琉璃门剖面图

Section of the Gate with Glazed Roof Tiles, West Tomb of Good Fortune

昌西陵隆恩殿香炉大样图

Detailed drawing of the Incense Burner of the Hall of Monumental Grace, West Tomb of Good Fortune

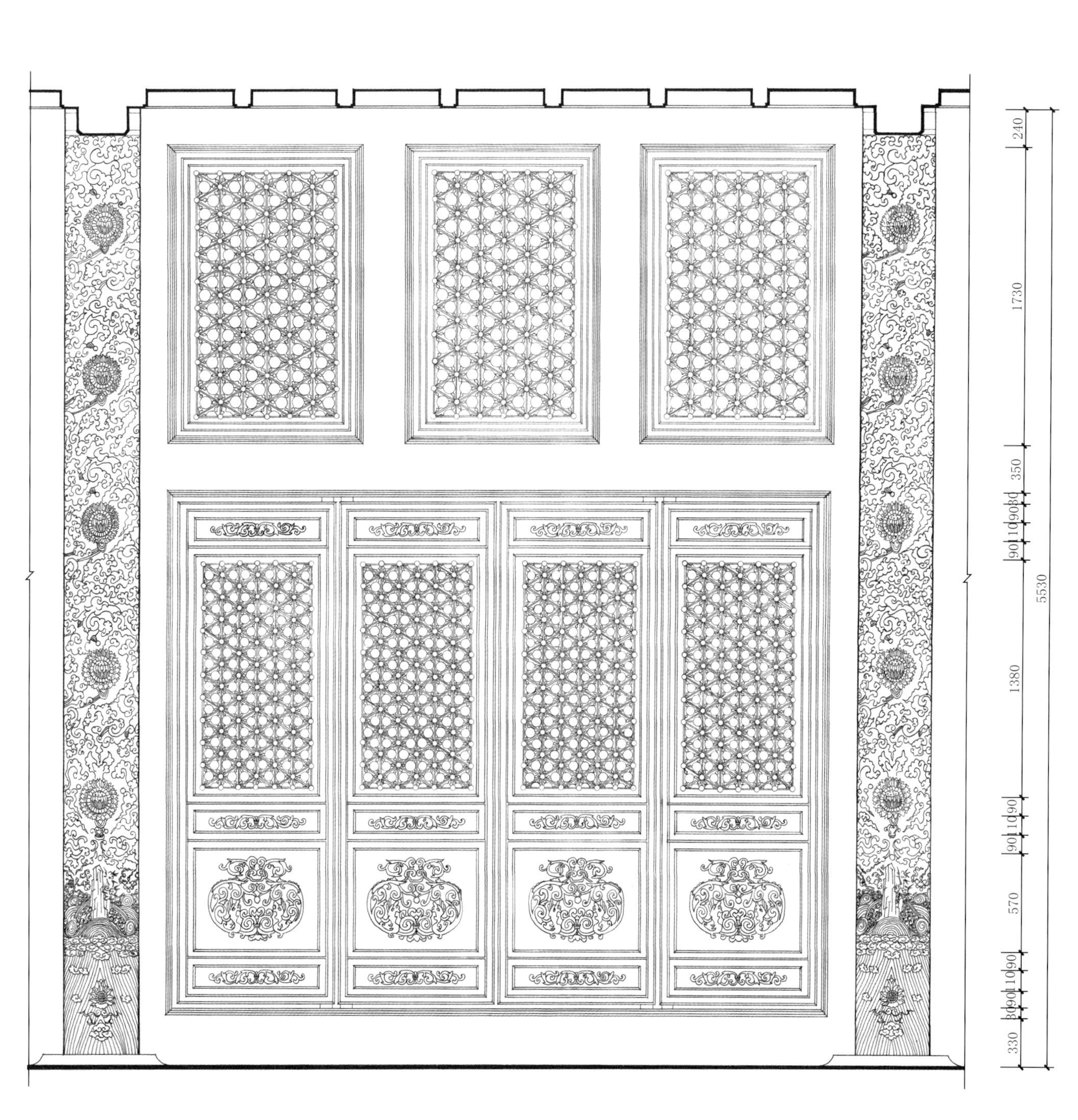

昌西陵隆恩殿隔扇大样图

Detailed drawing of the Two-Panel Lattice Door of the Hall of Monumental Grace, West Tomb of Good Fortune

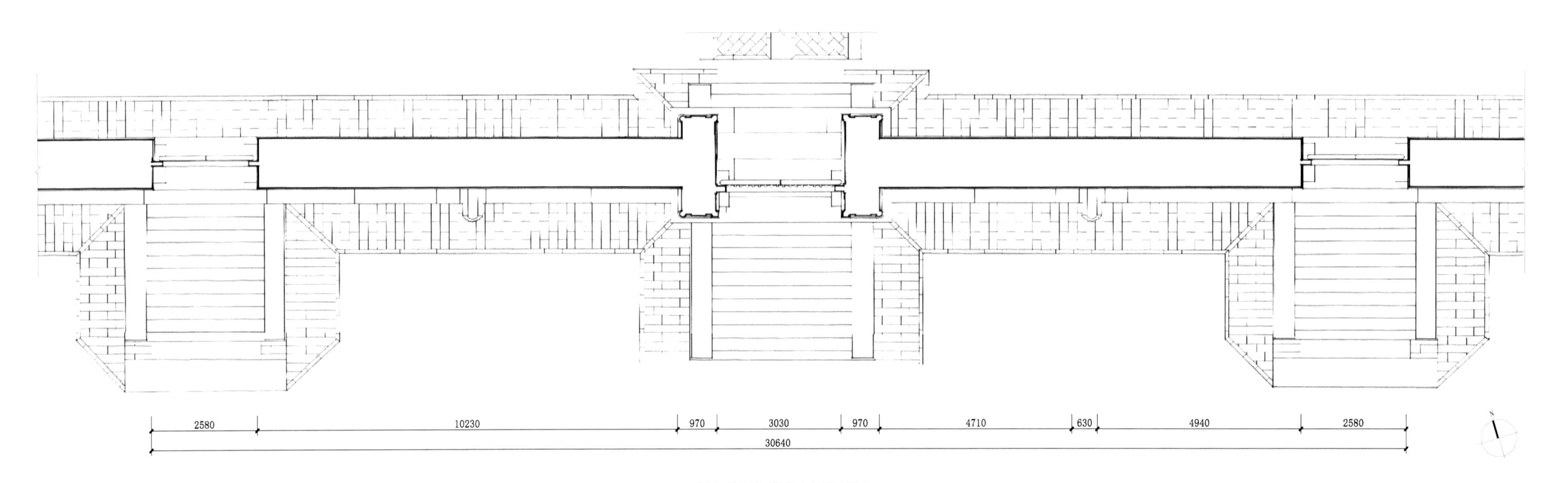

昌西陵琉璃花门平面图
Plan of the Gate with Glazed Roof Tiles,West Tomb of Good Fortune

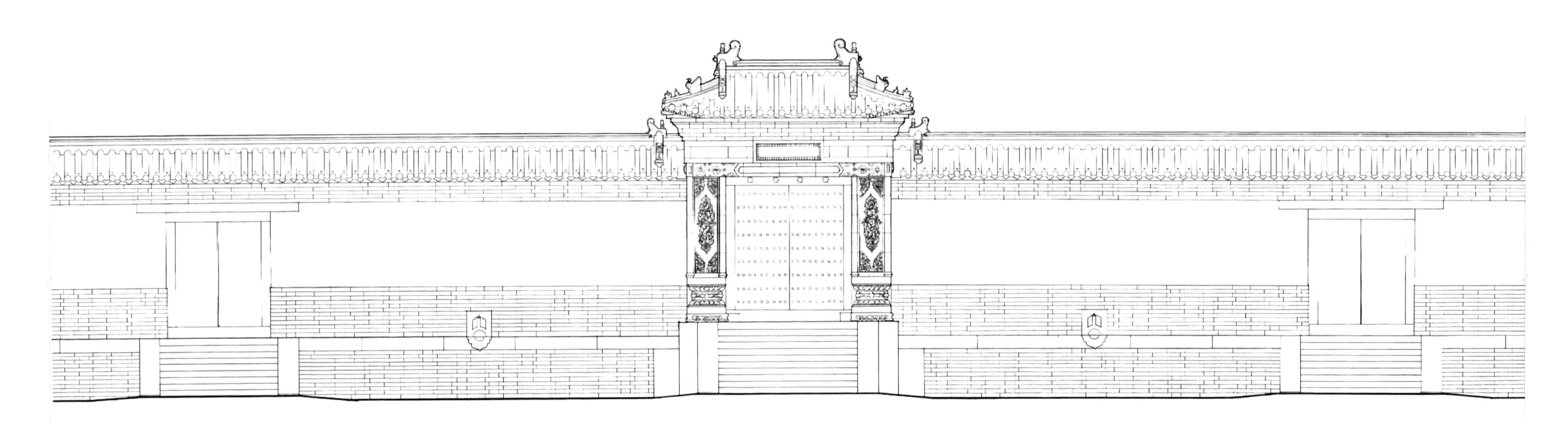

昌西陵琉璃花门正立面图
Front elevation of the Gate with Glazed Roof Tiles,West Tomb of Good Fortune

昌西陵石台五供正立面图

Front elevation of the Five Stone Ritual Vessels,West Tomb of Good Fortune

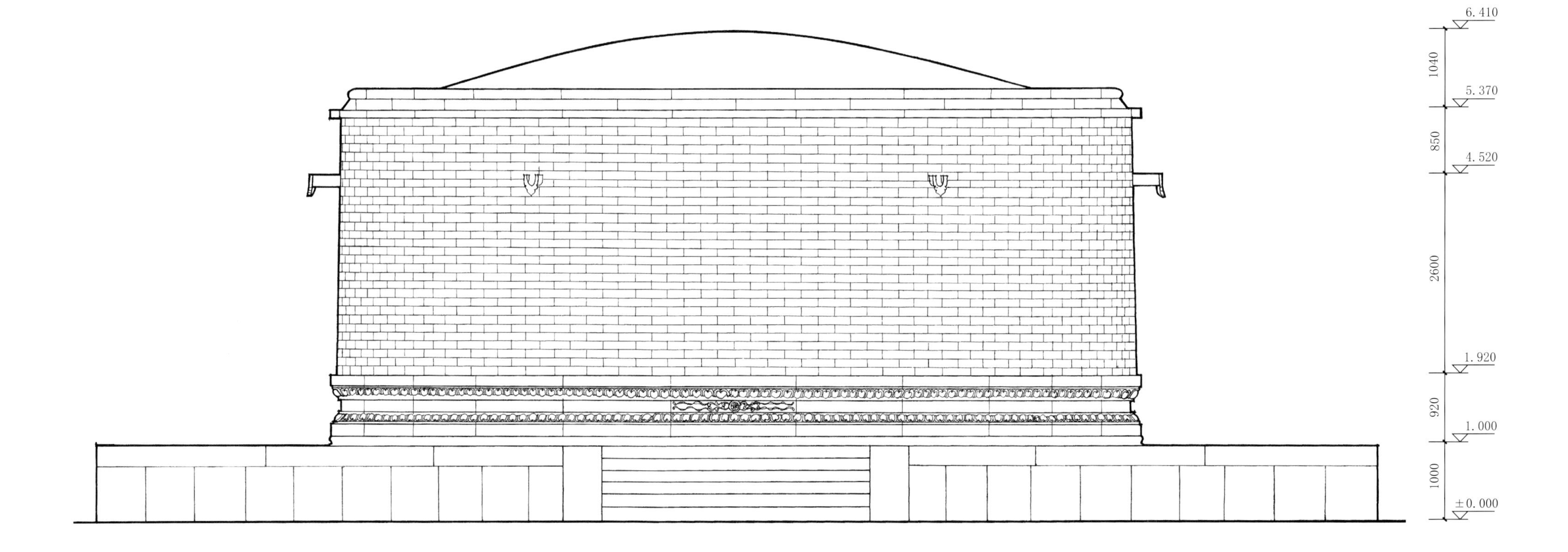

昌西陵宝顶正立面图

Front elevation of the Tumulus,West Tomb of Good Fortune

昌陵妃园寝

Imperial Consorts' Tombs affiliated with the Tomb of Good Fortune (Changling Feiyuanqin)

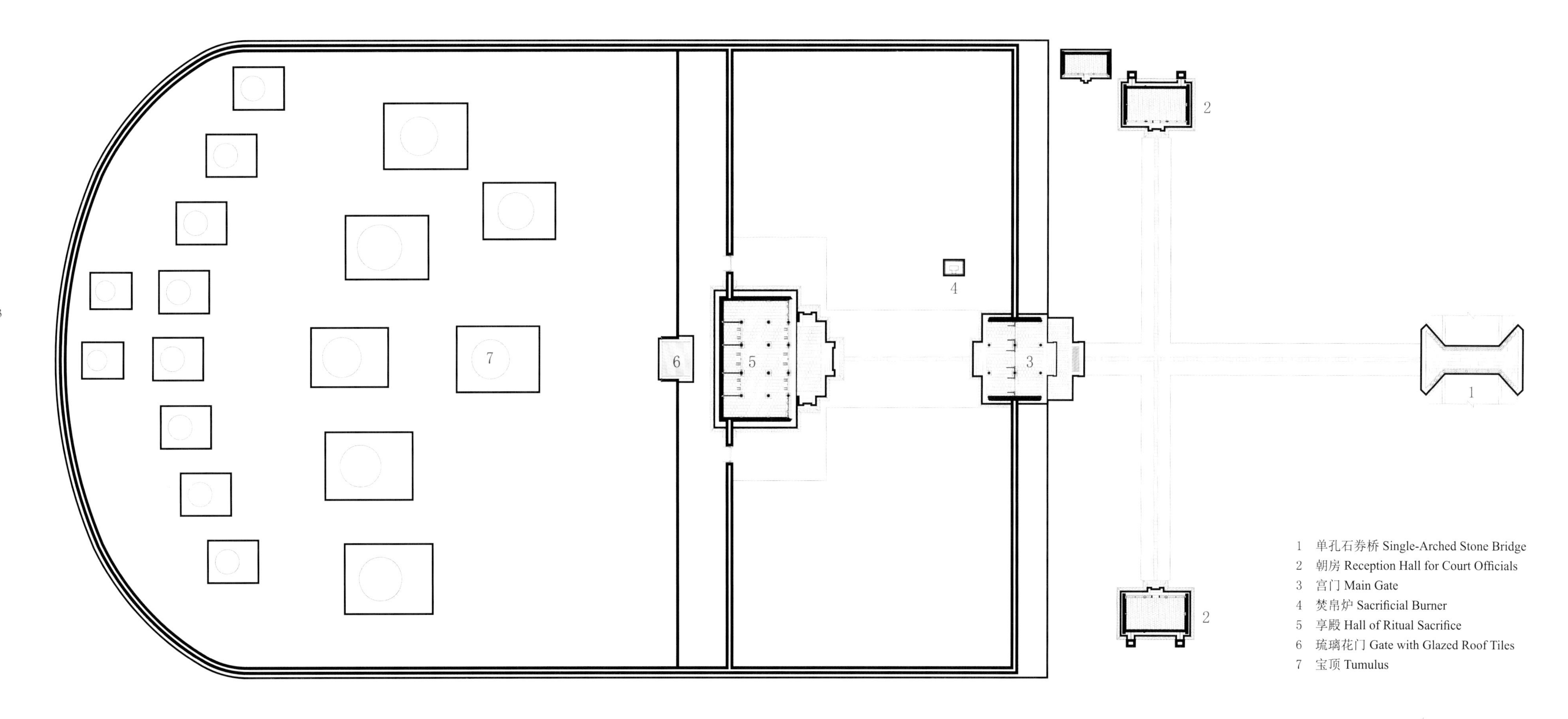

昌陵妃园寝组群平面图

Site plan of the building complex of the Imperial Consorts'Tombs affiliated with the Tomb of Good Fortune

慕 陵

Tomb of Veneration (Muling)

1 五孔石券桥 Five-Arched Stone Bridge
2 龙凤门 Dragon and Phoenix Gate
3 下马牌 Stela Marking the Place for Dismounting from One's Horse
4 神道碑亭 Pavilion for the Stela on the Spirit Way
5 神厨库门 Gate in the Culinary Courtyard for Sacrifices
6 神厨 Kitchen in the Culinary Courtyard for Sacrifices
7 神库 Repository in the Culinary Courtyard for Sacrifices
8 宰牲亭 Ritual Abattoir in the Culinary Courtyard for Sacrifices
9 井亭 Pavilion for the Well
10 三路三孔桥 Three-Way Three-Arch Bridges
11 朝房 Reception Hall for Court Officials
12 值房 Guard House
13 隆恩门 Gate of Monumental Grace
14 焚帛炉 Sacrificial Burner
15 配殿 Side Hall
16 隆恩殿 Hall of Monumental Grace
17 三路三孔桥 Three-Way Three-Arch Bridges
18 琉璃花门 Gate with Glazed Roof Tiles
19 石台五供 Five Stone Ritual Vessels
20 宝顶 Tumulus

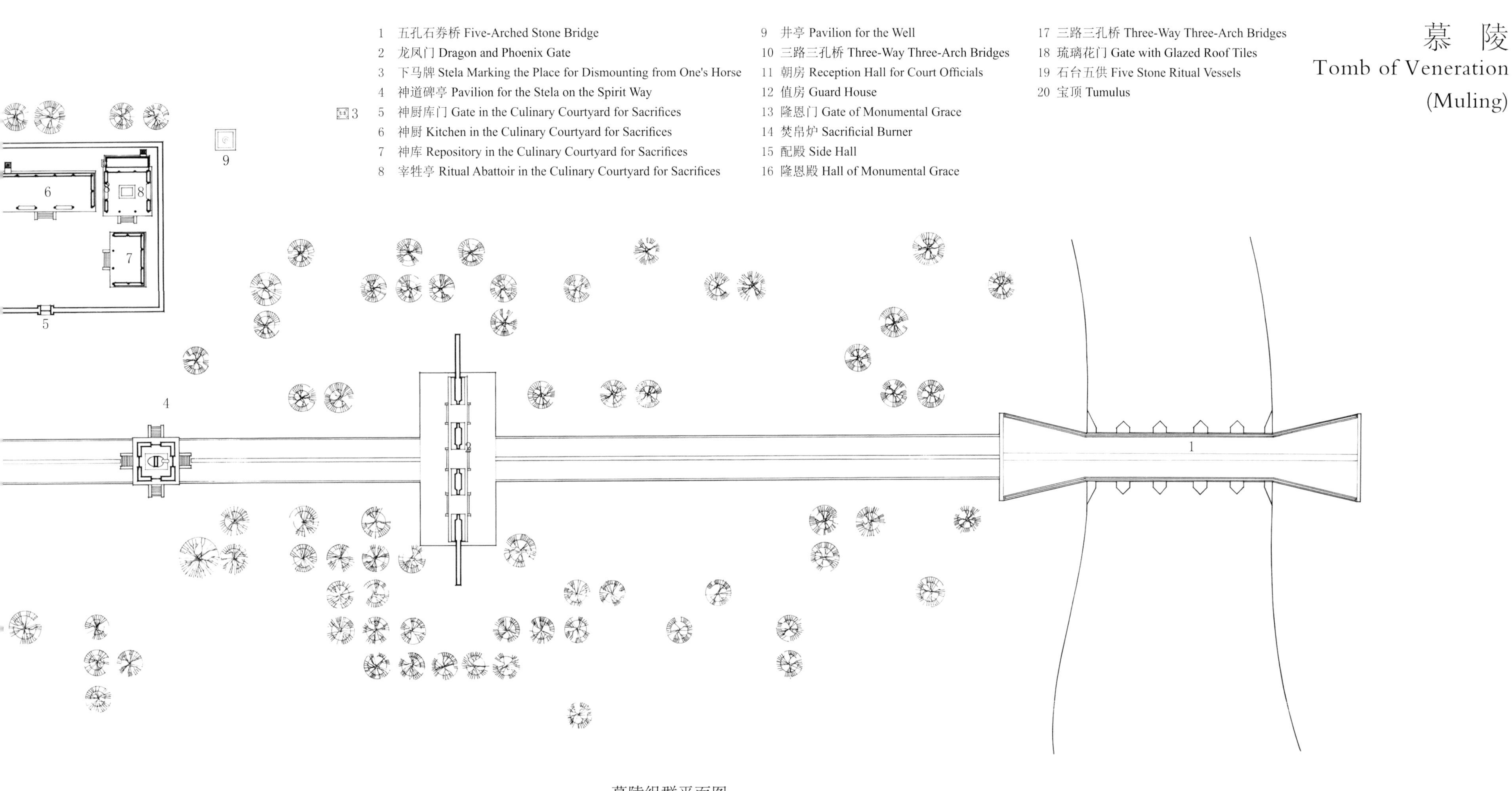

回3

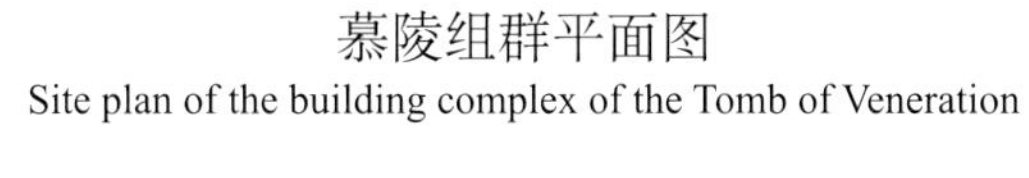

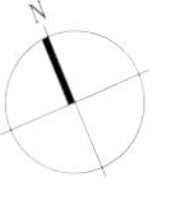

慕陵组群平面图
Site plan of the building complex of the Tomb of Veneration

回3

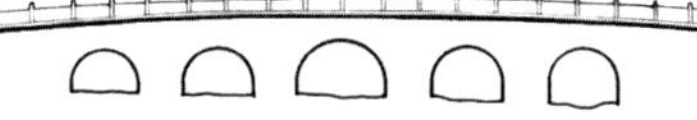

慕陵组群剖面图
Site section of the building complex of the Tomb of Veneration

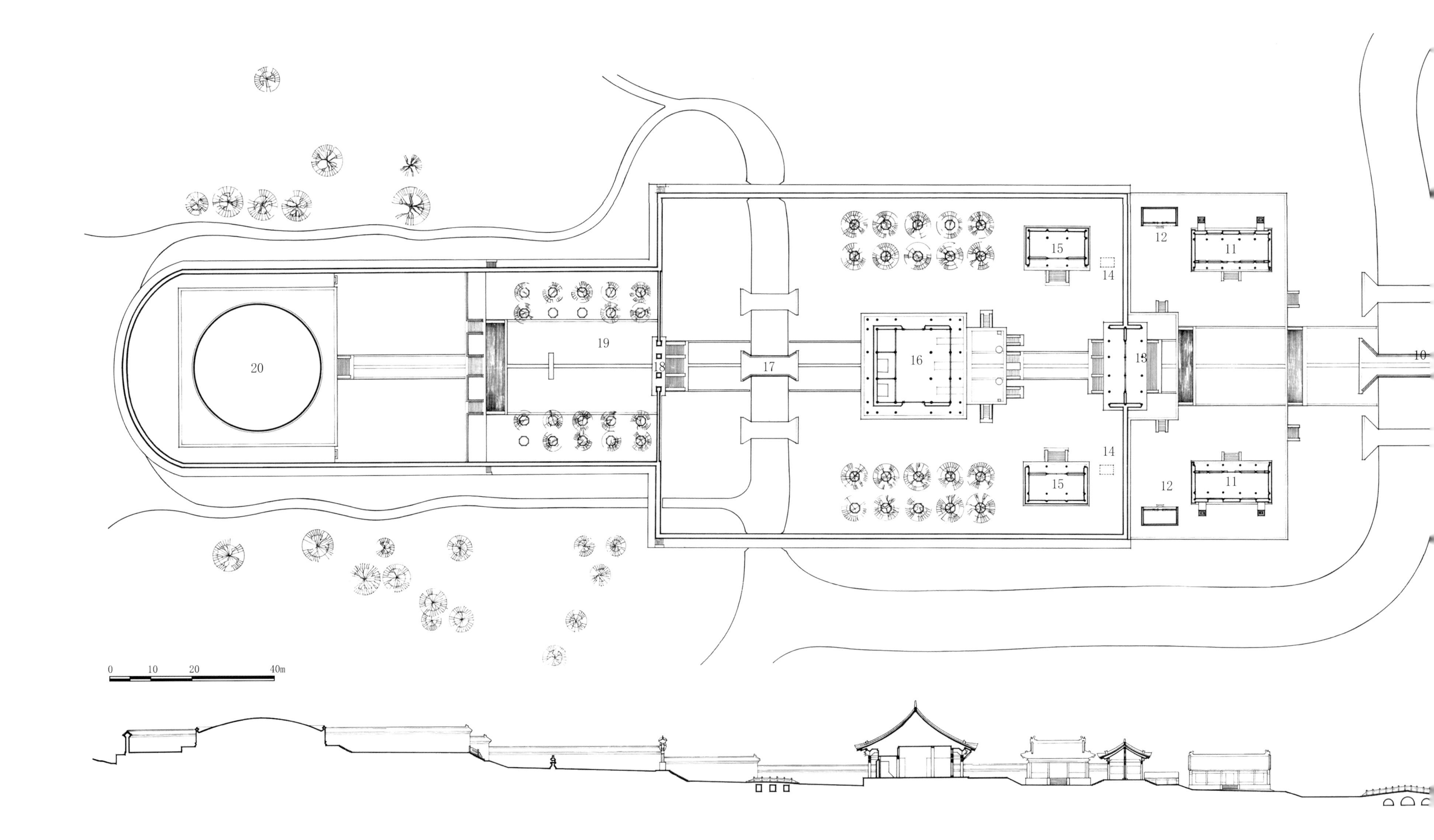

20
19
18
17
16
15
14
13
12
11
10
0
10
20
40m

慕陵龙凤门正立面图
Front elevation of the Dragon and Phoenix Gate,Tomb of Veneration

慕陵龙凤门横剖面图
Cross section of the Dragon and Phoenix Gate,Tomb of Veneration

慕陵龙凤门侧立面图
Side elevation of the Dragon and Phoenix Gate,Tomb of Veneration

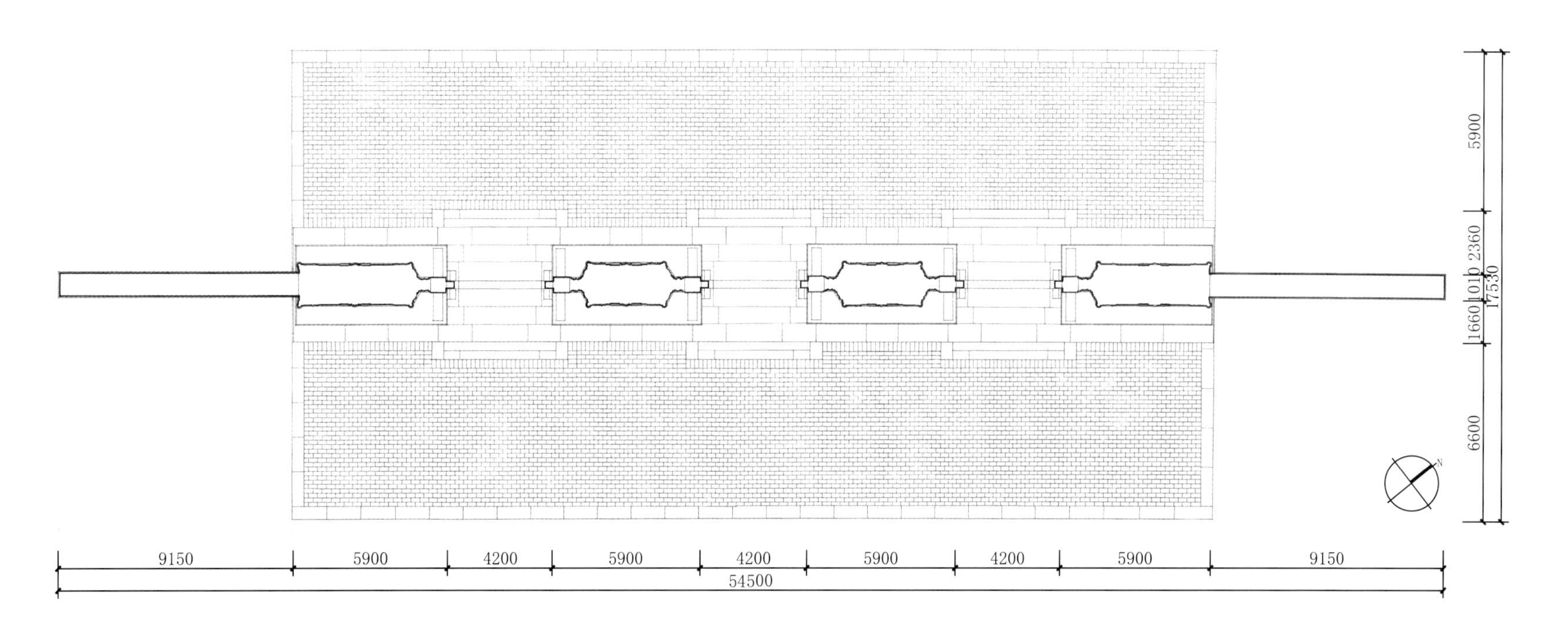

慕陵龙凤门平面图

Plan of the Dragon and Phoenix Gate,Tomb of Veneration

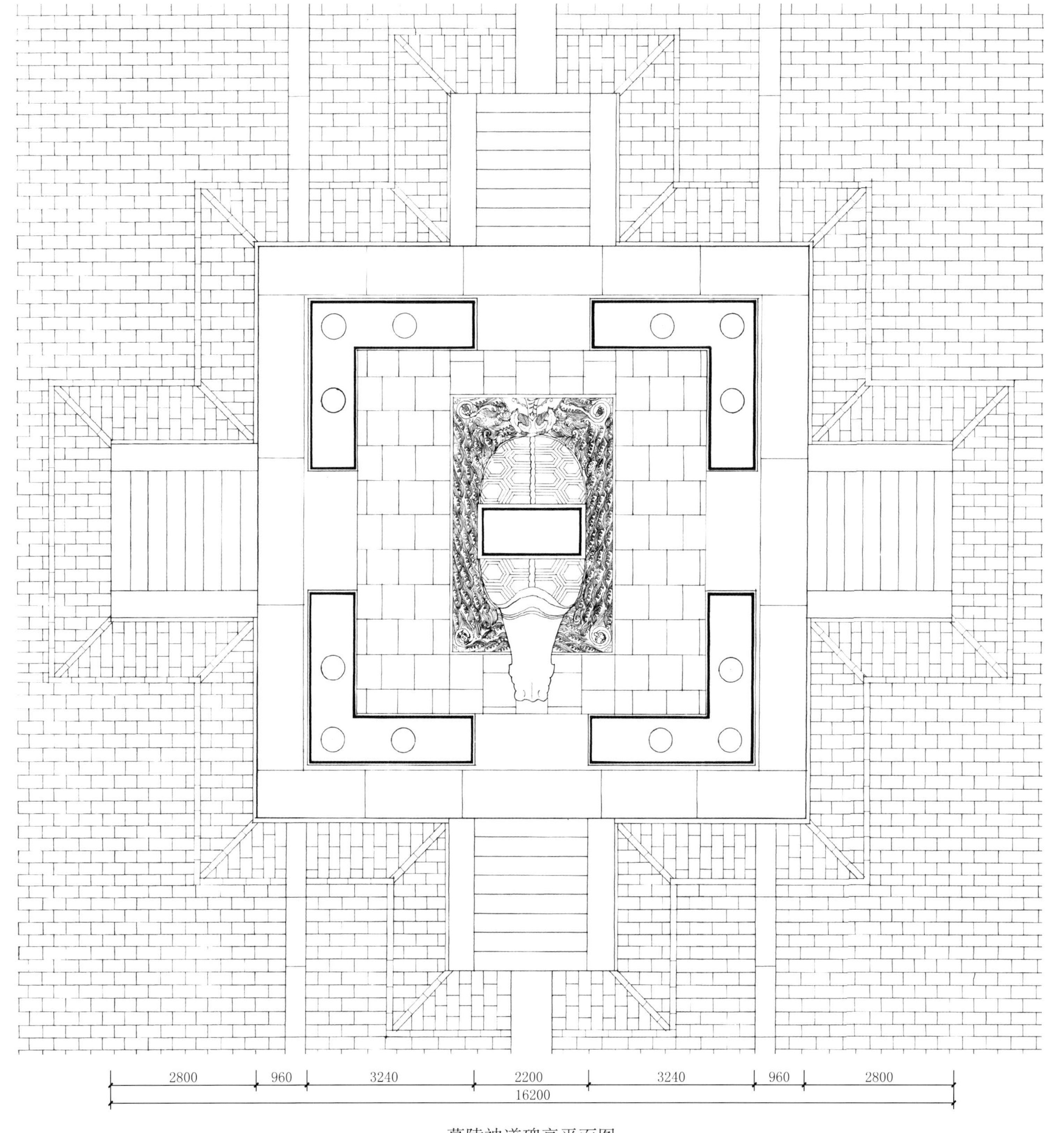

慕陵神道碑亭平面图

Plan of the Pavilion for the Stela on the Spirit Way,Tomb of Veneration

慕陵神道碑亭正立面图

Front elevation of the Pavilion for the Stela on the Spirit Way,Tomb of Veneration

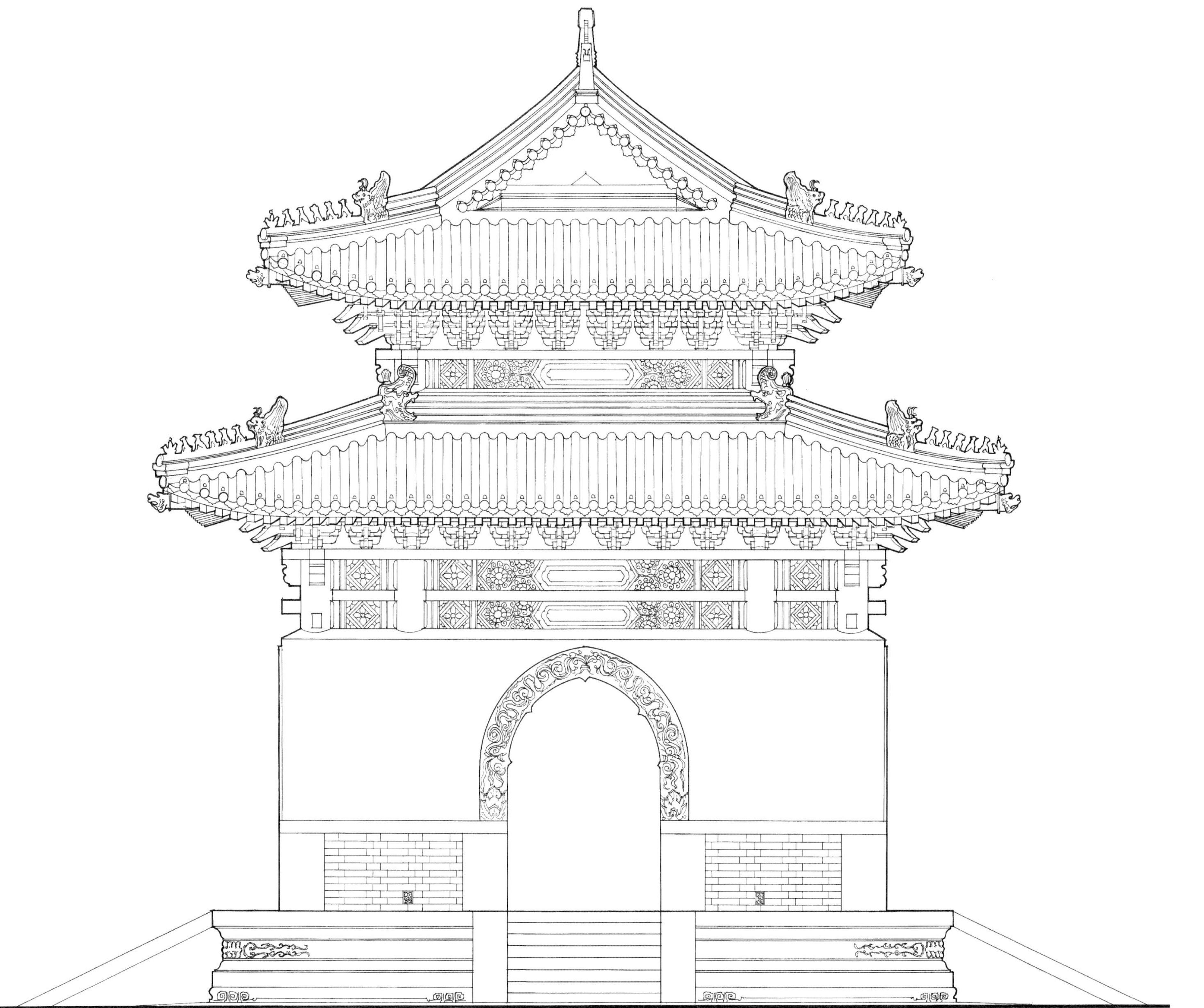

慕陵神道碑亭侧立面图

Side elevation of the Pavilion for the Stela on the Spirit Way,Tomb of Veneration

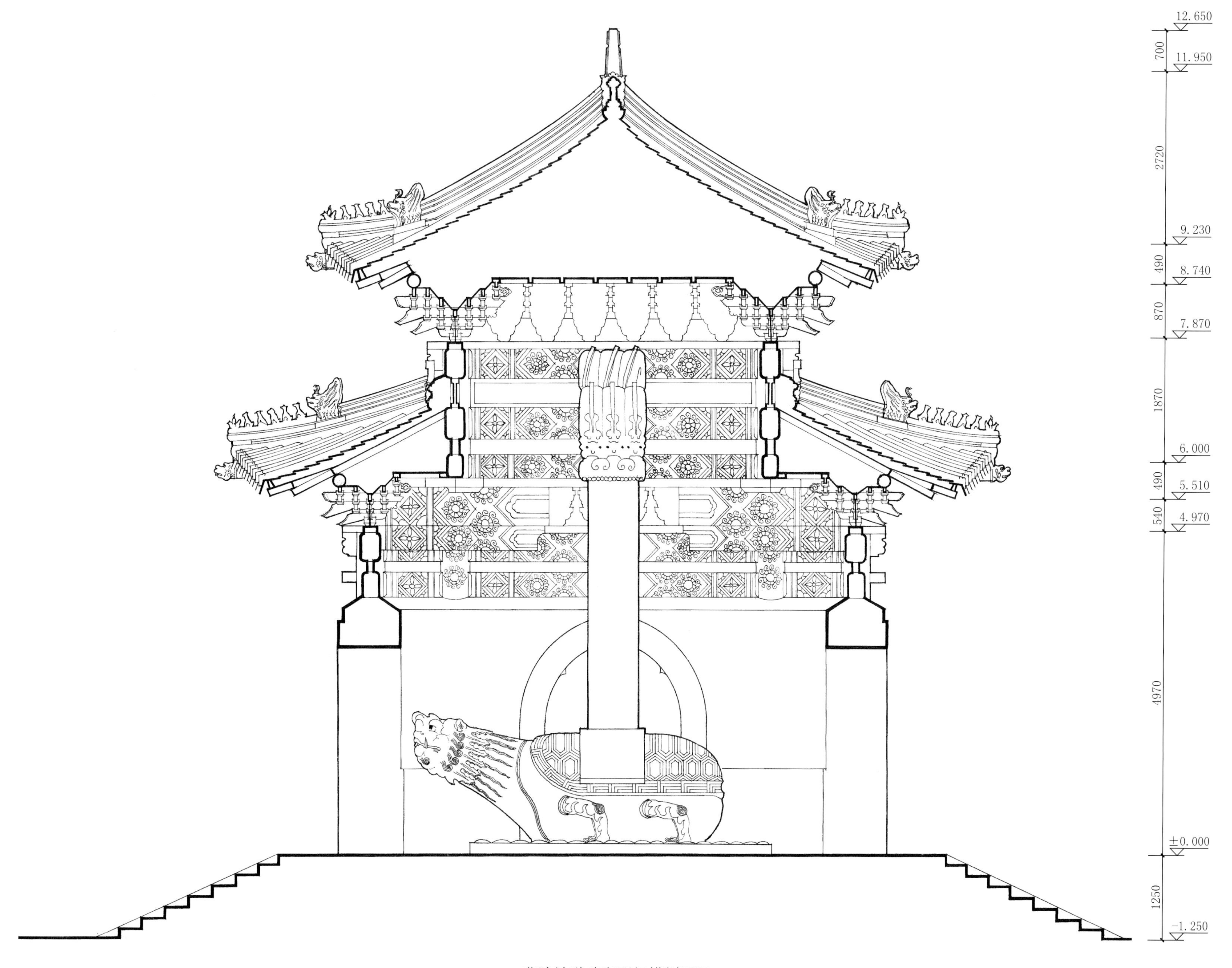

慕陵神道碑亭明间横剖面图

Cross section of the Central Chamber within the Pavilion for the Stela on the Spirit Way,Tomb of Veneration

慕陵神道碑亭纵剖面图

Longitudinal section of the Pavilion for the Stela on the Spirit Way,Tomb of Veneration

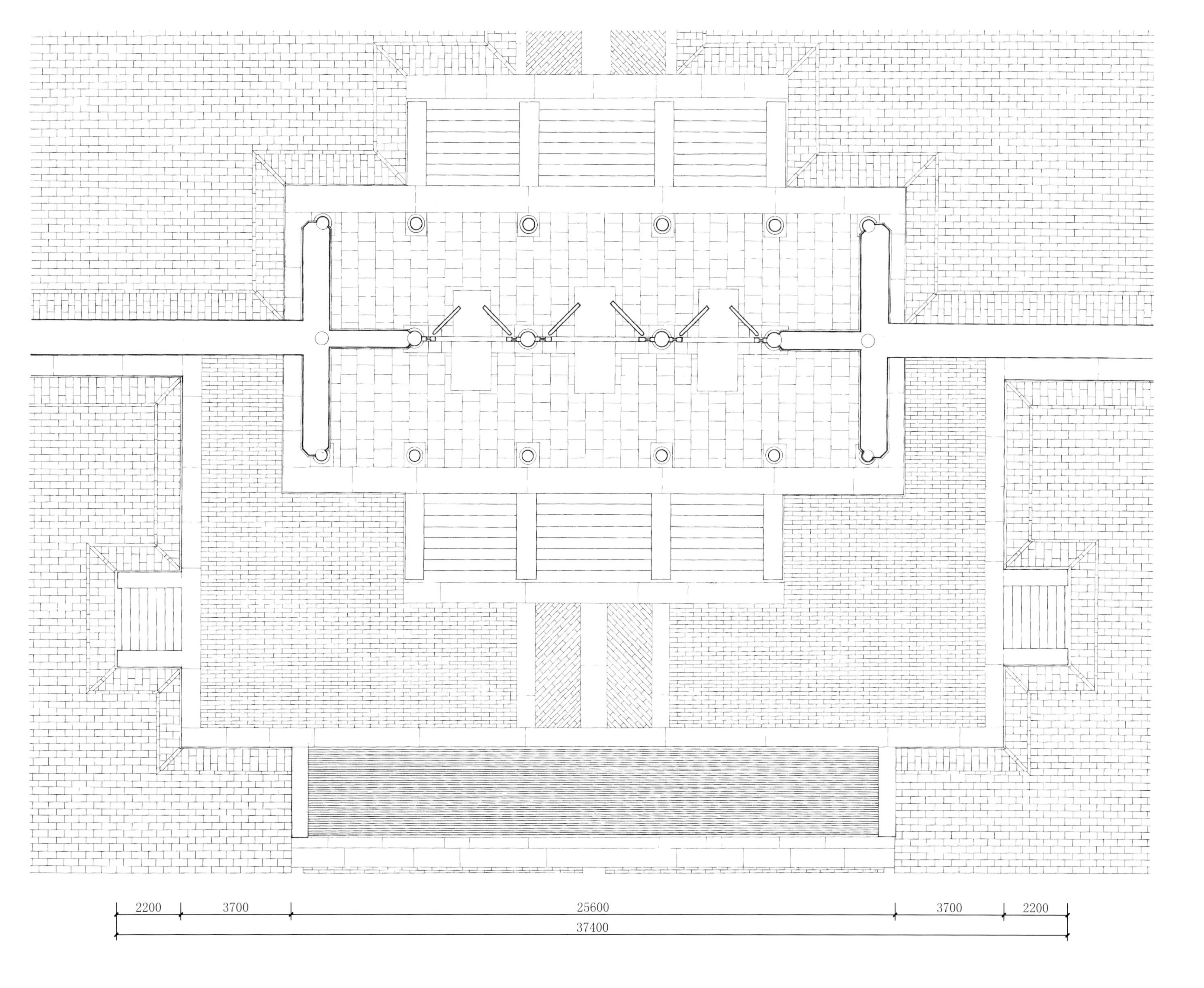

慕陵隆恩门平面图

Plan of the Gate of Monumental Grace,Tomb of Veneration

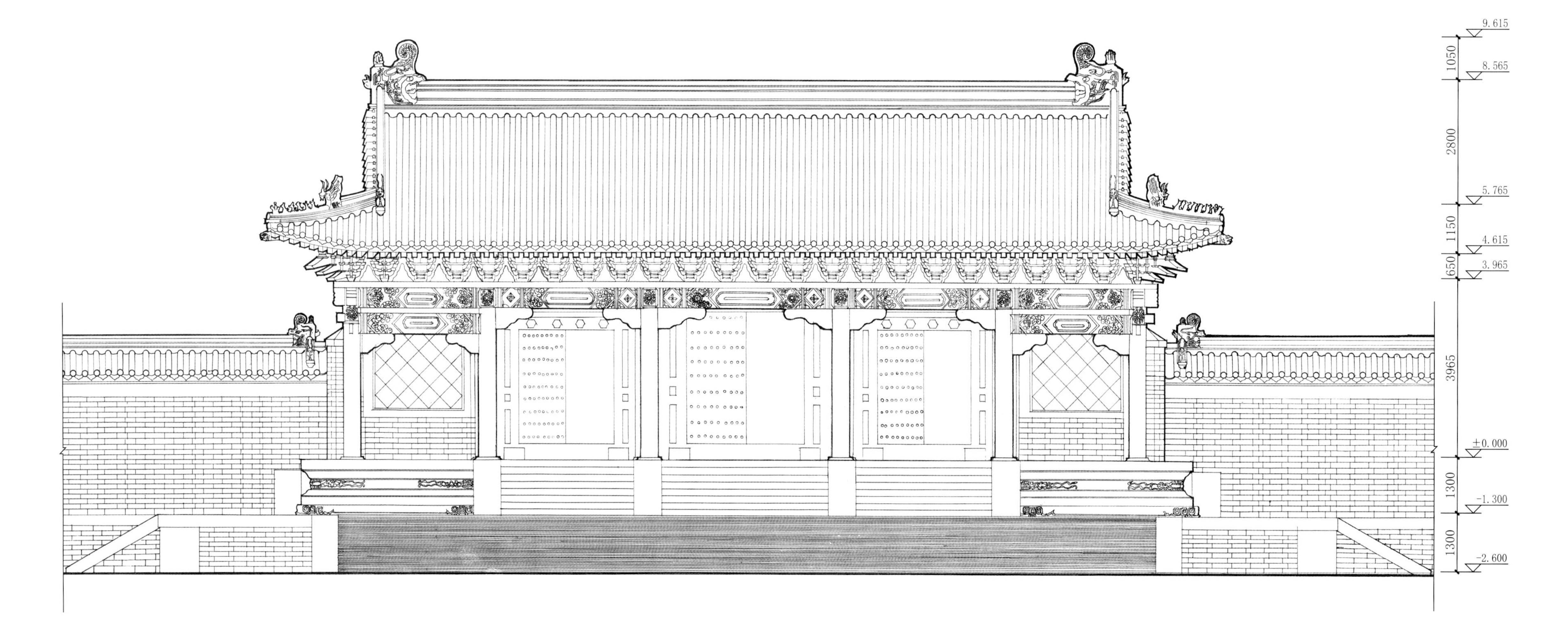

慕陵隆恩门正立面图
Front elevation of the Gate of Monumental Grace,Tomb of Veneration

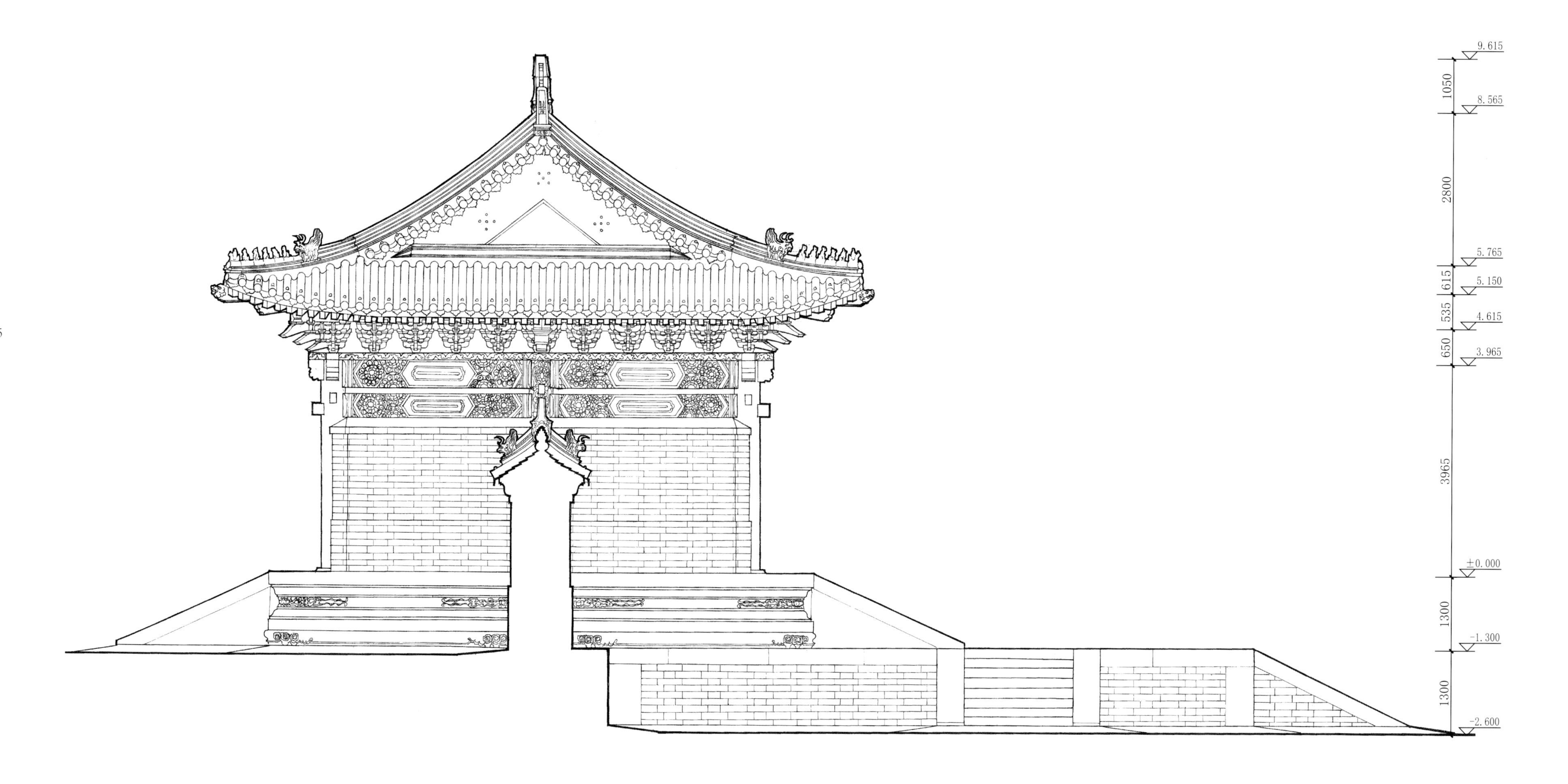

慕陵隆恩门侧立面图

Side elevation of the Gate of Monumental Grace,Tomb of Veneration

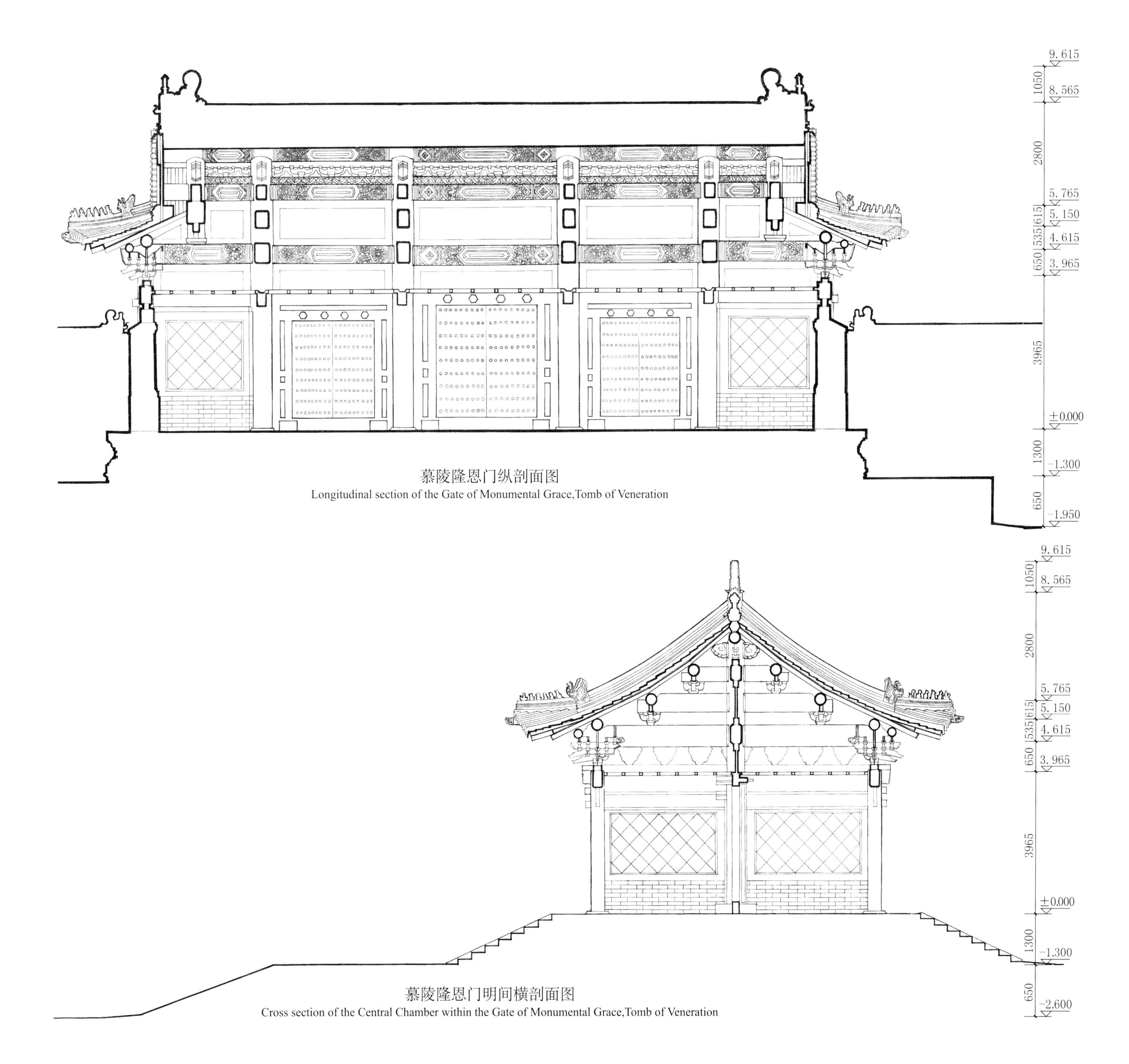

慕陵隆恩门纵剖面图
Longitudinal section of the Gate of Monumental Grace,Tomb of Veneration

慕陵隆恩门明间横剖面图
Cross section of the Central Chamber within the Gate of Monumental Grace,Tomb of Veneration

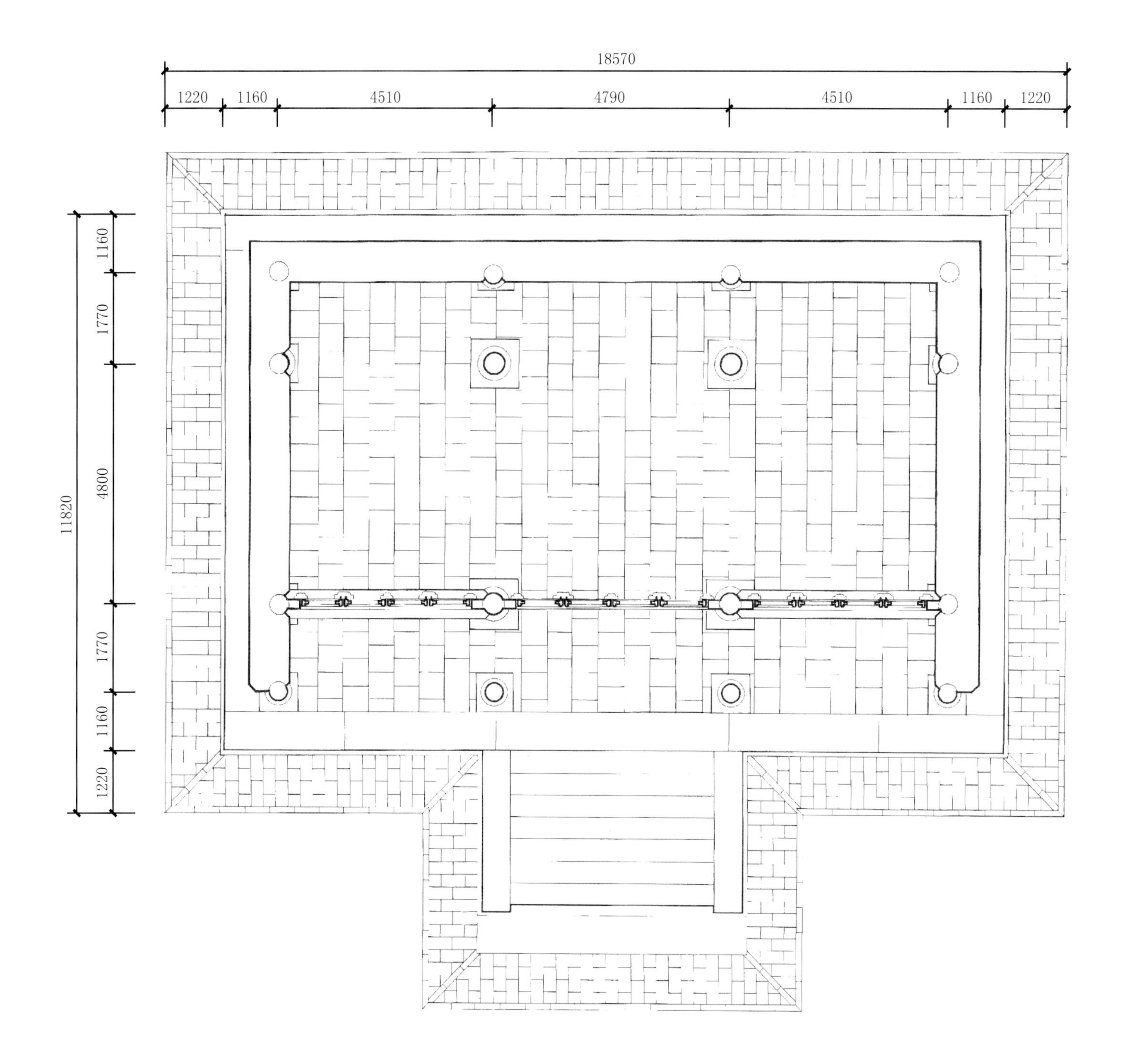

慕陵配殿平面图
Plan of the Side Hall,Tomb of Veneration

慕陵配殿正立面图
Front elevation of the Side Hall,Tomb of Veneration

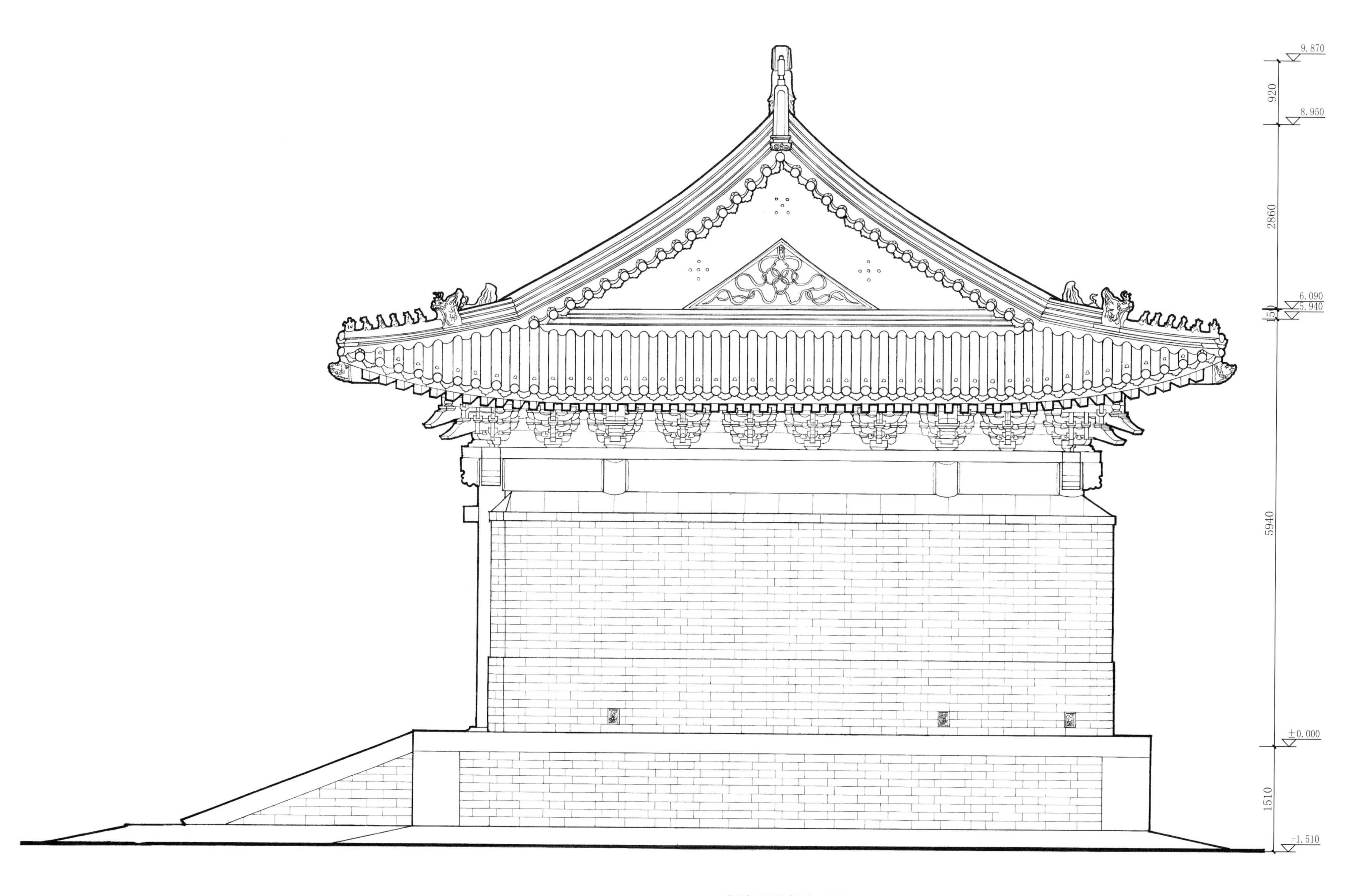

慕陵配殿侧立面图

Side elevation of the Side Hall,Tomb of Veneration

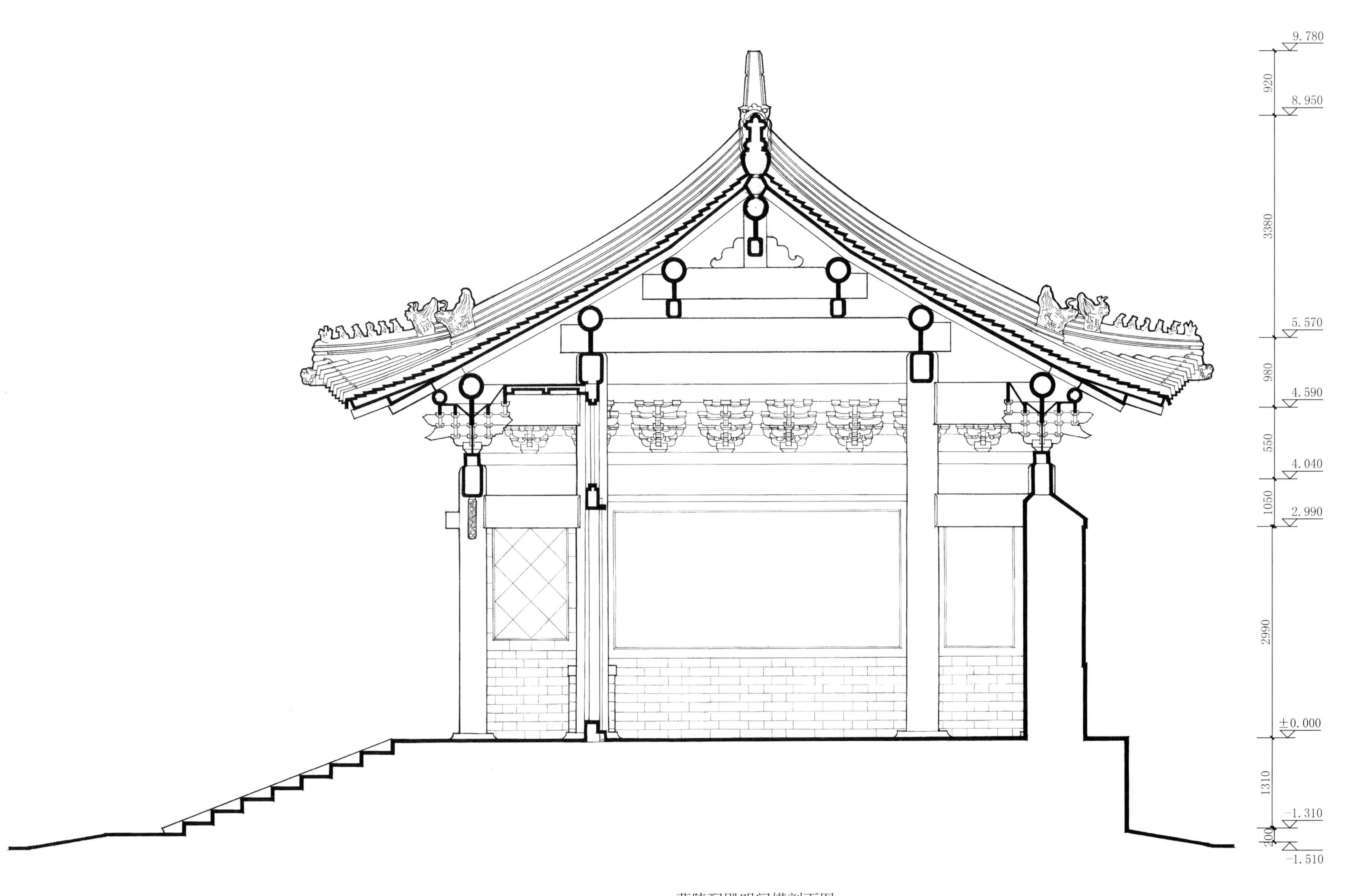

慕陵配殿明间横剖面图

Cross section of the Central Chamber within the Side Hall,Tomb of Veneration

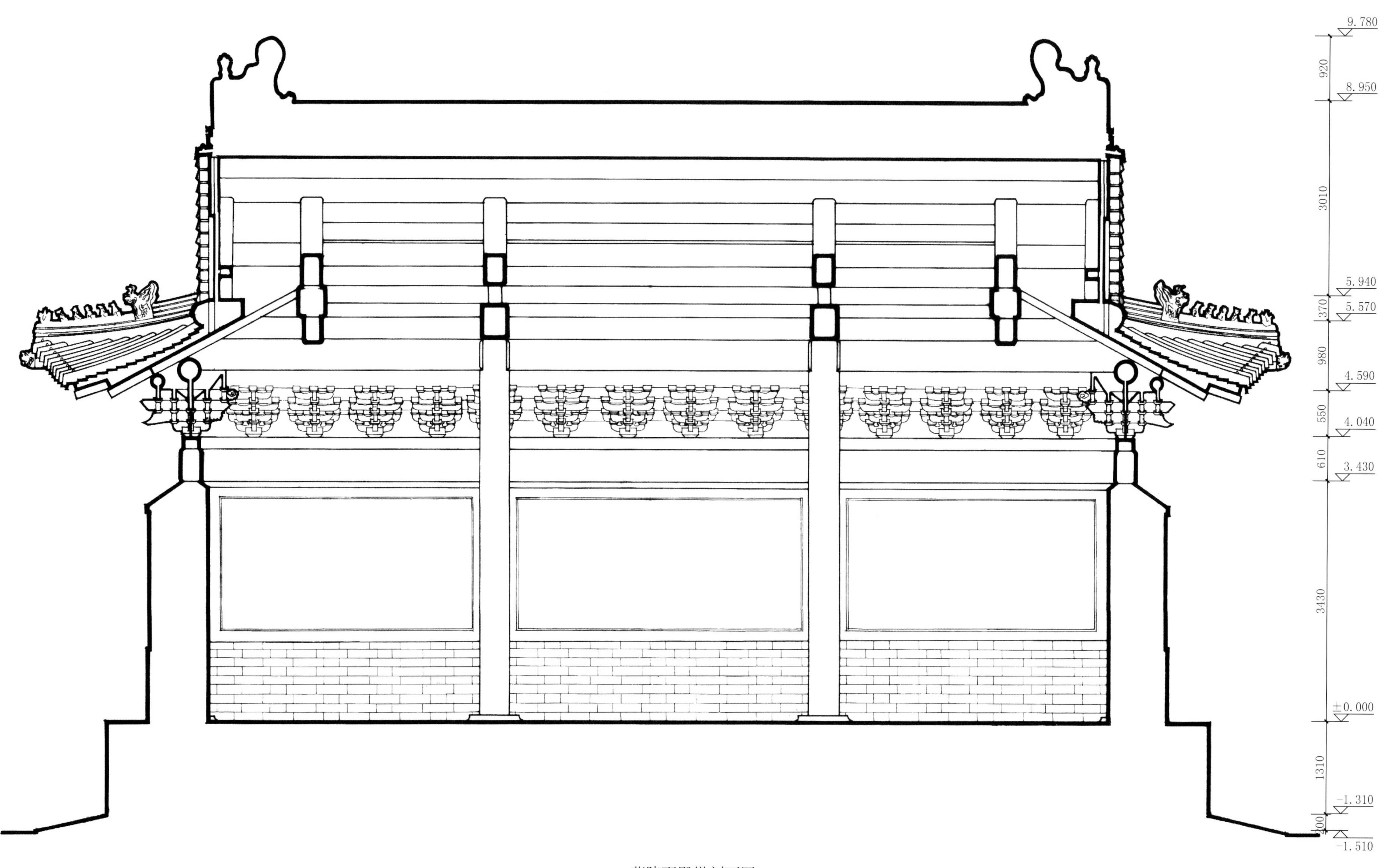

慕陵配殿纵剖面图

Longitudinal section of the Side Hall,Tomb of Veneration

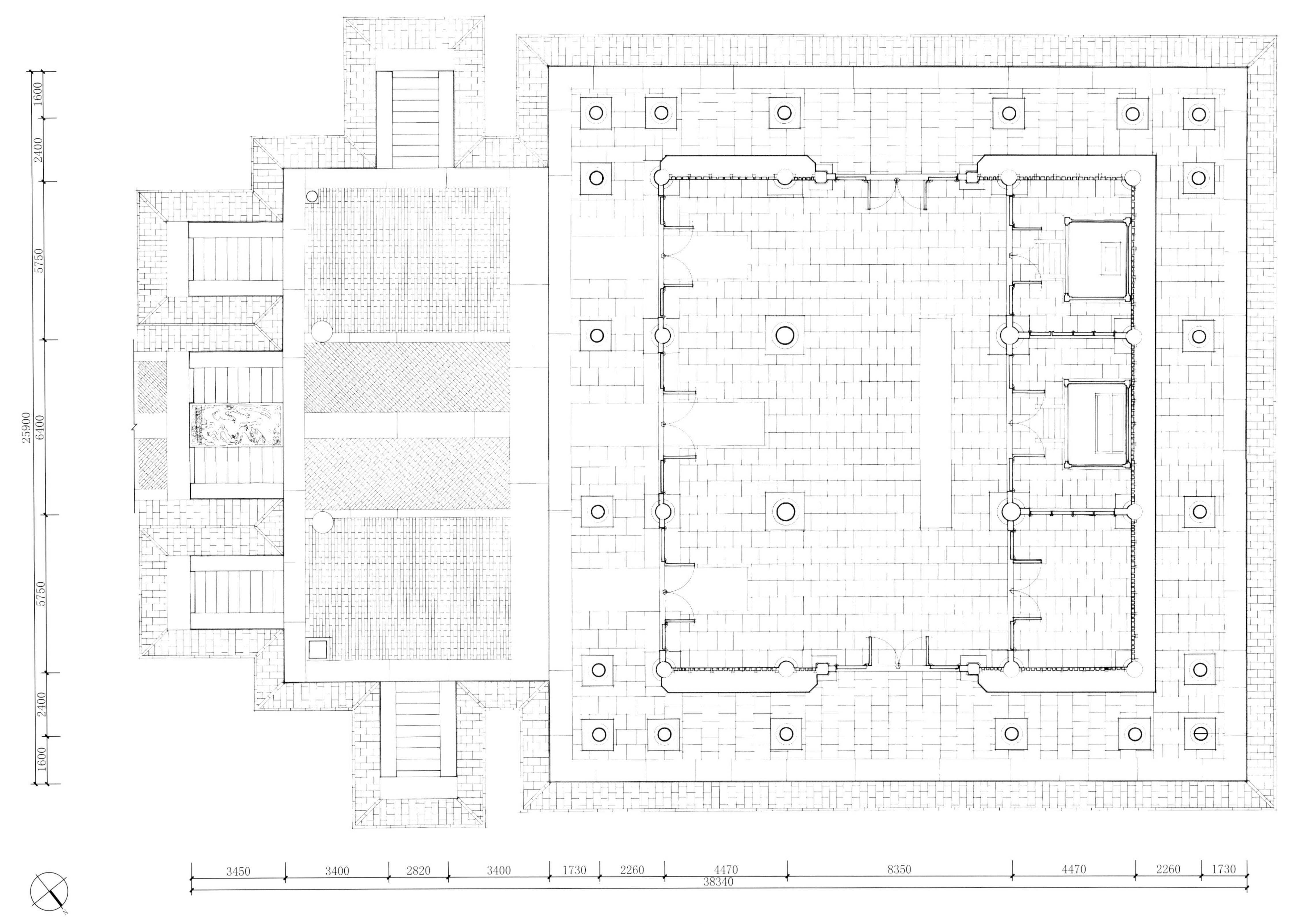

慕陵隆恩殿平面图
Plan of the Hall of Monumental Grace,Tomb of Veneration

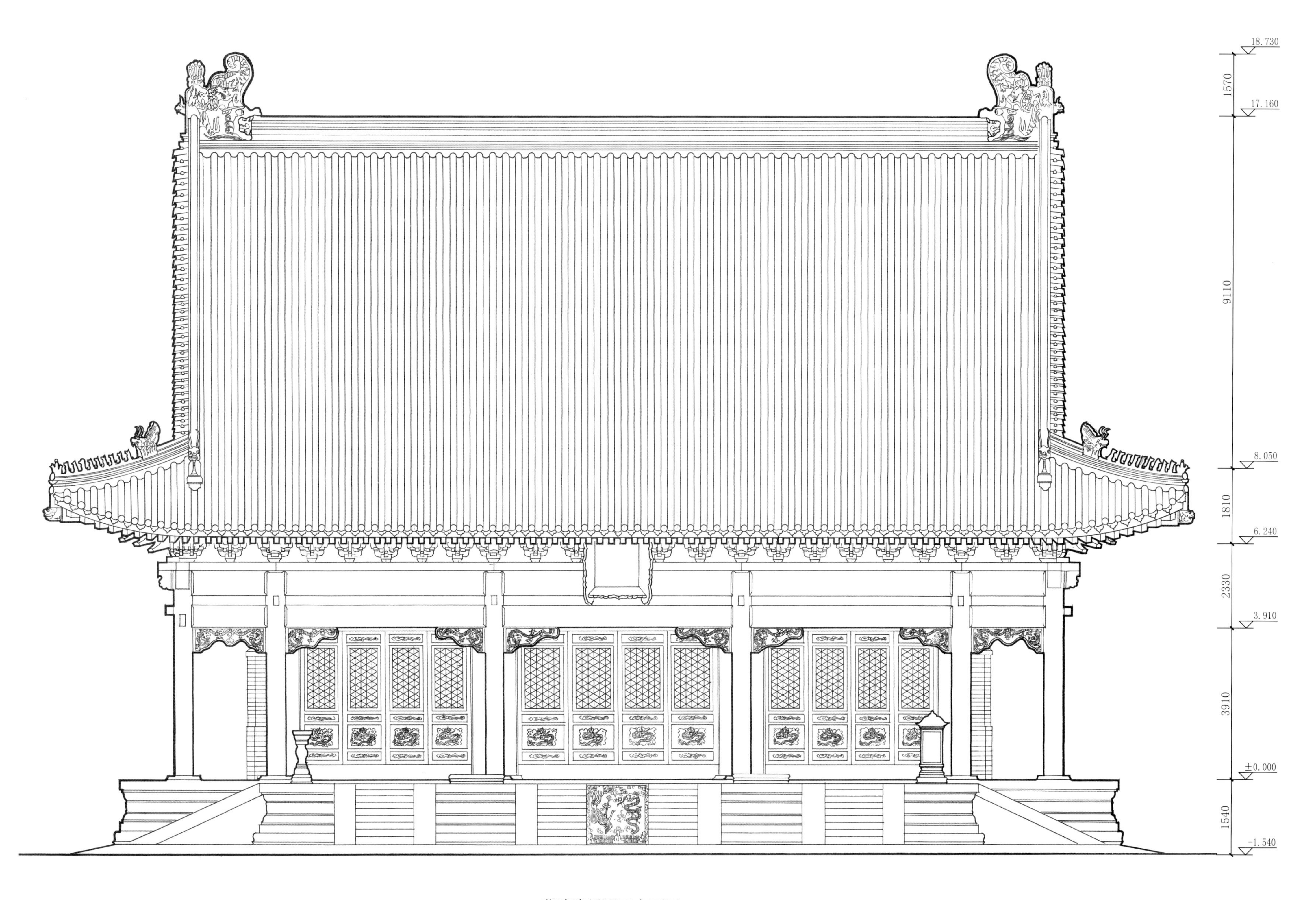

慕陵隆恩殿正立面图

Front elevation of the Hall of Monumental Grace,Tomb of Veneration

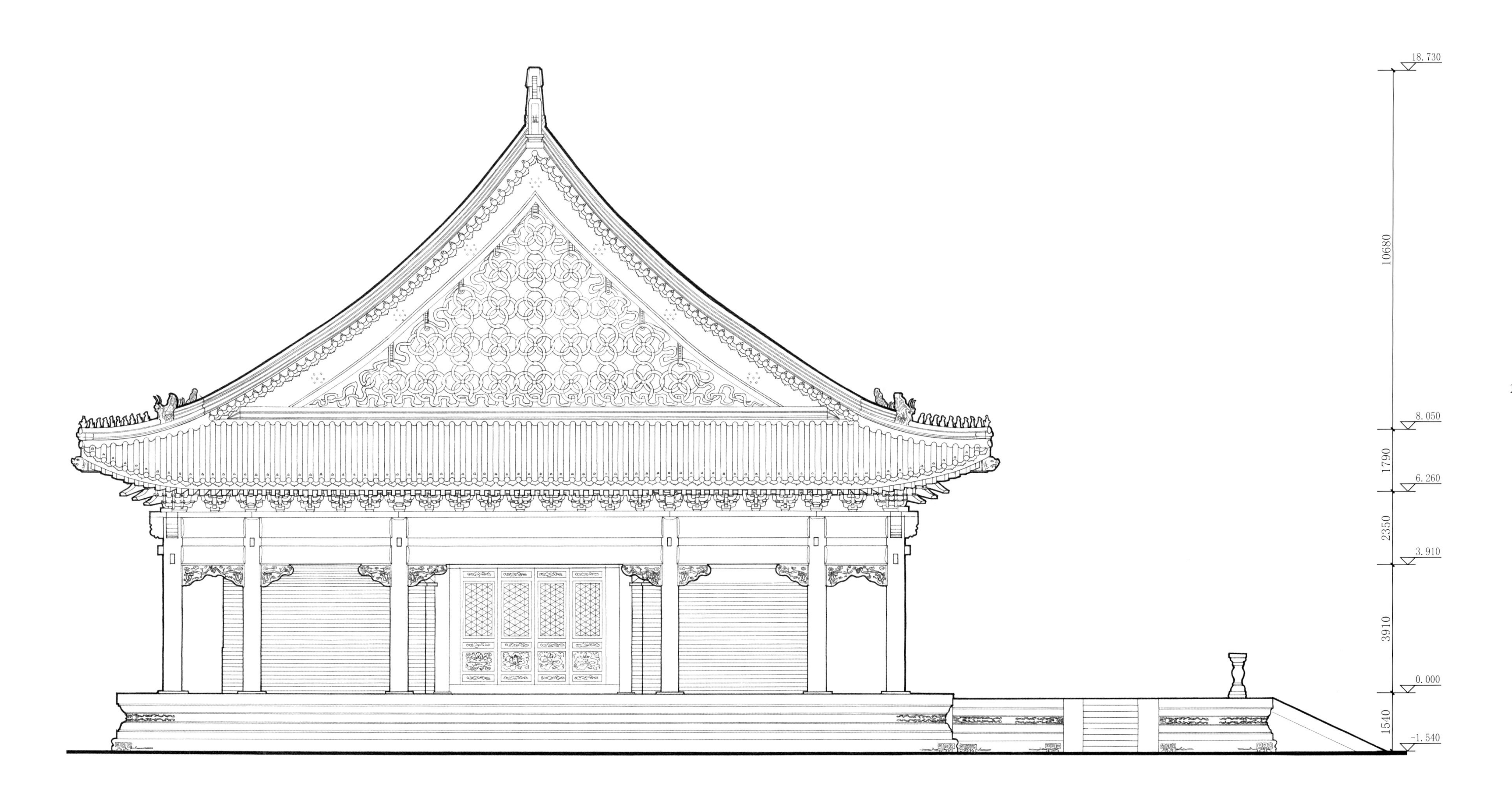

慕陵隆恩殿侧立面图

Side elevation of the Hall of Monumental Grace,Tomb of Veneration

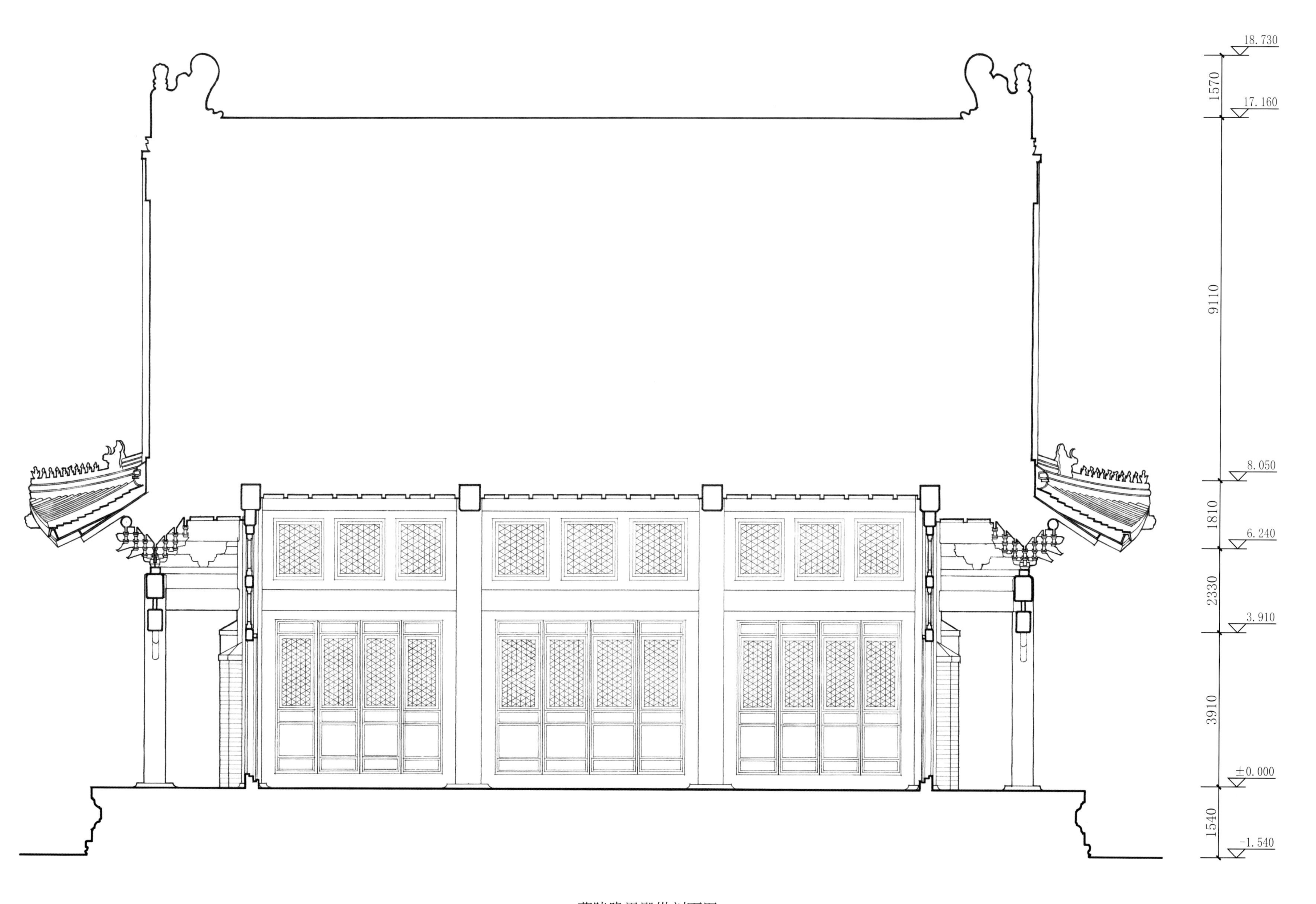

慕陵隆恩殿纵剖面图

Longitudinal section of the Hall of Monumental Grace,Tomb of Veneration

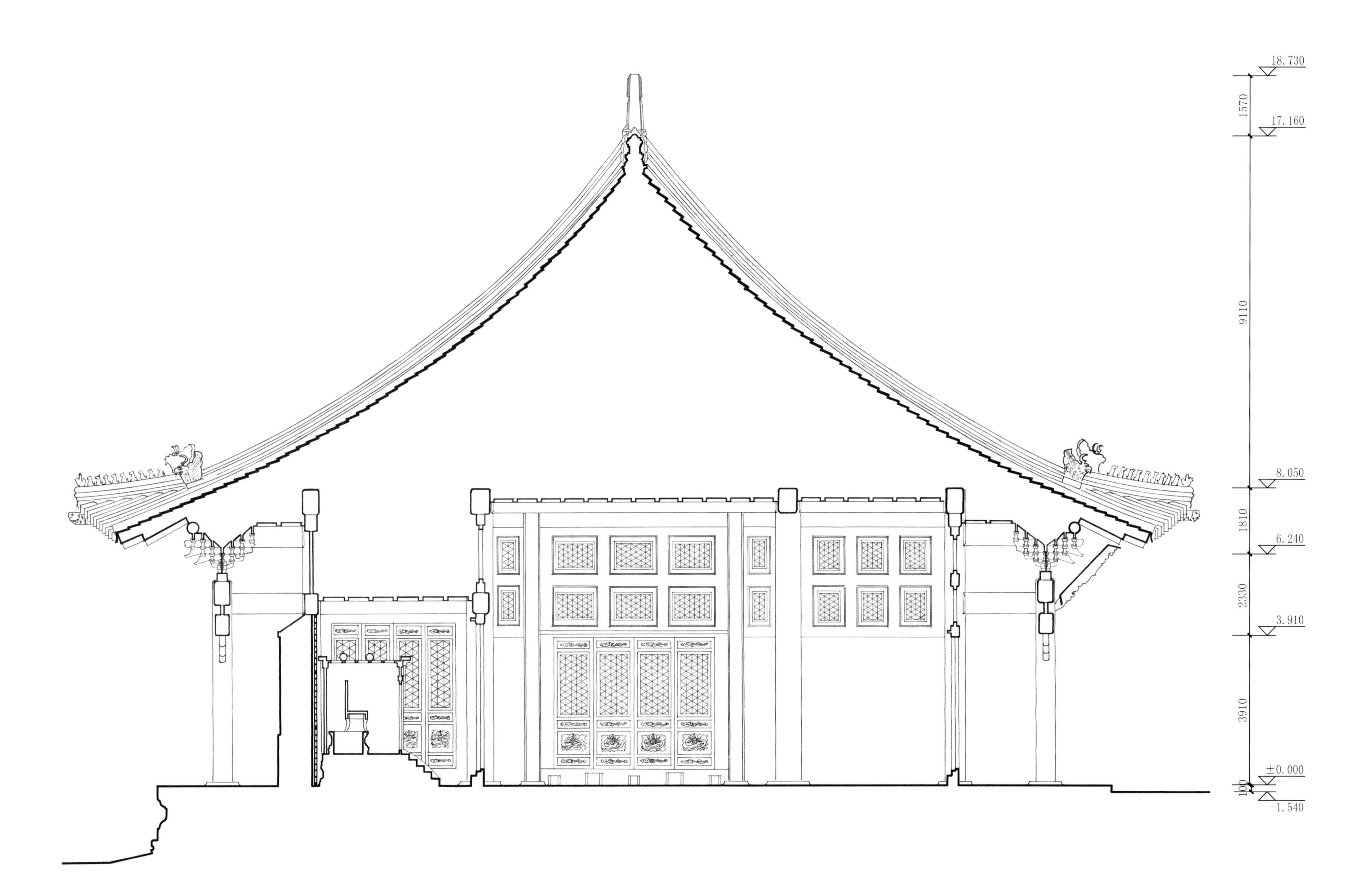

慕陵隆恩殿明间横剖面图

Cross section of the Central Chamber within the Hall of Monumental Grace,Tomb of Veneration

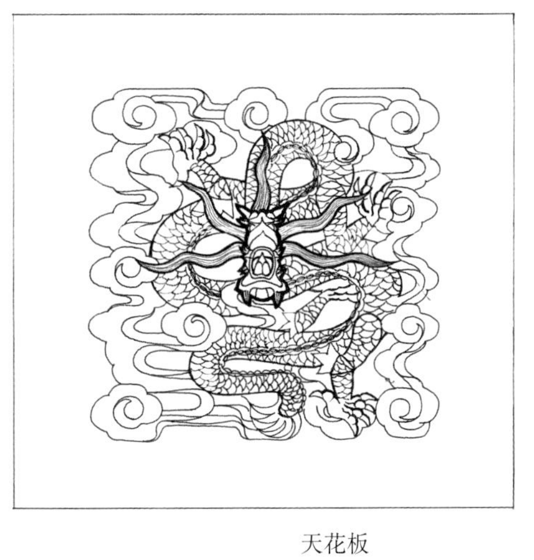

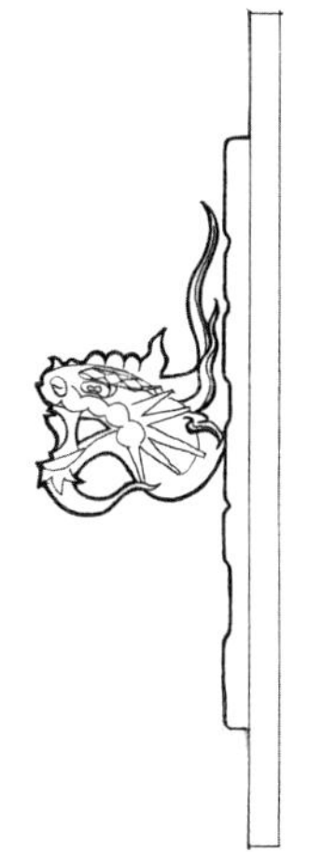

天花板

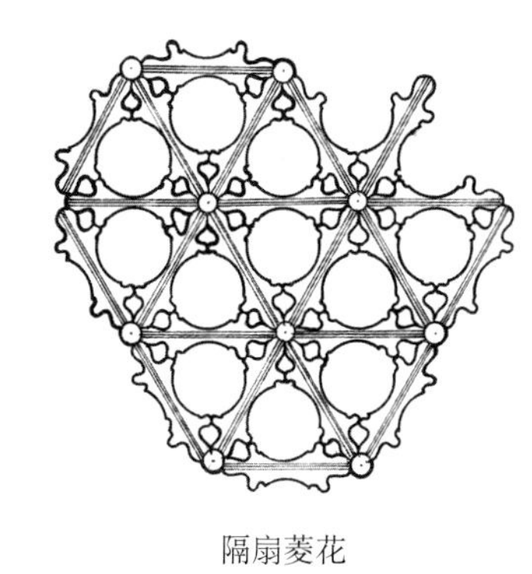

隔扇菱花

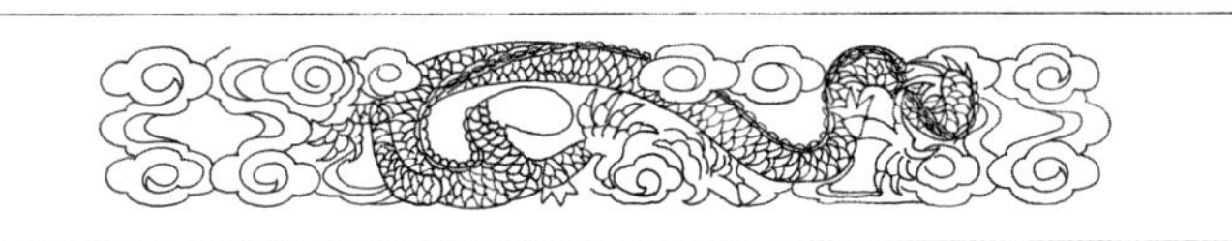

绦环板

裙板

雀替

丹陛

慕陵隆恩殿构件大样图

Detailed drawing of the structural component of the Hall of Monumental Grace,Tomb of Veneration

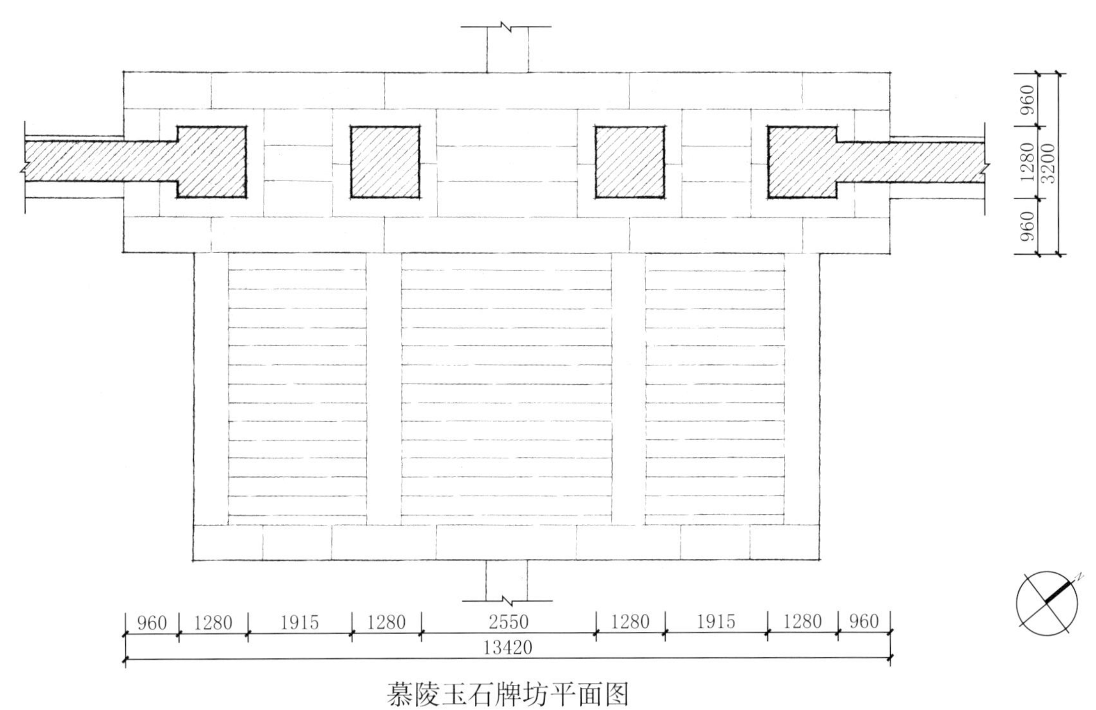

慕陵玉石牌坊平面图

Plan of the Jade Memorial Gateway, Tomb of Veneration

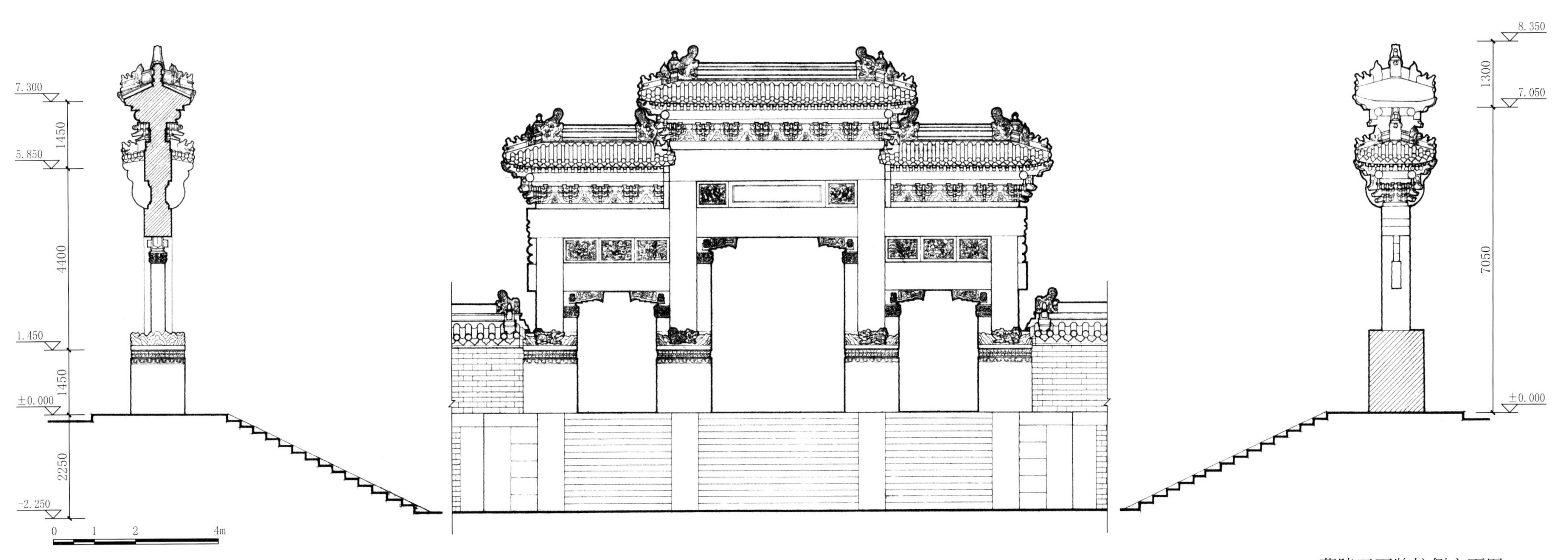

慕陵玉石牌坊剖立面图

Sectional elevation of the Jade Memorial Gateway, Tomb of Veneration

慕陵玉石牌坊正立面图

Front elevation of the Jade Memorial Gateway, Tomb of Veneration

慕陵玉石牌坊侧立面图

Side elevation of the Jade Memorial Gateway, Tomb of Veneration

慕陵石台五供正立面图

Front elevation of the Five Stone Ritual Vessels,Tomb of Veneration

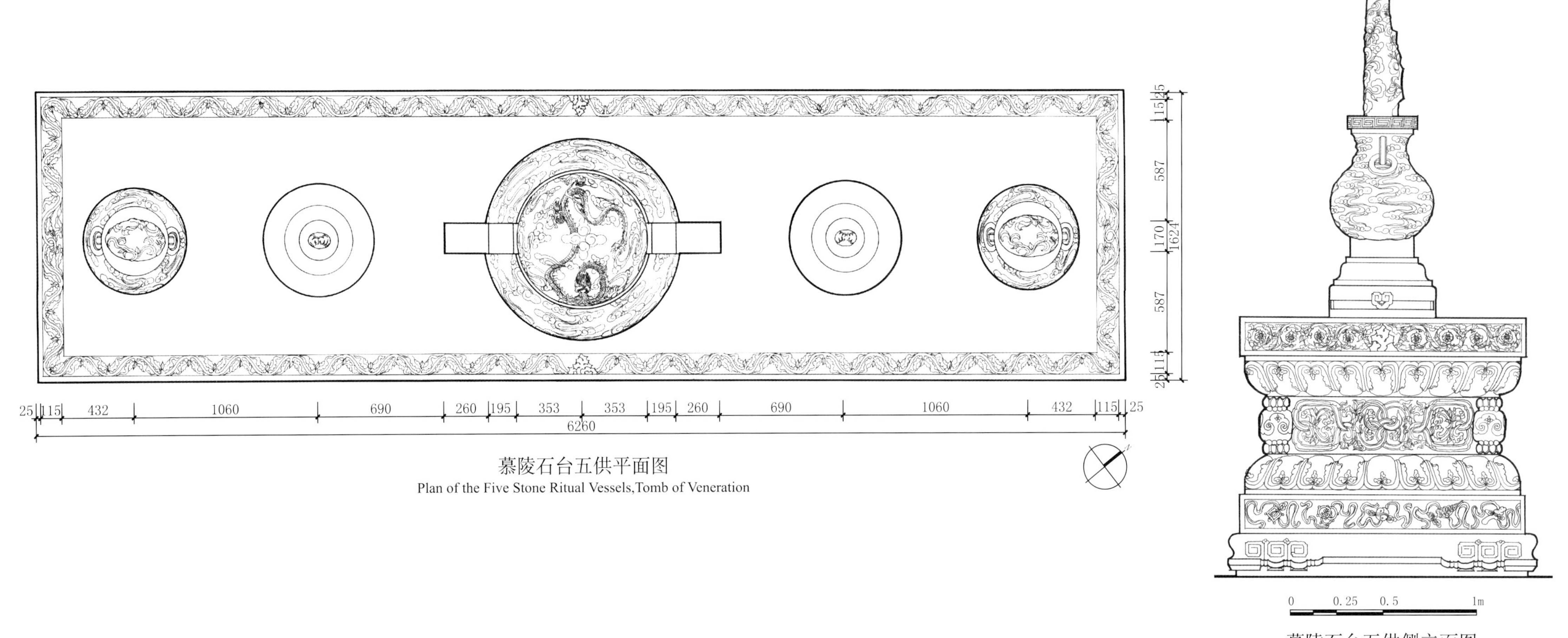

慕陵石台五供平面图

Plan of the Five Stone Ritual Vessels,Tomb of Veneration

慕陵石台五供侧立面图

Side elevation of the Five Stone Ritual Vessels,Tomb of Veneration

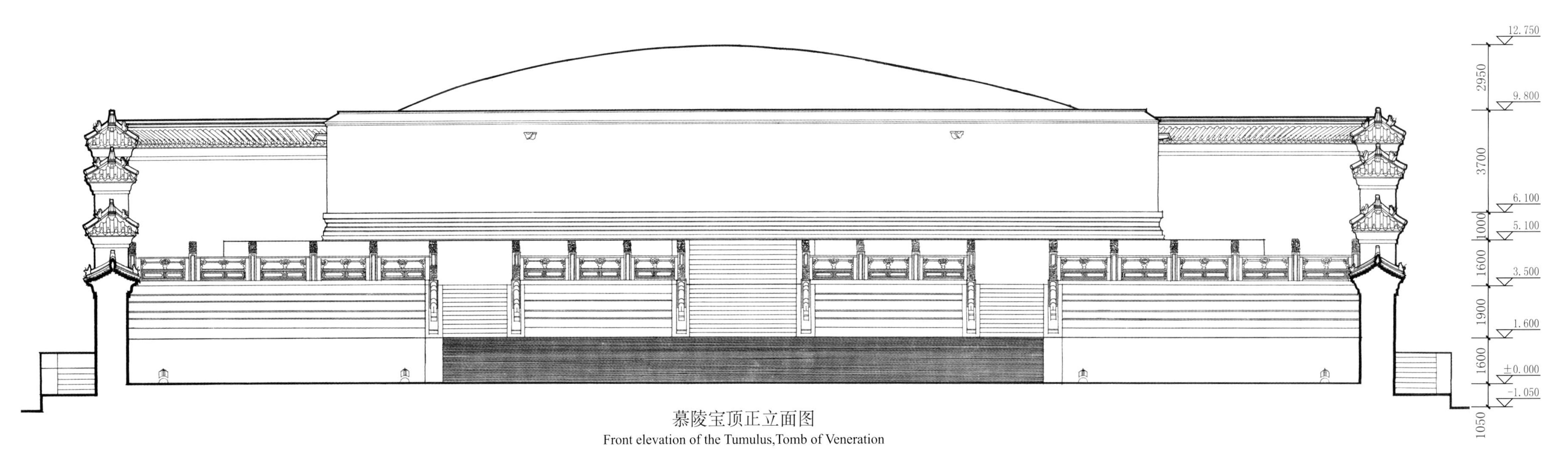

慕陵宝顶正立面图

Front elevation of the Tumulus,Tomb of Veneration

慕东陵
East Tomb of Veneration (Mu Dongling)

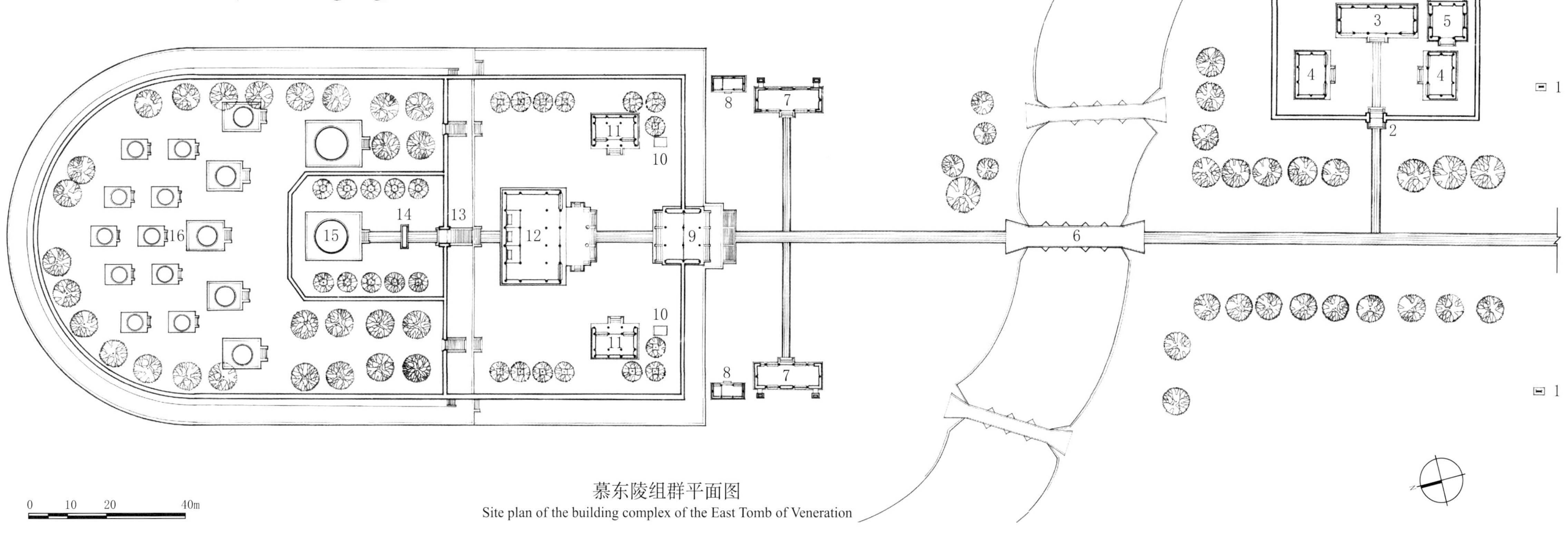

慕东陵组群平面图
Site plan of the building complex of the East Tomb of Veneration

1 下马牌 Stela Marking the Place for Dismounting from One's Horse
2 神厨库门 Gate in the Culinary Courtyard for Sacrifices
3 神厨 Kitchen in the Culinary Courtyard for Sacrifices
4 神库 Repository in the Culinary Courtyard for Sacrifices
5 宰牲亭 Ritual Abattoir in the Culinary Courtyard for Sacrifices
6 五孔石平桥 Five-Arched Flat Stone Bridge
7 朝房 Reception Hall for Court Officials
8 值房 Guard House
9 隆恩门 Gate of Monumental Grace
10 焚帛炉 Sacrificial Burner
11 配殿 Side Hall
12 隆恩殿 Hall of Monumental Grace
13 琉璃花门 Gate with Glazed Roof Tiles
14 石台五供 Five Stone Ritual Vessels
15 后陵宝顶 Tumulus of the Queen
16 妃嫔宝顶 Tumulus of the Imperial Consorts

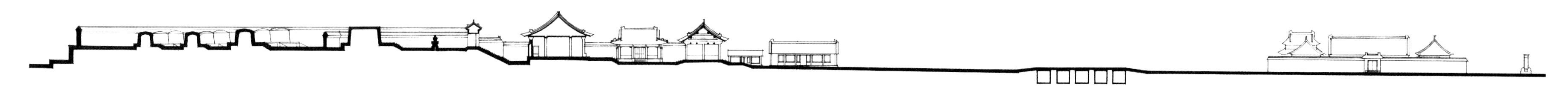

慕东陵组群剖面图
Site section of the building complex of the East Tomb of Veneration

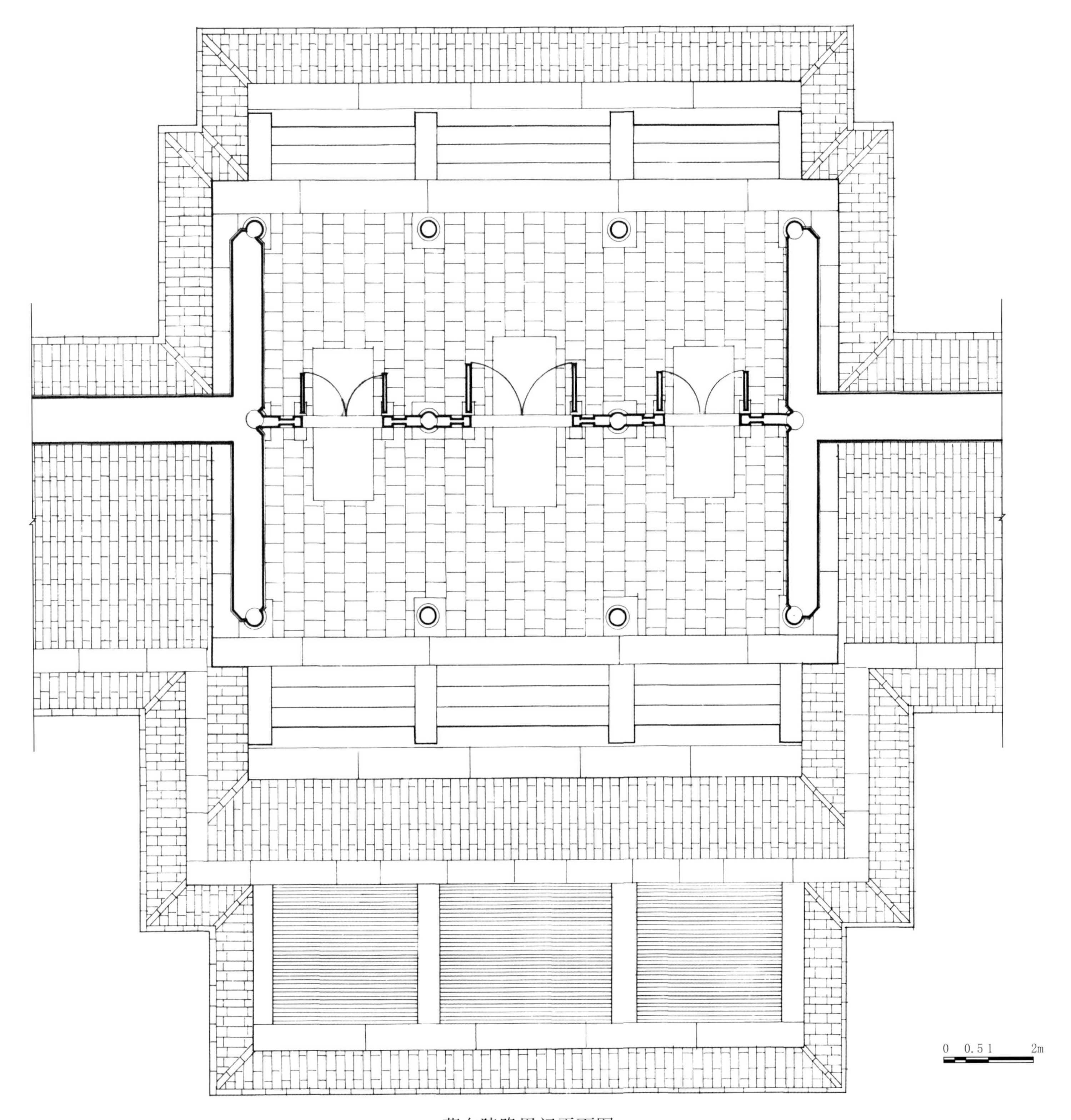

慕东陵隆恩门平面图

Plan of the Gate of Monumental Grace,East Tomb of Veneration

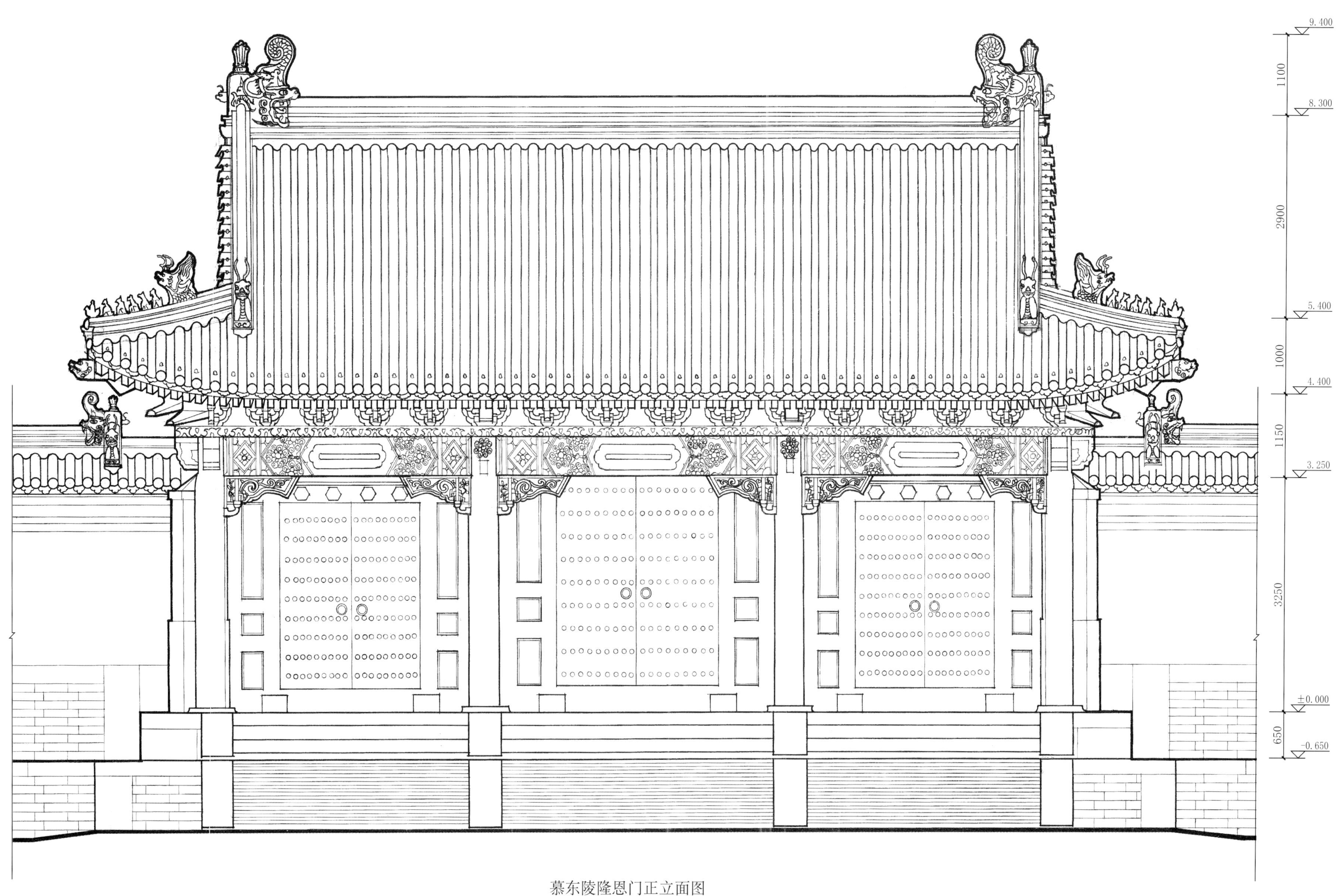

慕东陵隆恩门正立面图

Front elevation of the Gate of Monumental Grace,East Tomb of Veneration

慕东陵隆恩门侧立面图
Side elevation of the Gate of Monumental Grace,East Tomb of Veneration

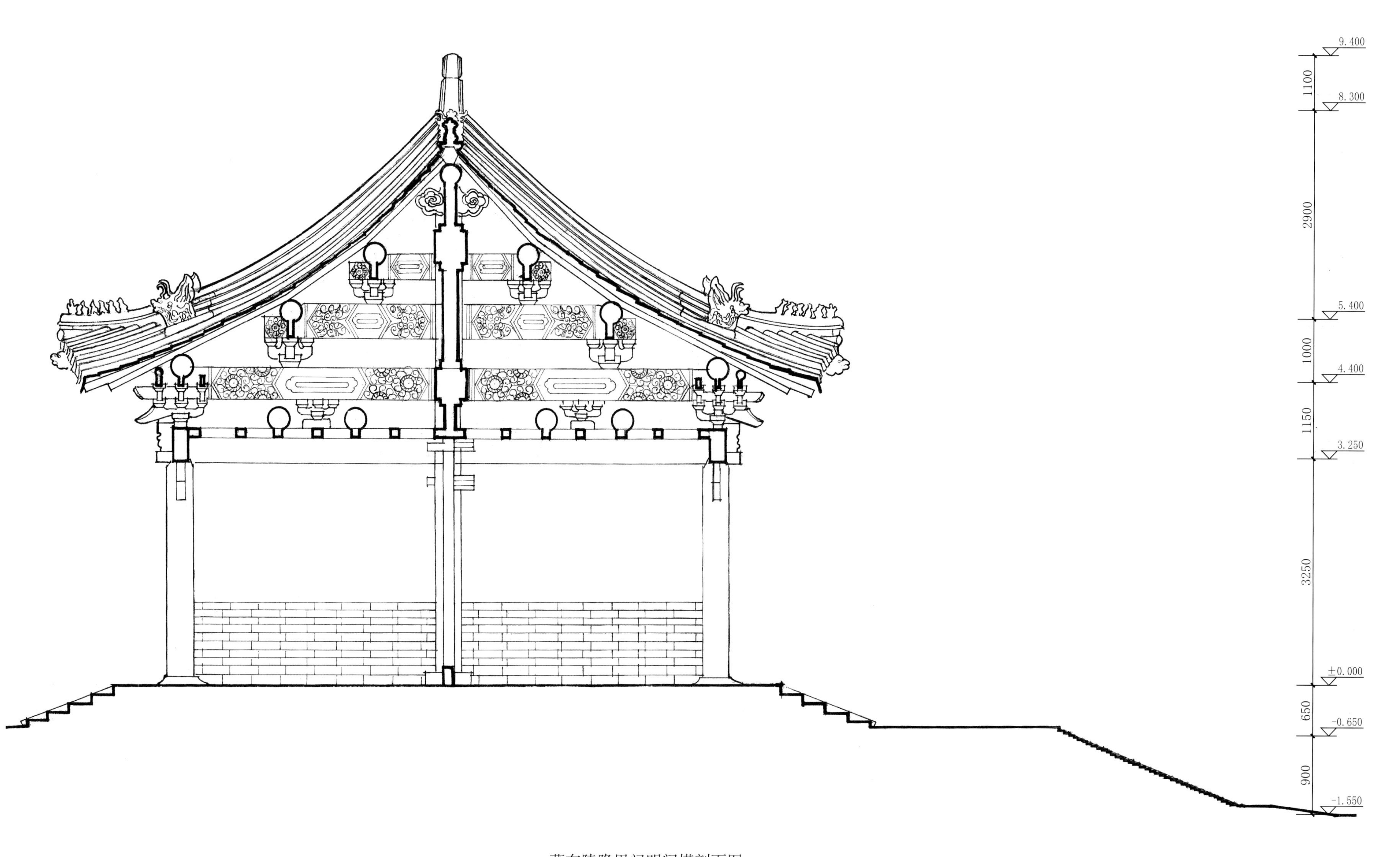

慕东陵隆恩门明间横剖面图

Cross section of the Central Chamber within the Gate of Monumental Grace,East Tomb of Veneration

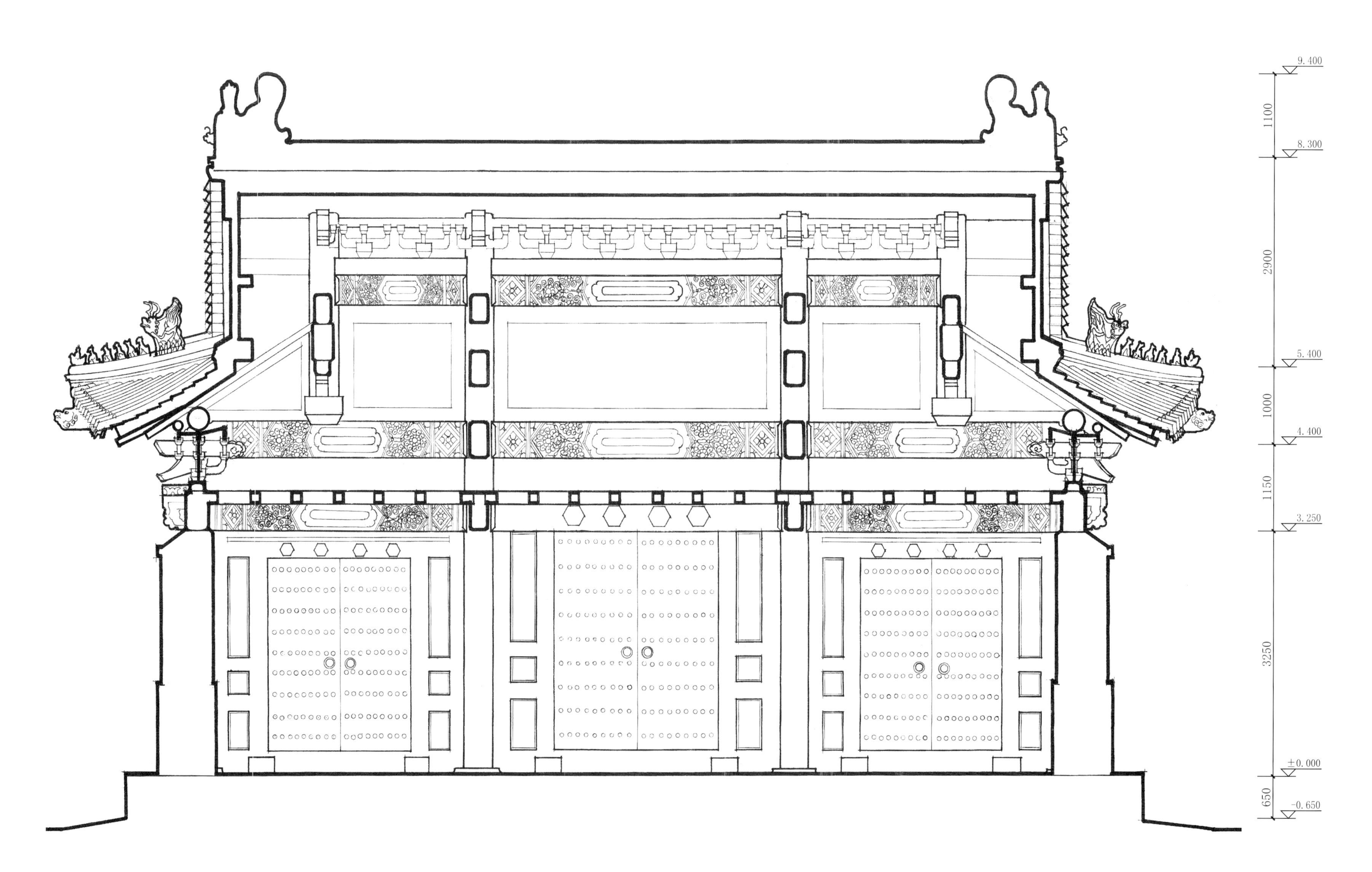

慕东陵隆恩门纵剖面图

Longitudinal section of the Gate of Monumental Grace,East Tomb of Veneration

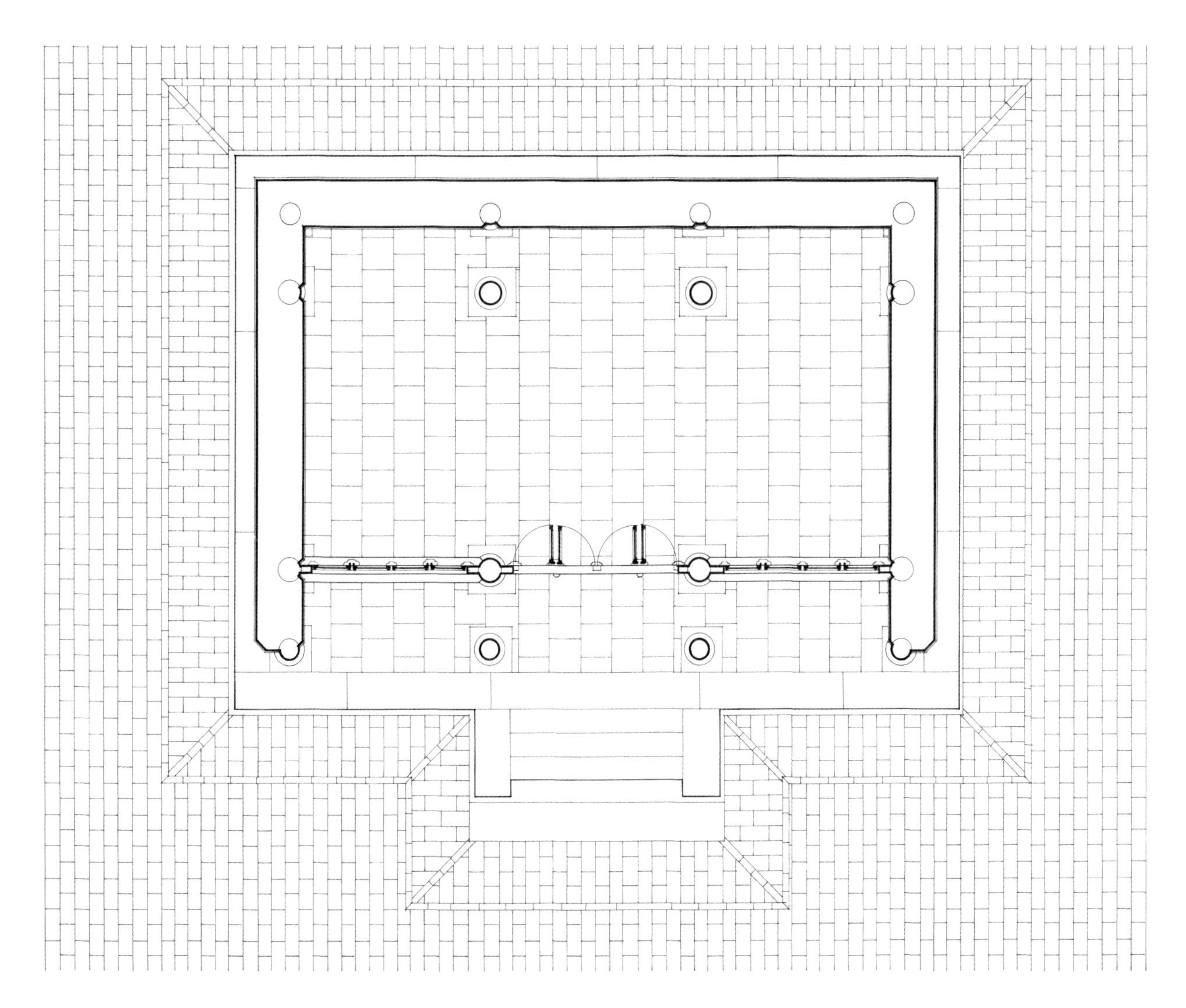

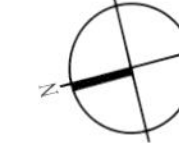

0 0.5 1 2m

慕东陵配殿平面图
Plan of the Side Hall,East Tomb of Veneration

慕东陵配殿正立面图

Front elevation of the Side Hall,East Tomb of Veneration

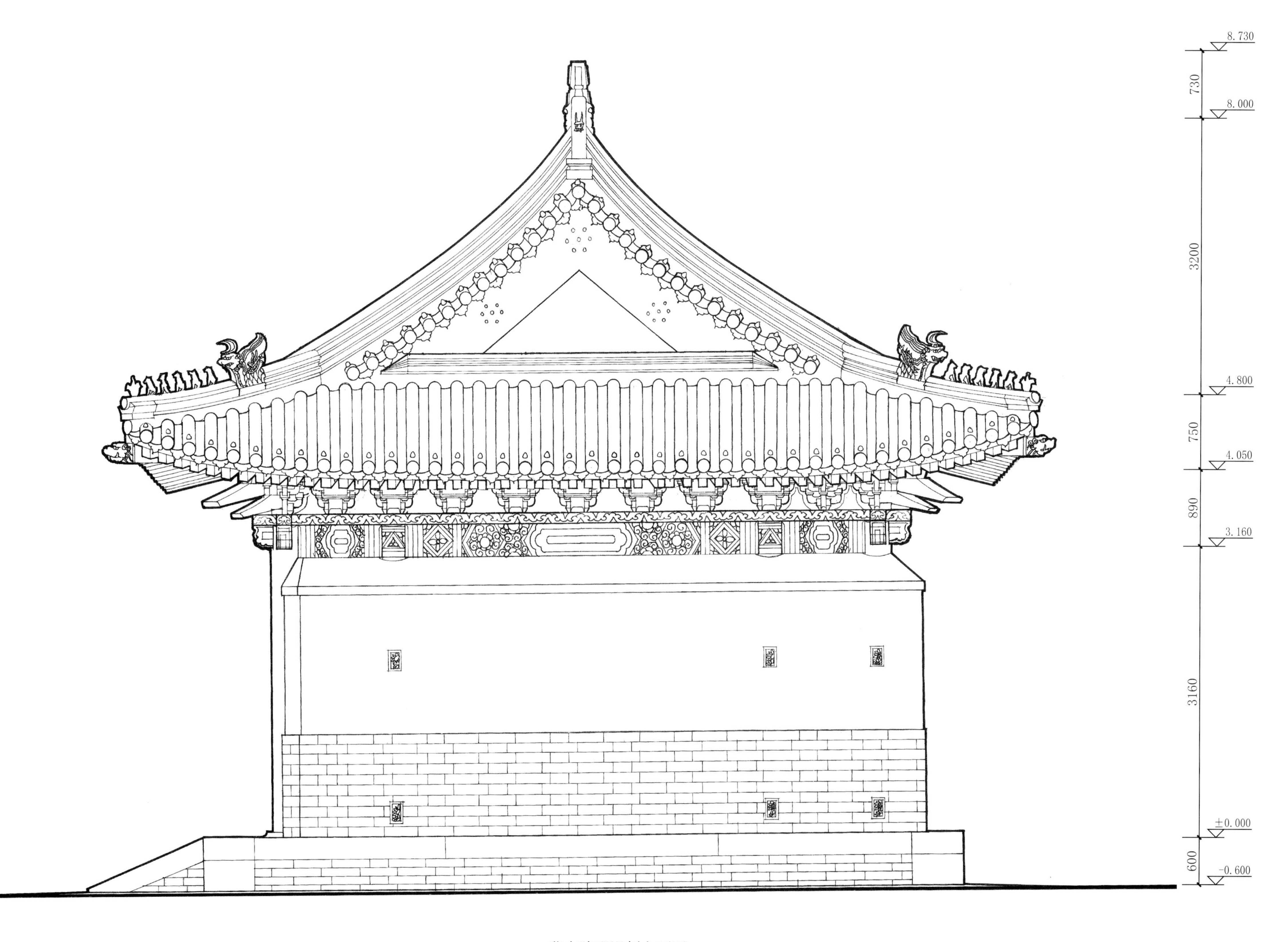

慕东陵配殿侧立面图
Side elevation of the Side Hall,East Tomb of Veneration

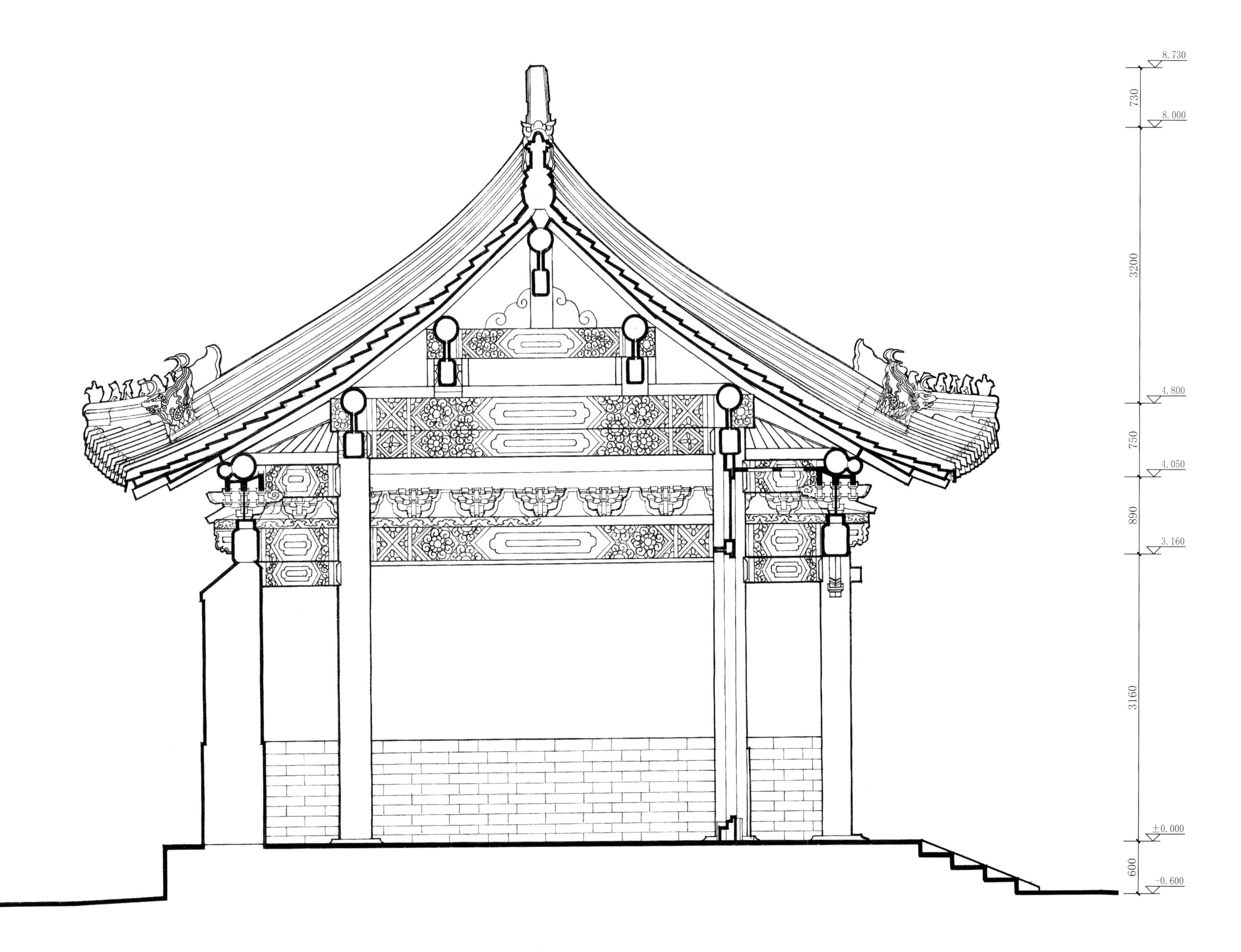

慕东陵配殿明间横剖面图
Cross section of the Central Chamber within the Side Hall,East Tomb of Veneration

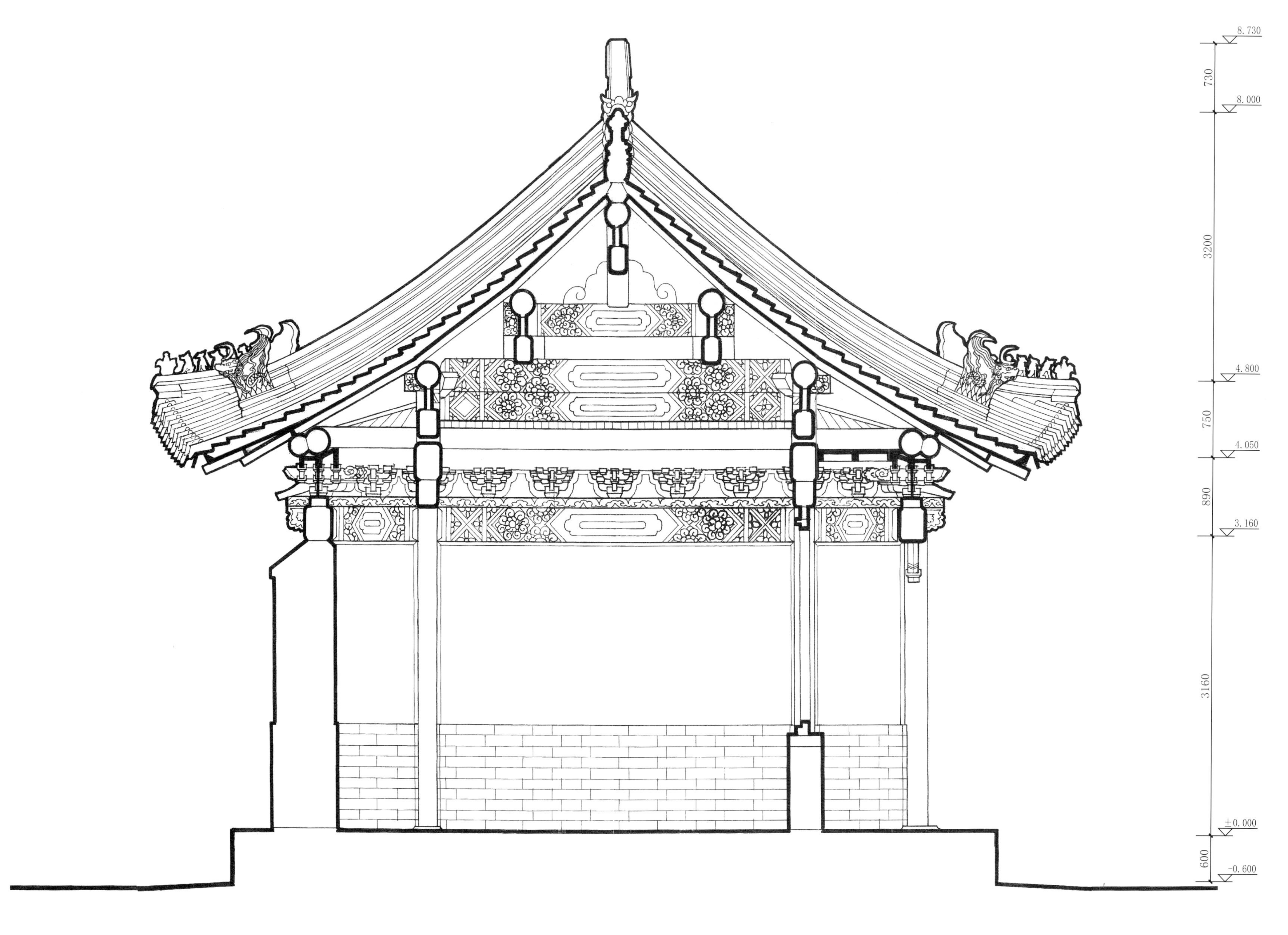

慕东陵配殿次间横剖面图

Cross section of the Side Chamber within the Side Hall,East Tomb of Veneration

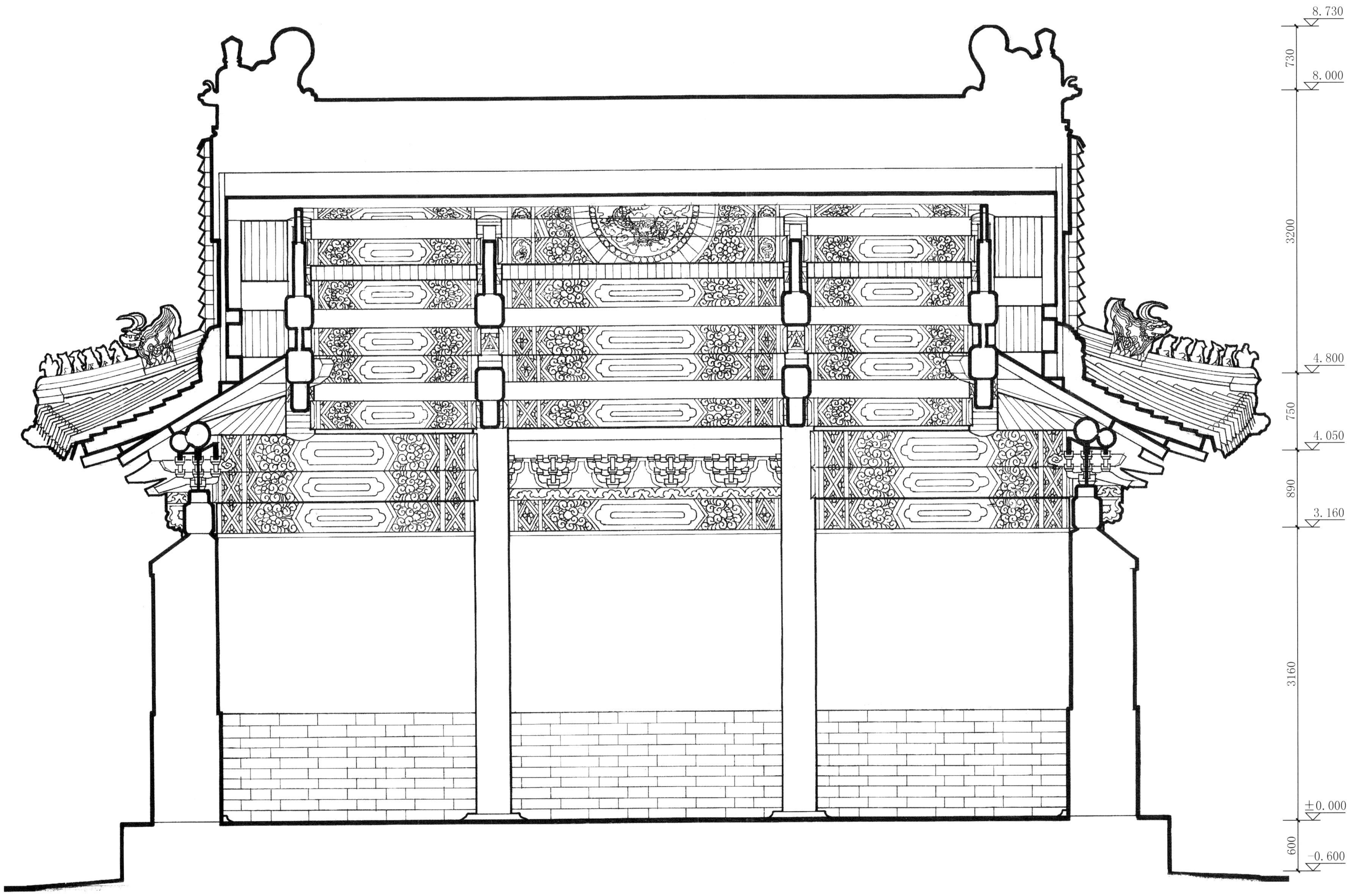

8.730

730

8.000

3200

4.800

750

4.050

890

3.160

3160

±0.000

600

-0.600

慕东陵配殿纵剖面图

Longitudinal section of the Side Hall,East Tomb of Veneration

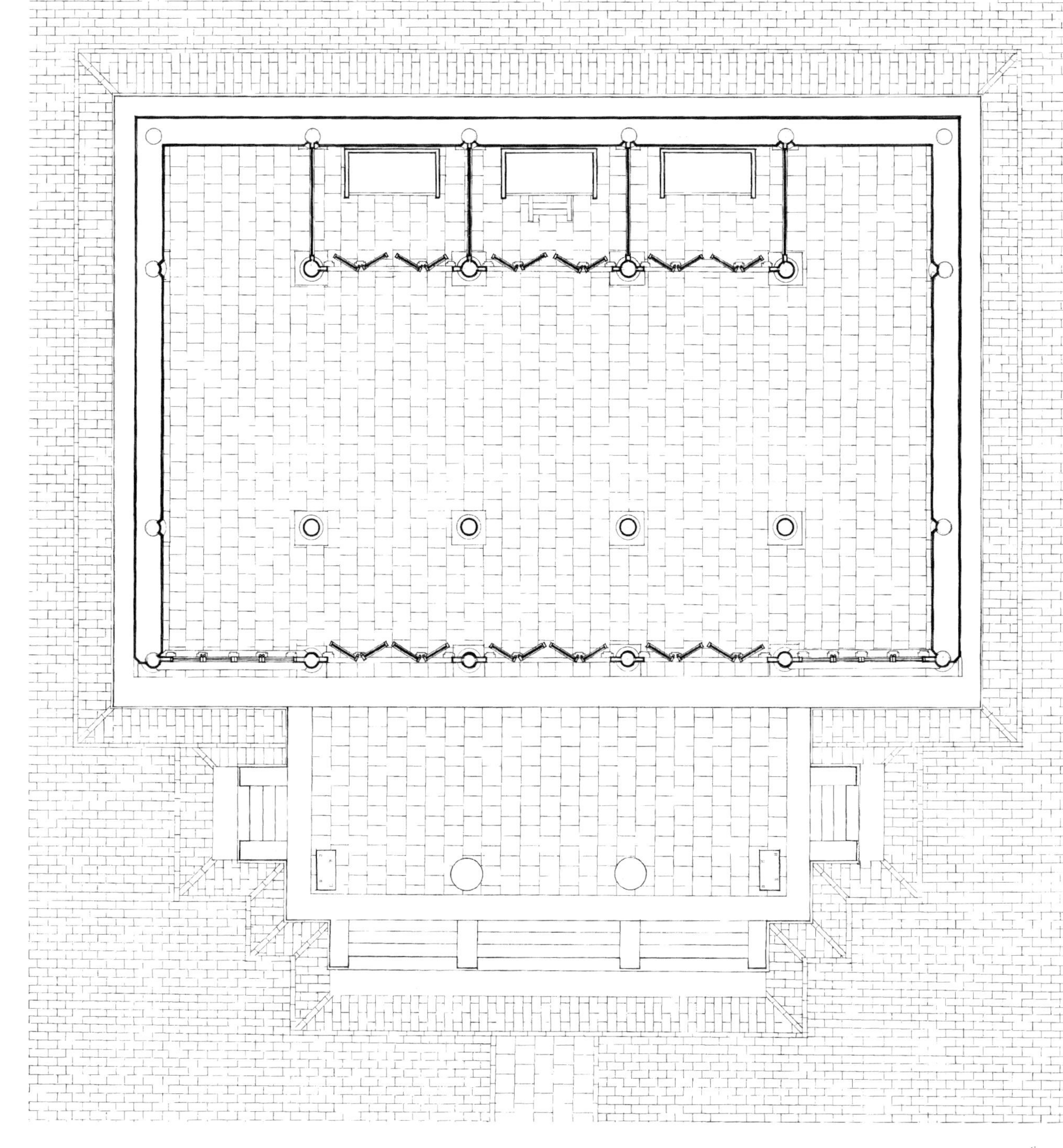

慕东陵隆恩殿平面图

Plan of the Hall of Monumental Grace,East Tomb of Veneration

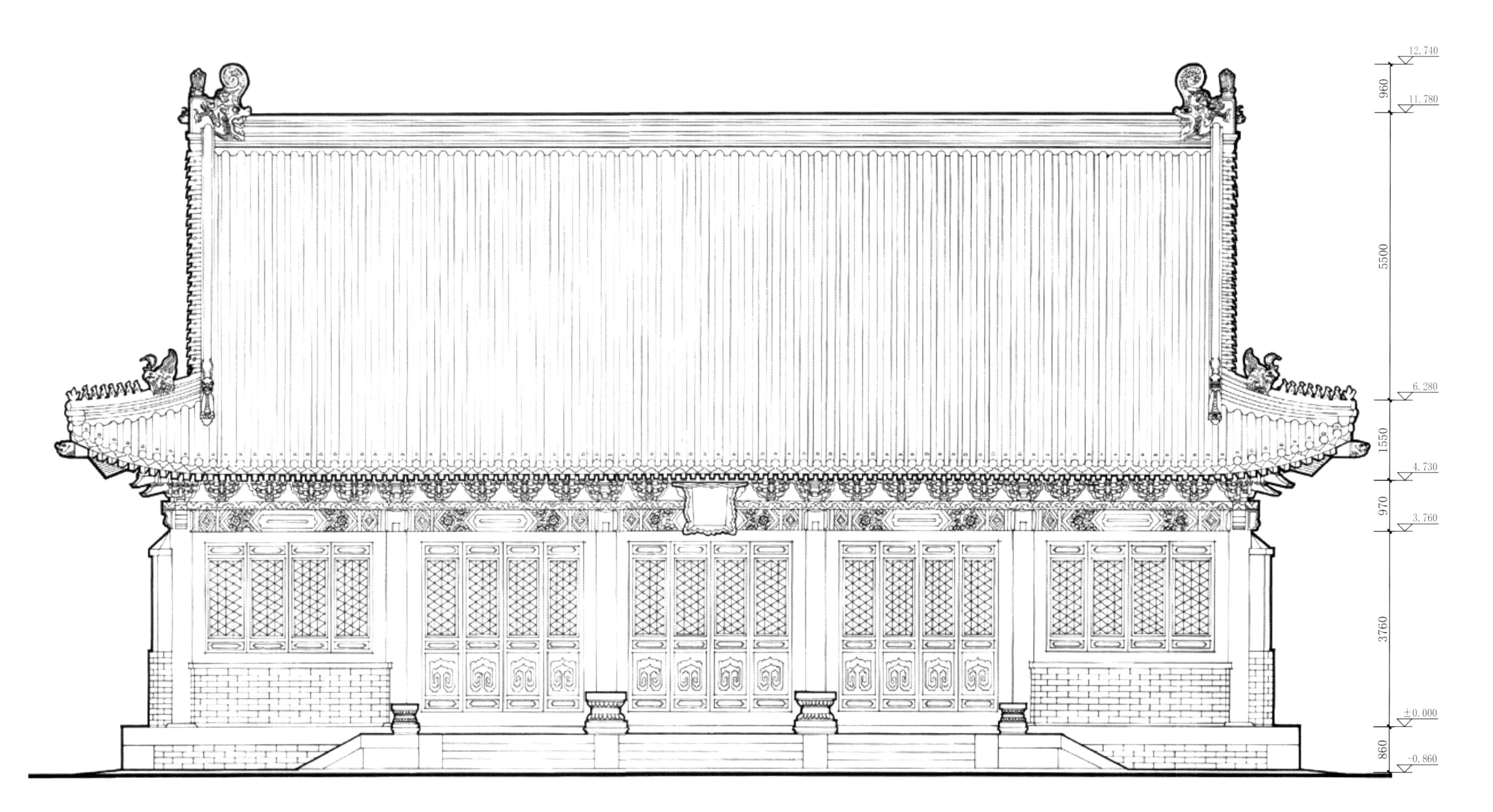

慕东陵隆恩殿正立面图
Front elevation of the Hall of Monumental Grace,East Tomb of Veneration

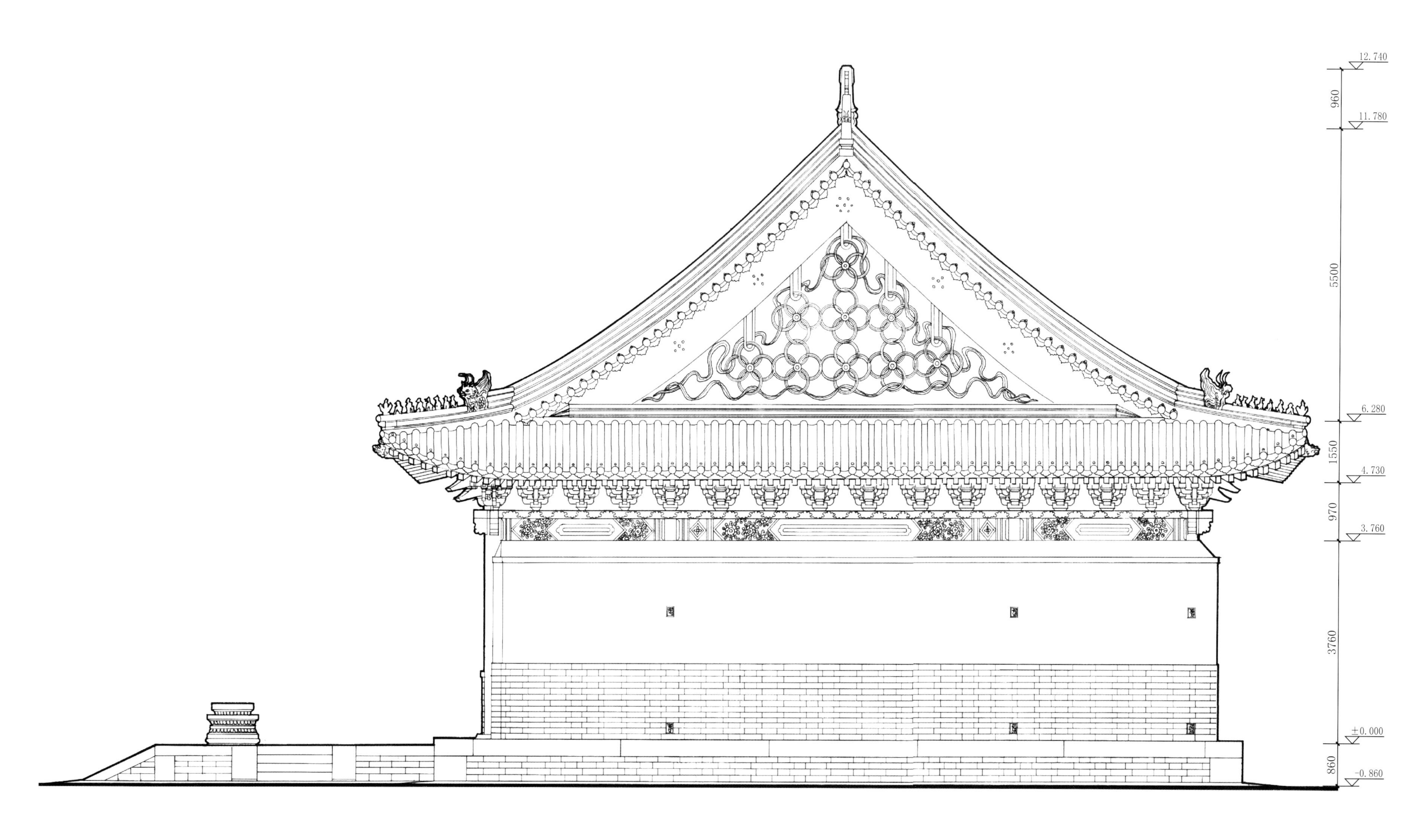

慕东陵隆恩殿侧立面图

Side elevation of the Hall of Monumental Grace,East Tomb of Veneration

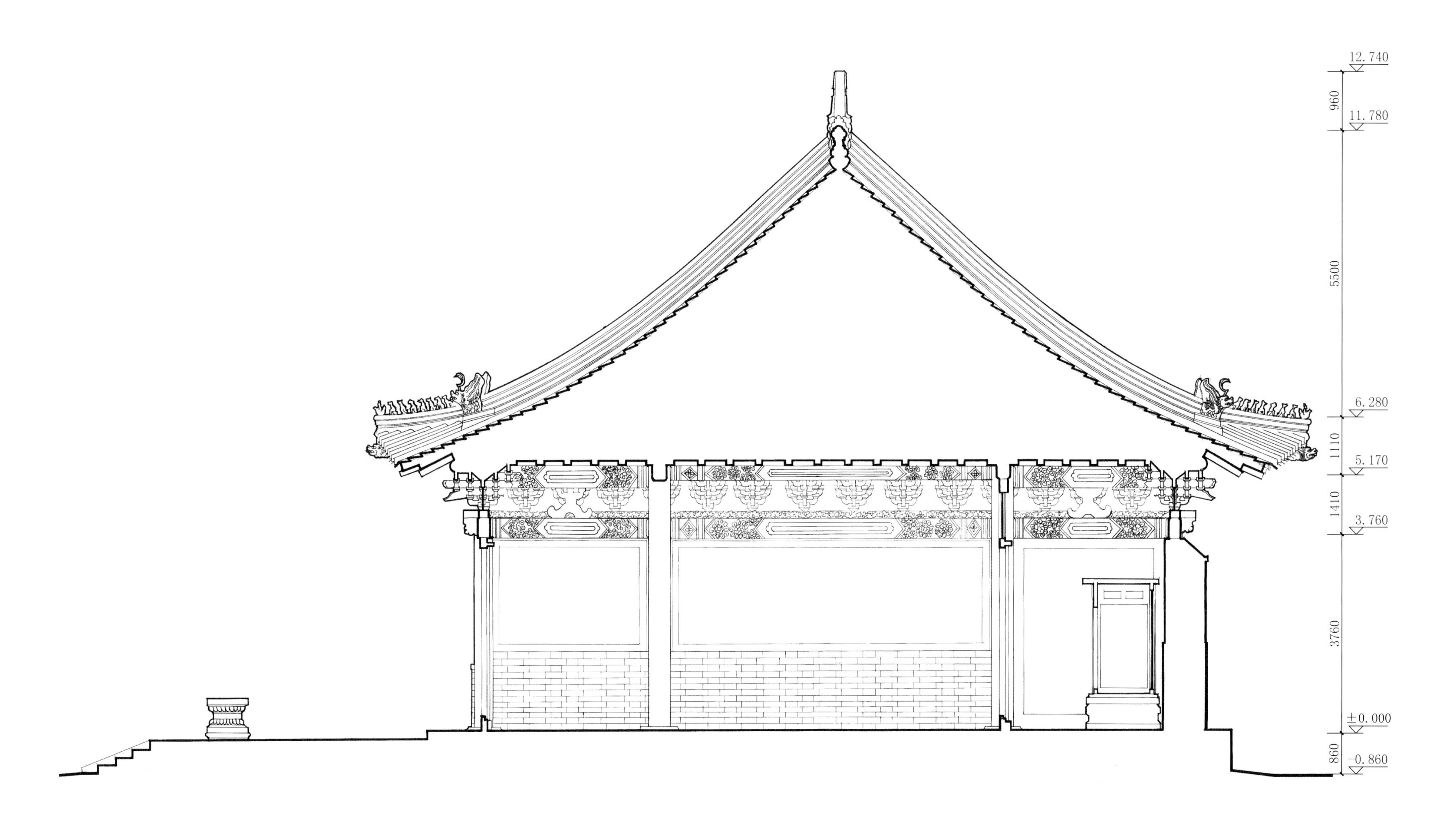

慕东陵隆恩殿明间横剖面图

Cross section of the Central Chamber within the Hall of Monumental Grace,East Tomb of Veneration

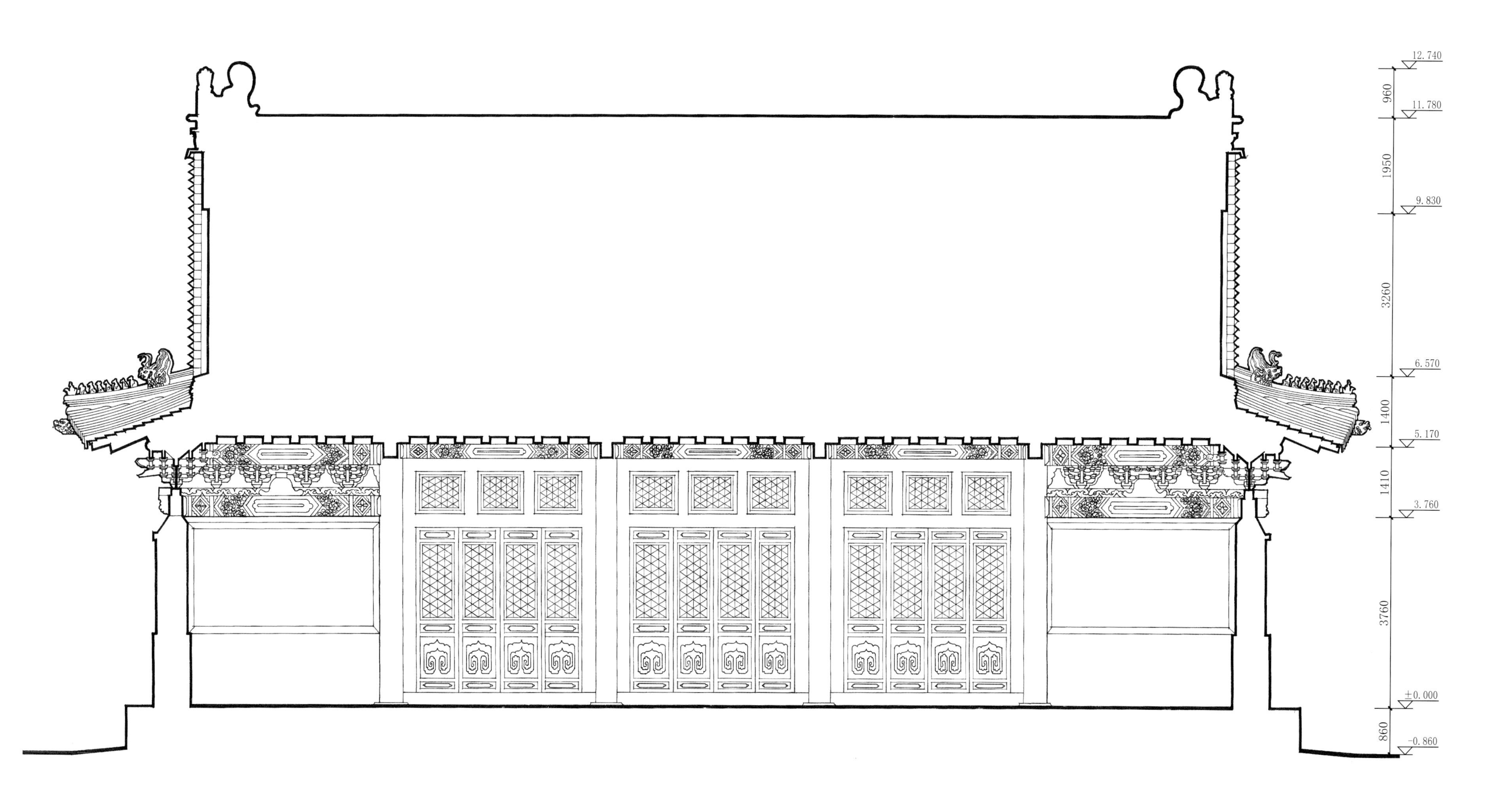

慕东陵隆恩殿纵剖面图

Longitudinal section of the Hall of Monumental Grace,East Tomb of Veneration

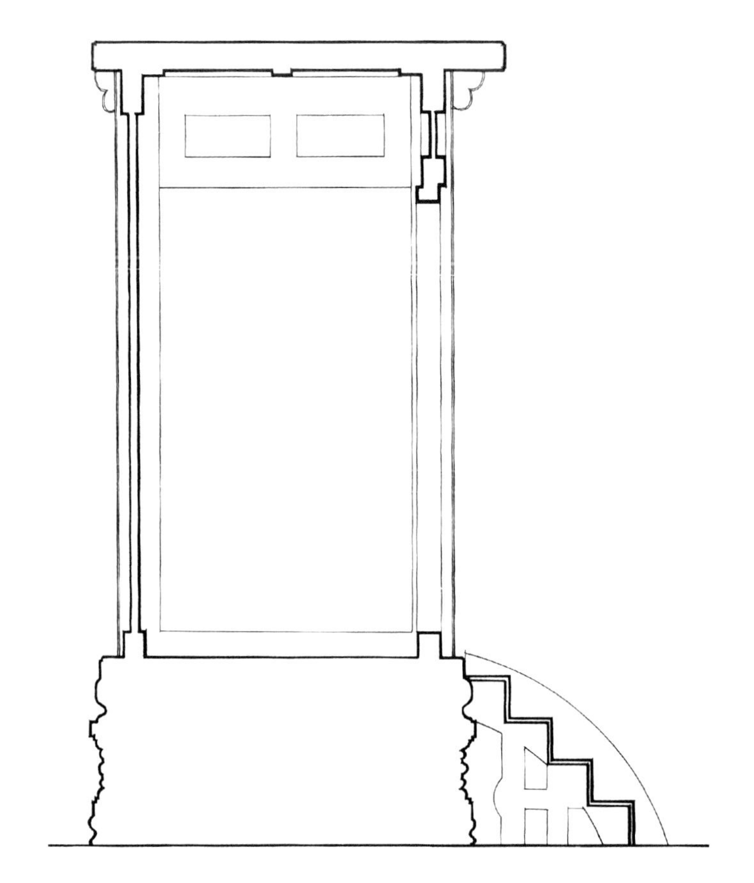

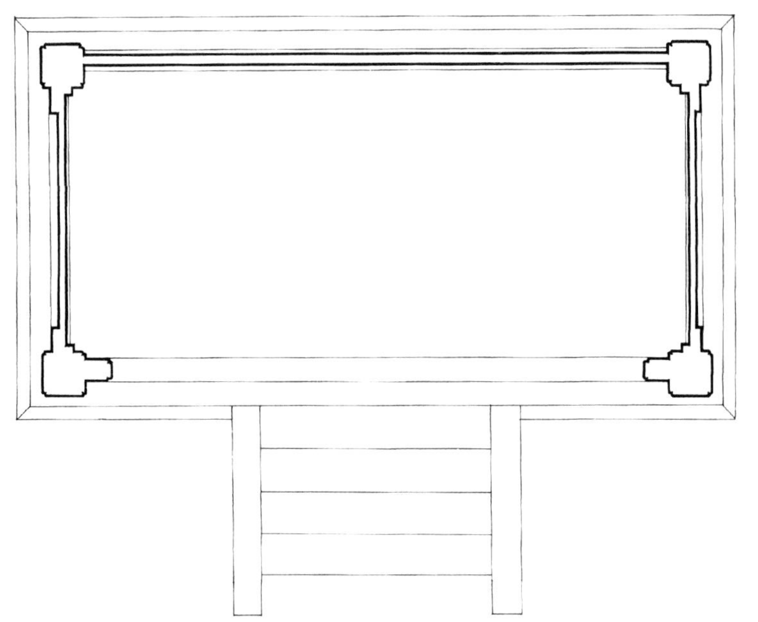

0 0.25 0.5 1m

慕东陵隆恩殿暖阁大样图

Detailed drawing of the Warming Room of the Hall of Monumental Grace,East Tomb of Veneration

0 1 2 4m

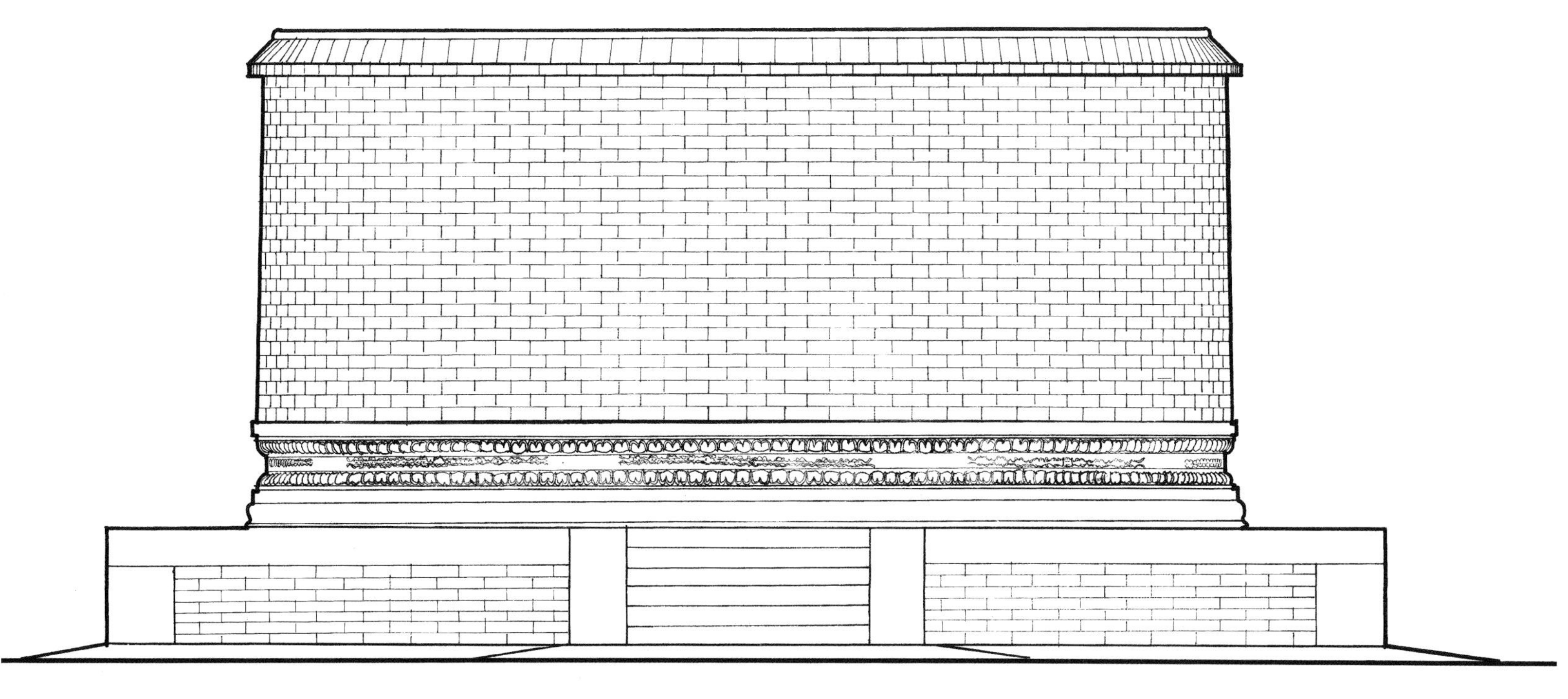

慕东陵后宝顶正立面图
Front elevation of the Queen's Tumulus,East Tomb of Veneration

N
0 1 2 4m

慕东陵后宝顶平面图
Plan of the Queen's Tumulus,East Tomb of Veneration

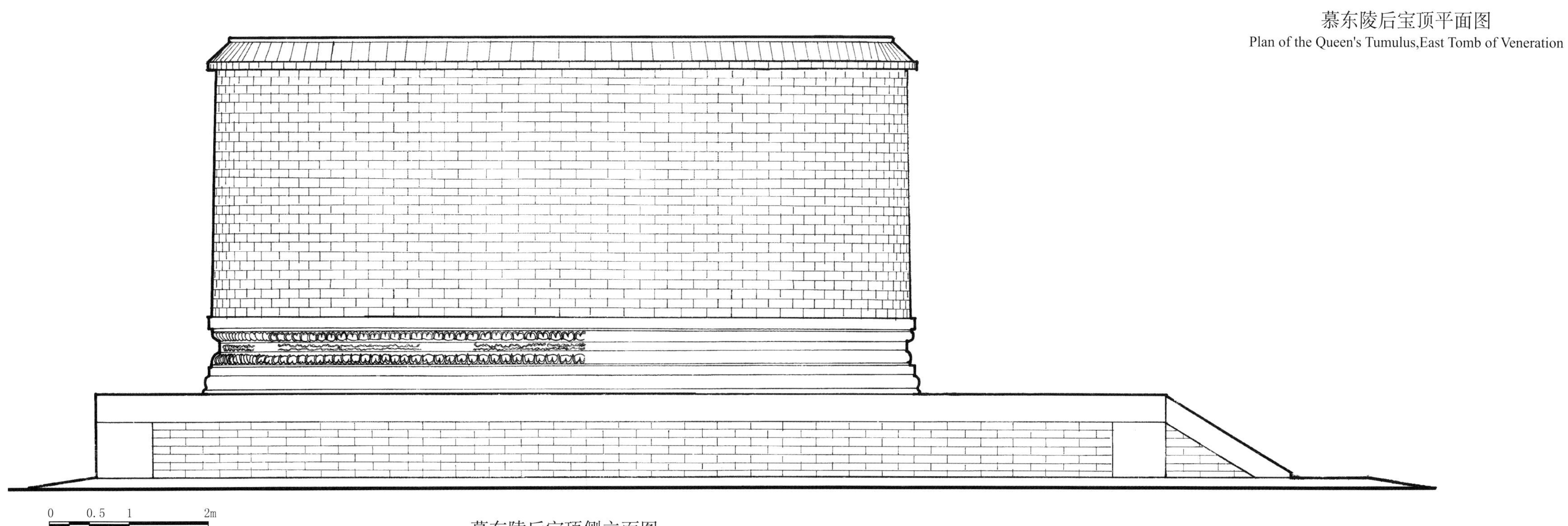

慕东陵后宝顶侧立面图
Side elevation of the Queen's Tumulus,East Tomb of Veneration

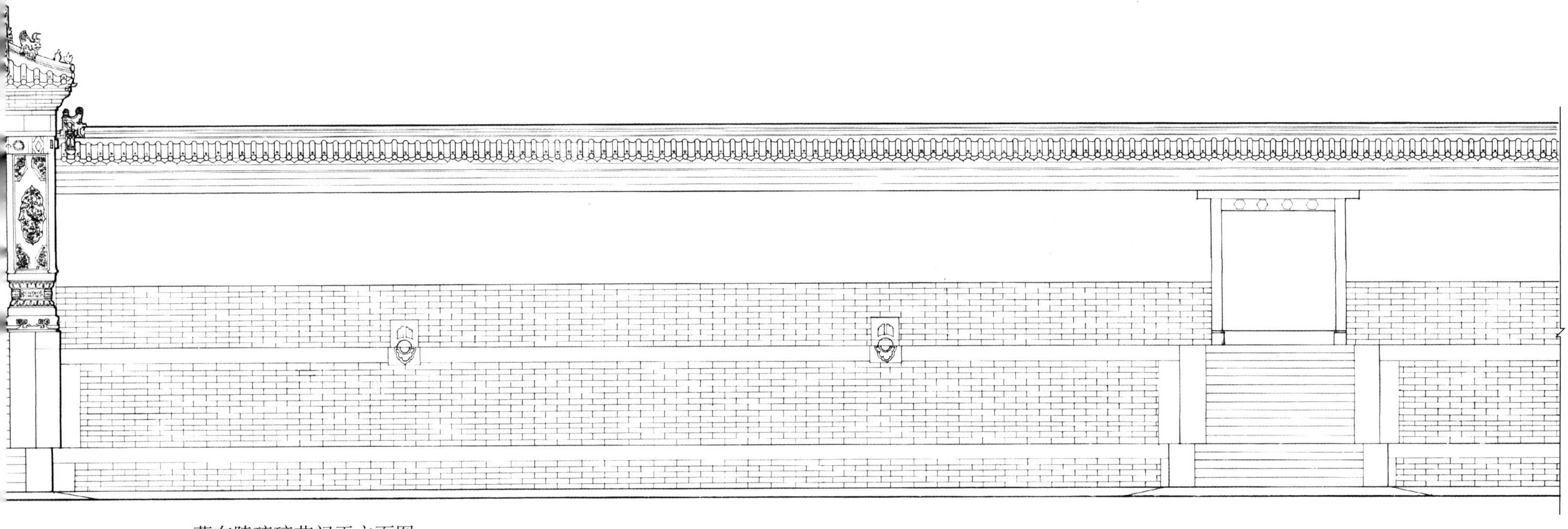

慕东陵琉璃花门正立面图
Front elevation of the Gate with Glazed Roof Tiles,East Tomb of Veneration

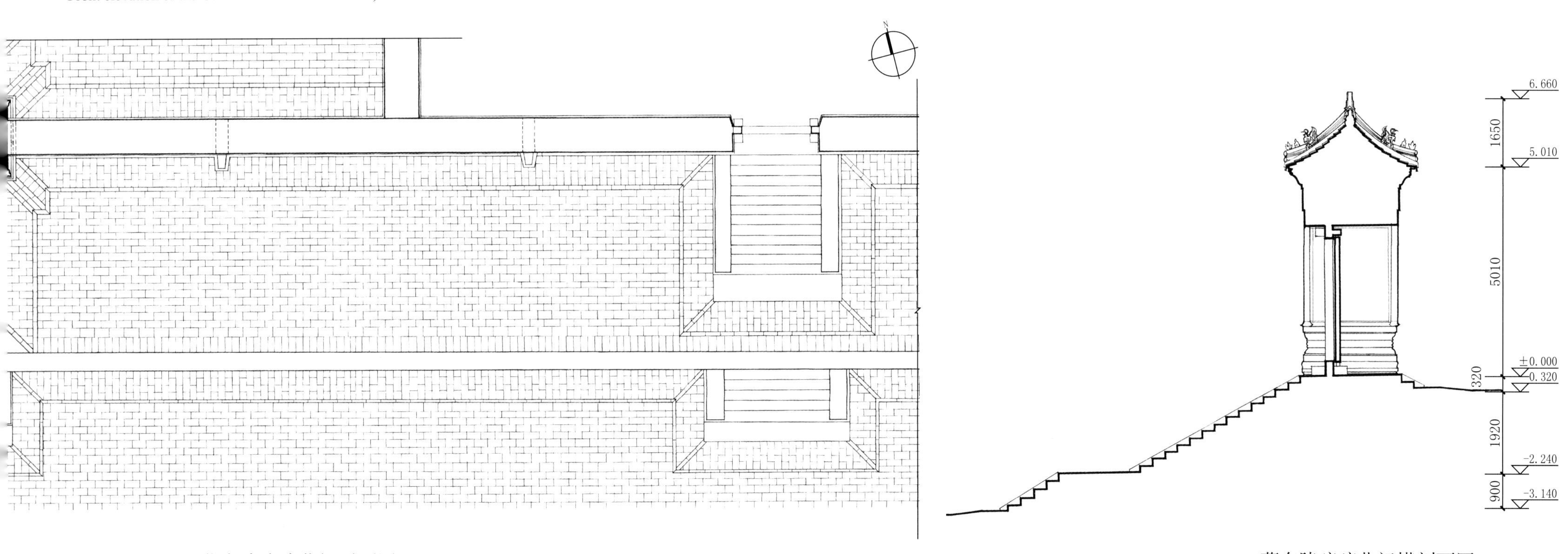

慕东陵琉璃花门平面图
Plan of the Gate with Glazed Roof Tiles,East Tomb of Veneration

慕东陵琉璃花门横剖面图
Cross section of the Gate with Glazed Roof Tiles,East Tomb of Veneration

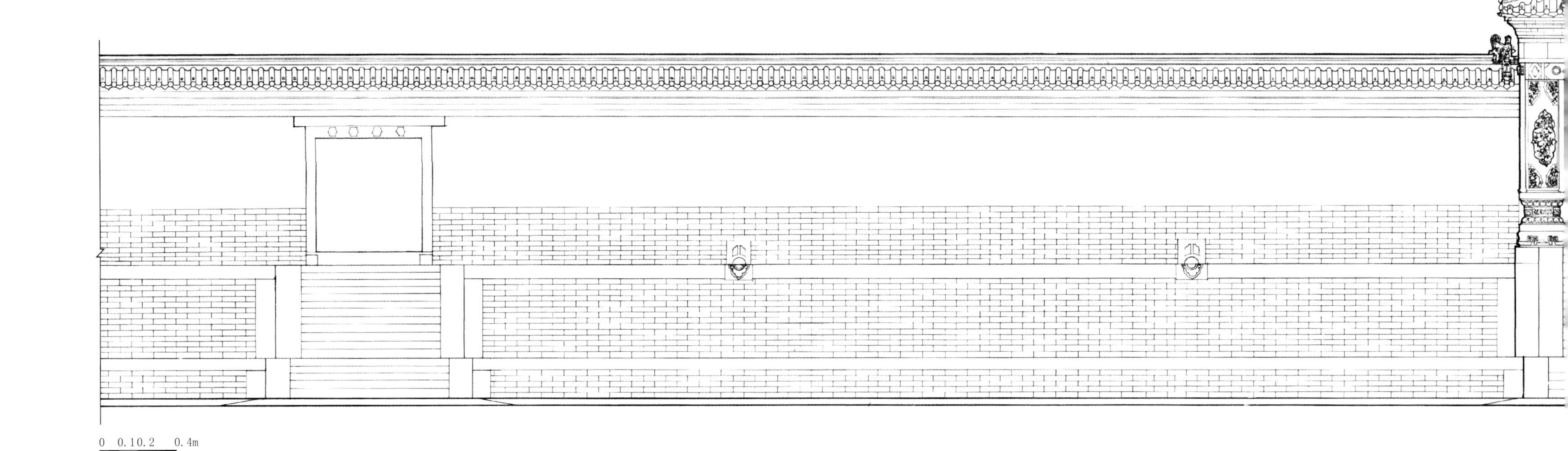

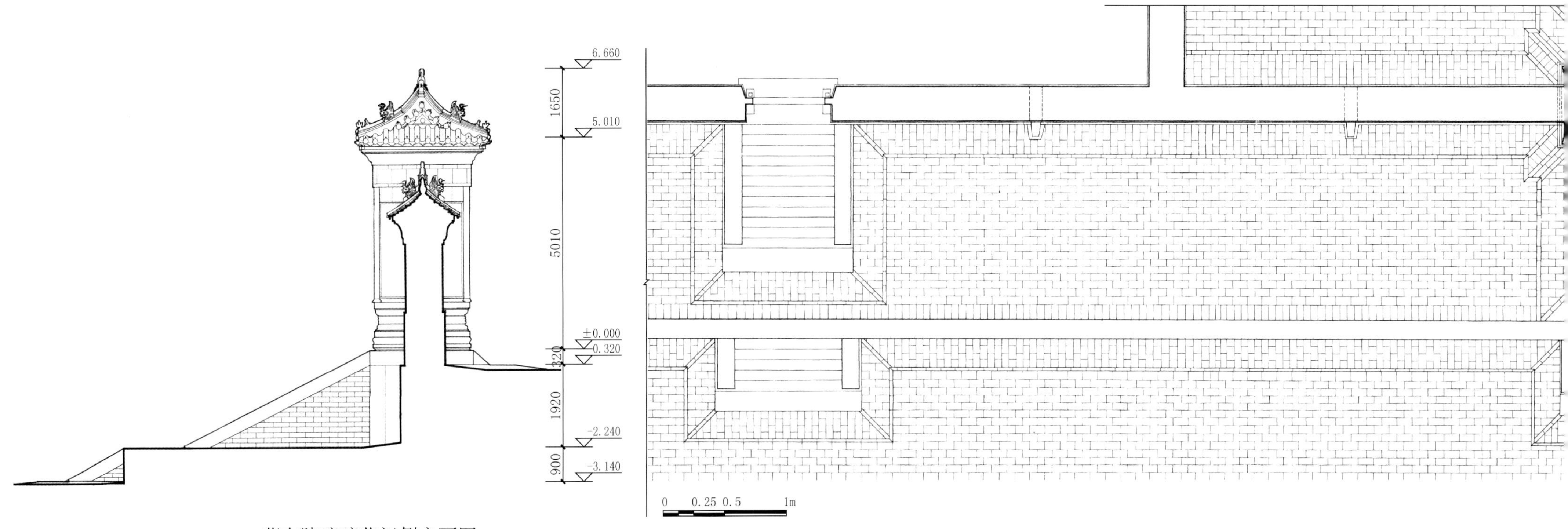

慕东陵琉璃花门侧立面图

Side elevation of the Gate with Glazed Roof Tiles,East Tomb of Veneration

崇 陵

Tomb of Serendipity (Chonglings)

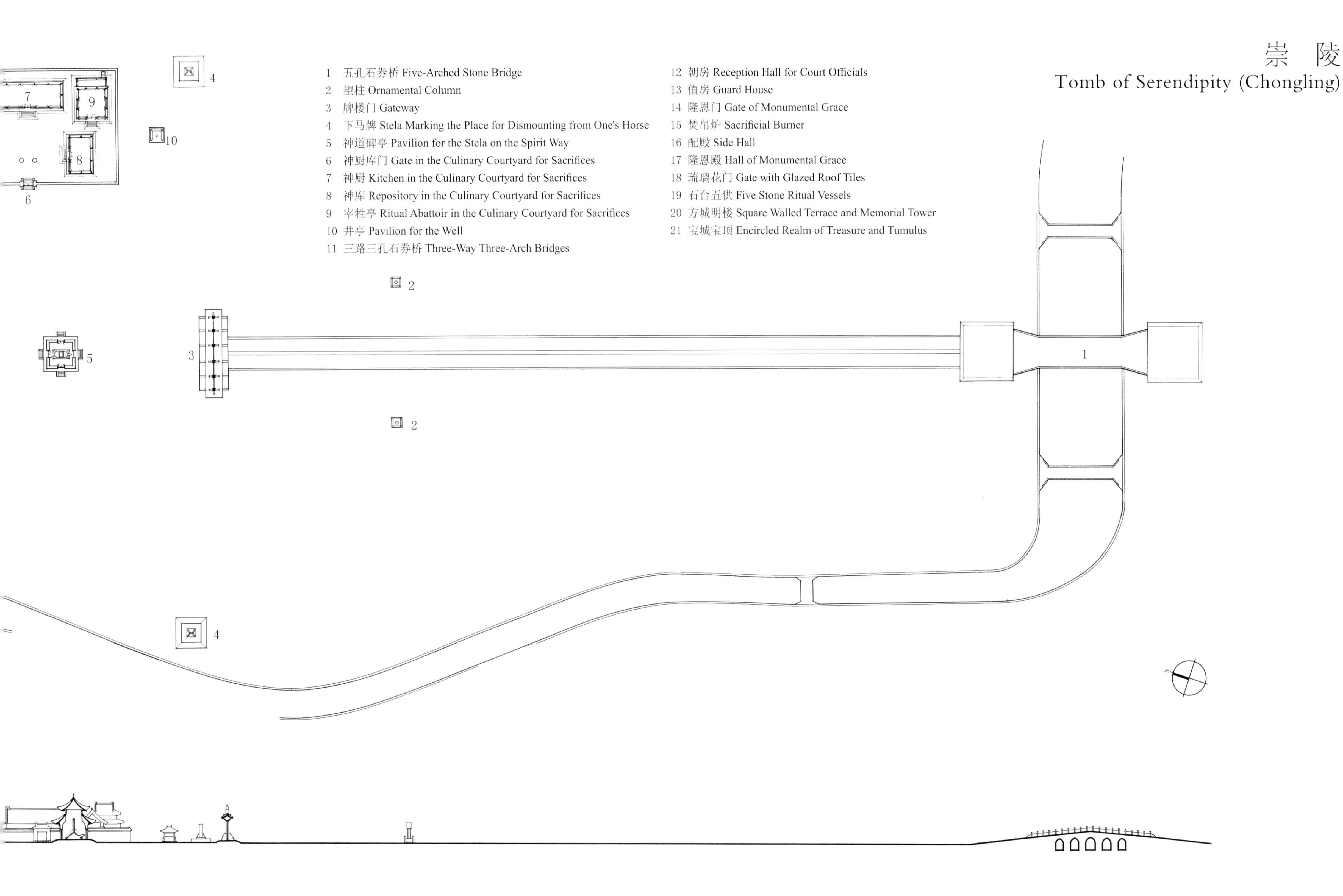

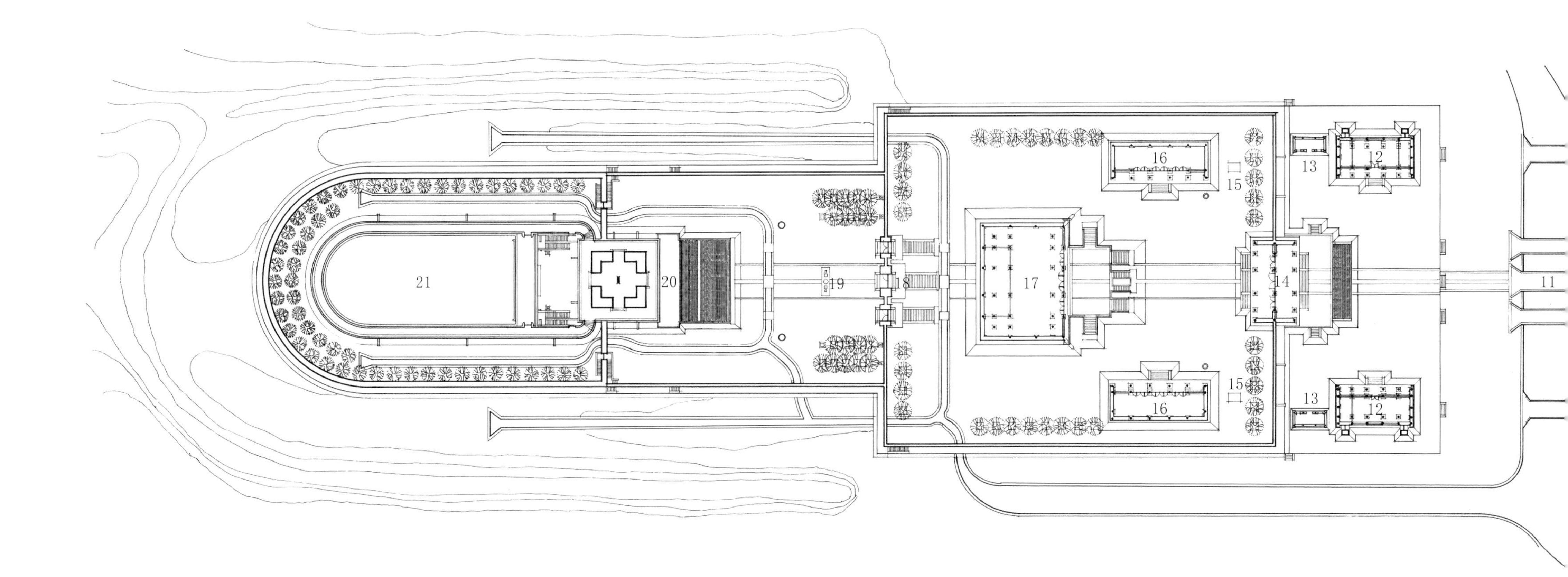

崇陵组群平面图

Site plan of the building complex of the Tomb of Serendipity

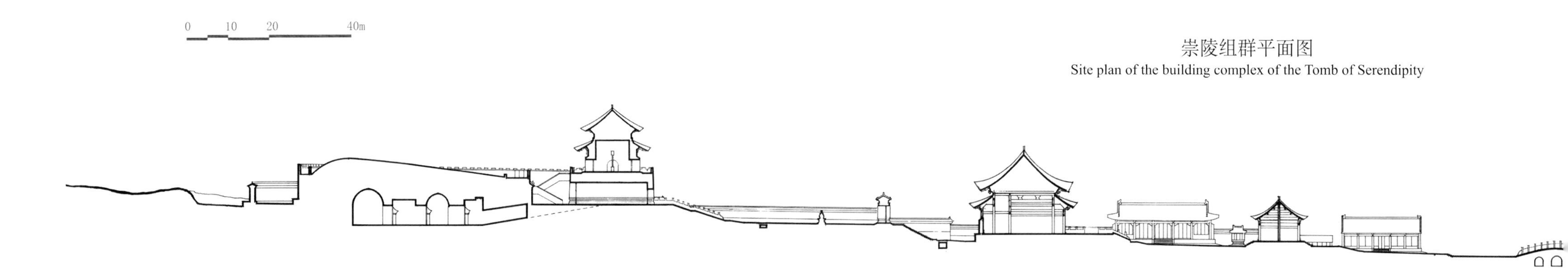

崇陵组群剖面图

Site section of the building complex of the Tomb of Serendipity

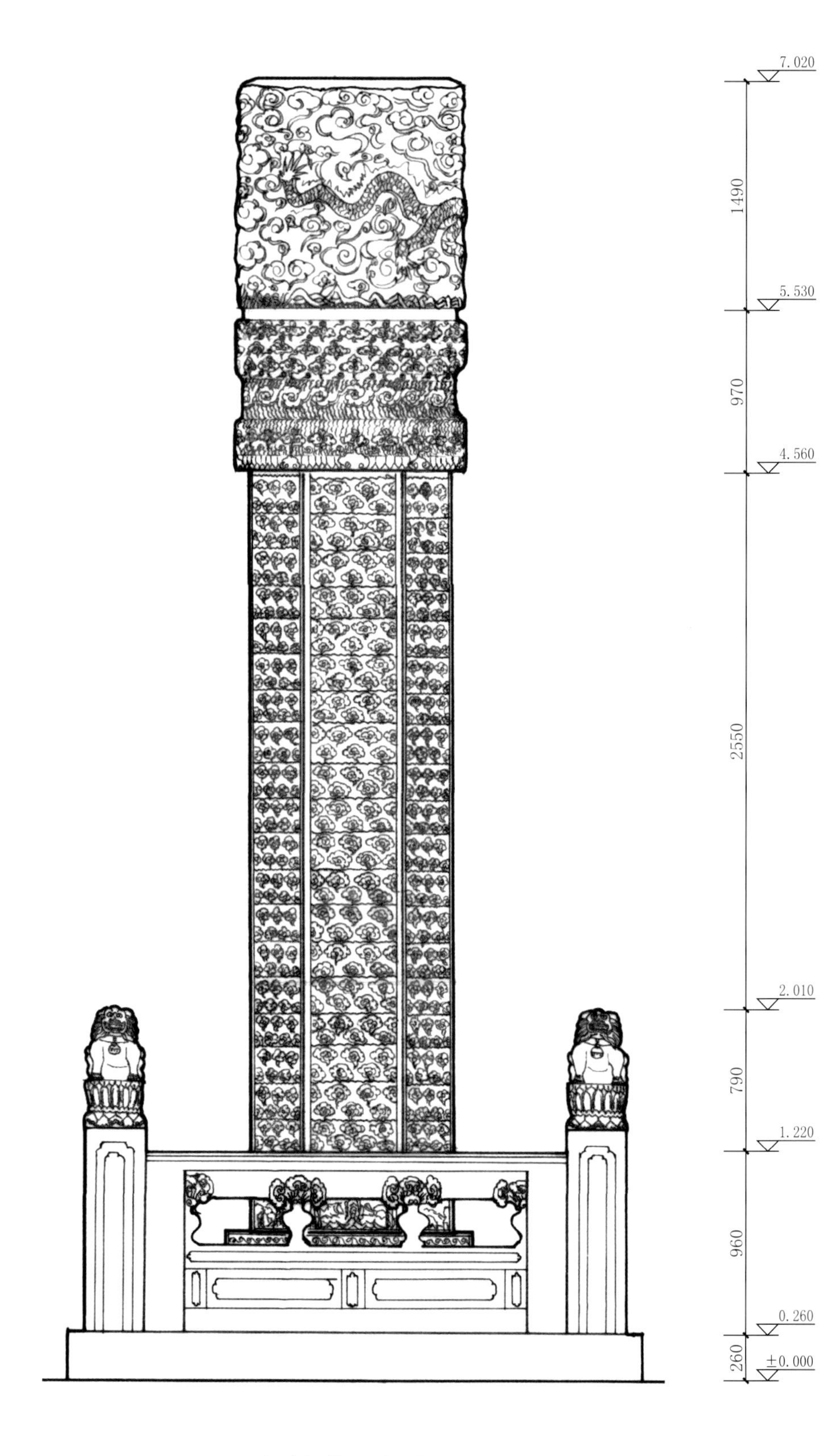

崇陵望柱正立面图
Front elevation of the Ornamental Column,Tomb of Serendipity

3200
100 90 90 100
1125 570 1125
3200

崇陵望柱平面图
Plan of the Ornamental Column,Tomb of Serendipity

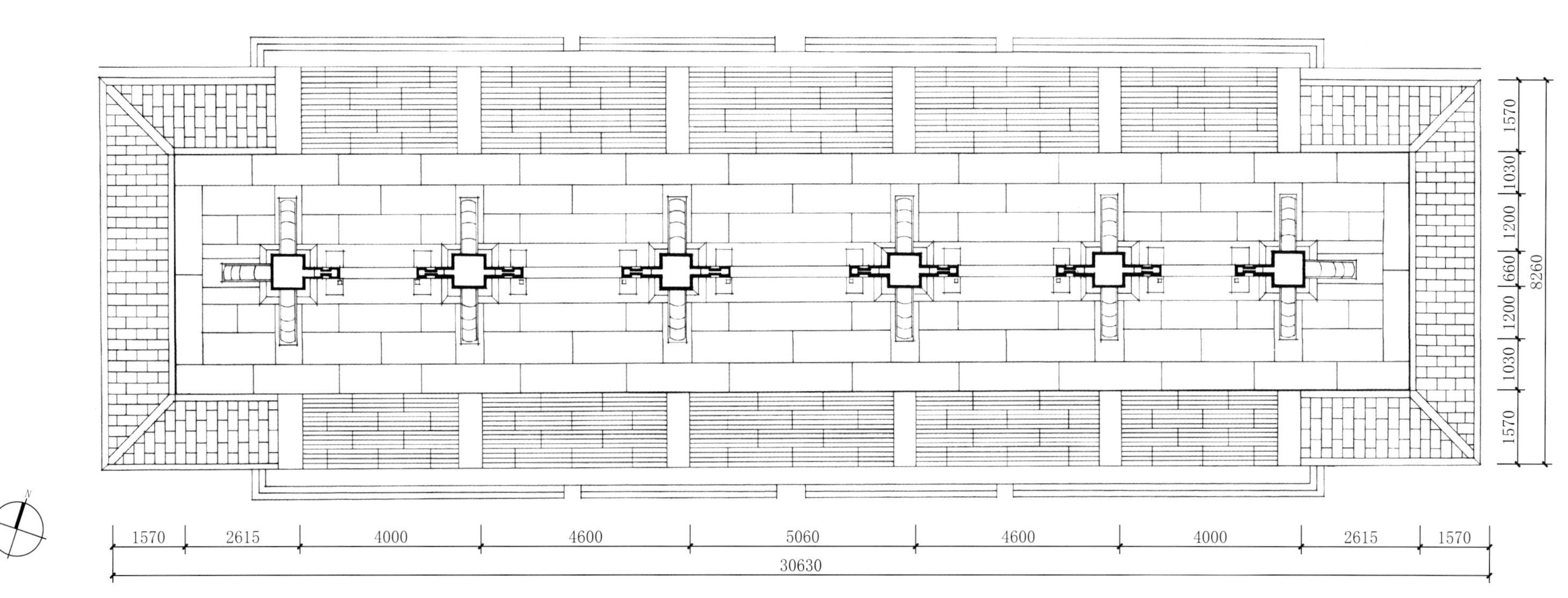

崇陵牌楼门平面图
Plan of the Gateway,Tomb of Serendipity

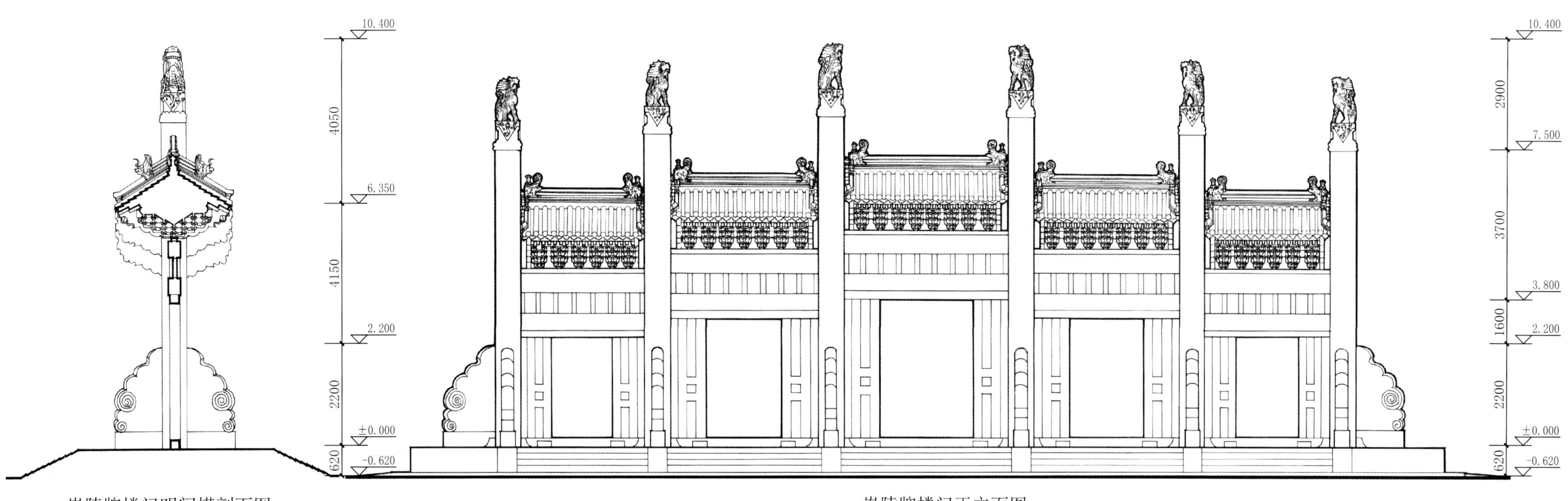

崇陵牌楼门明间横剖面图
Cross section of the Central Chamber within the Gateway,Tomb of Serendipity

崇陵牌楼门正立面图
Front elevation of the Gateway,Tomb of Serendipity

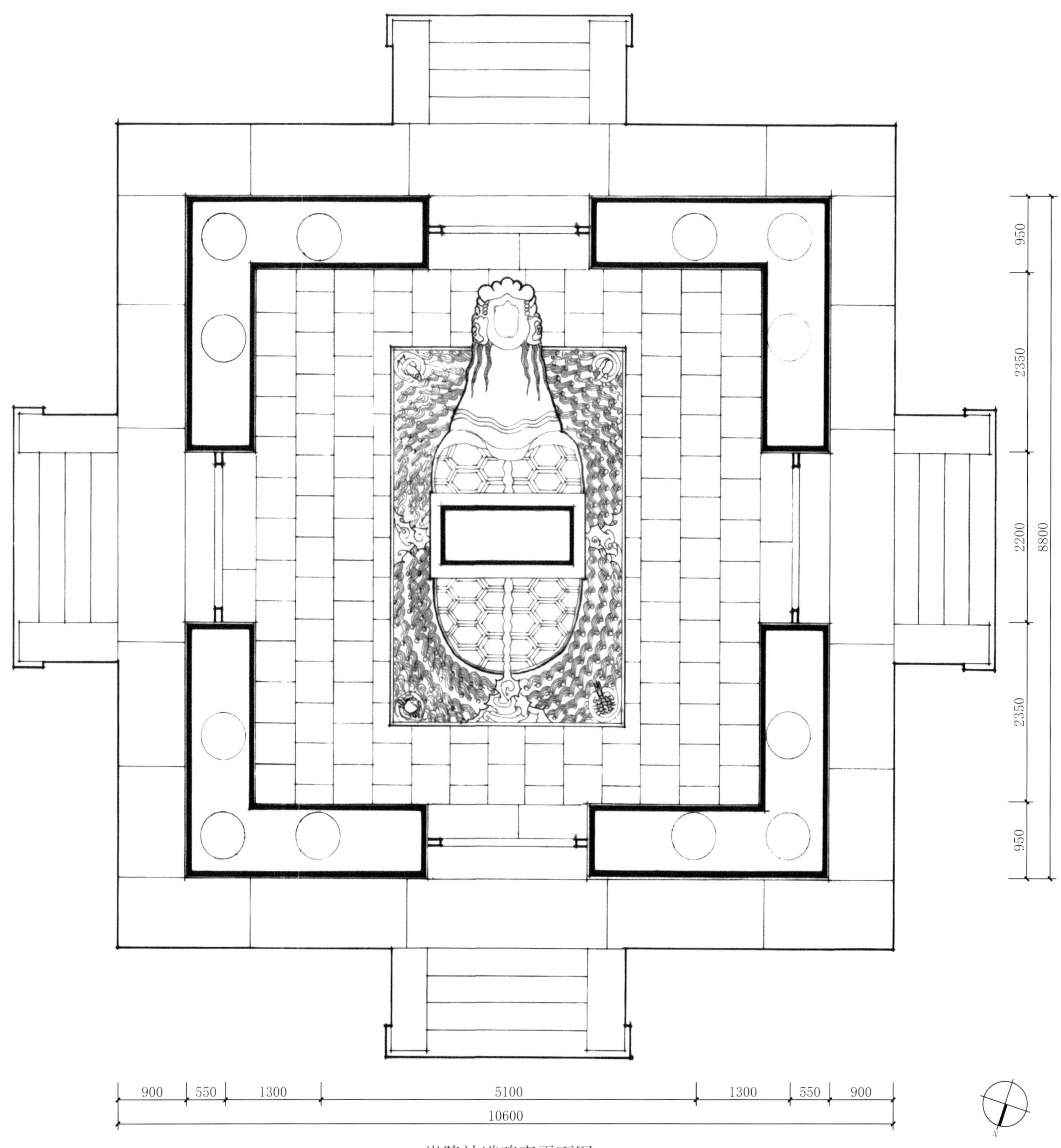

崇陵神道碑亭平面图

Plan of the Pavilion for the Stela on the Spirit Way,Tomb of Serendipity

崇陵神道碑亭正立面图

Front elevation of the Pavilion for the Stela on the Spirit Way,Tomb of Serendipity

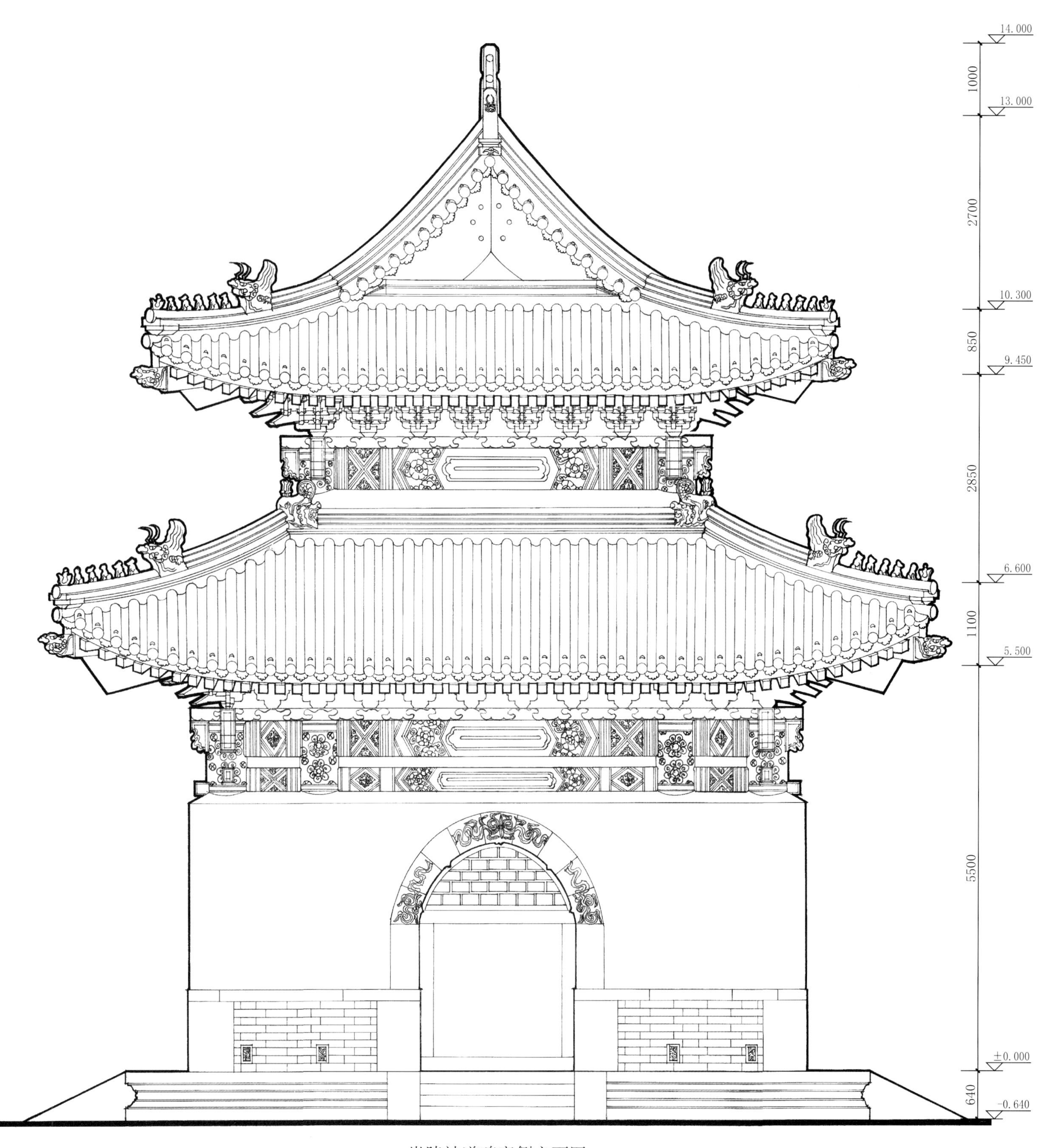

崇陵神道碑亭侧立面图
Side elevation of the Pavilion for the Stela on the Spirit Way,Tomb of Serendipity

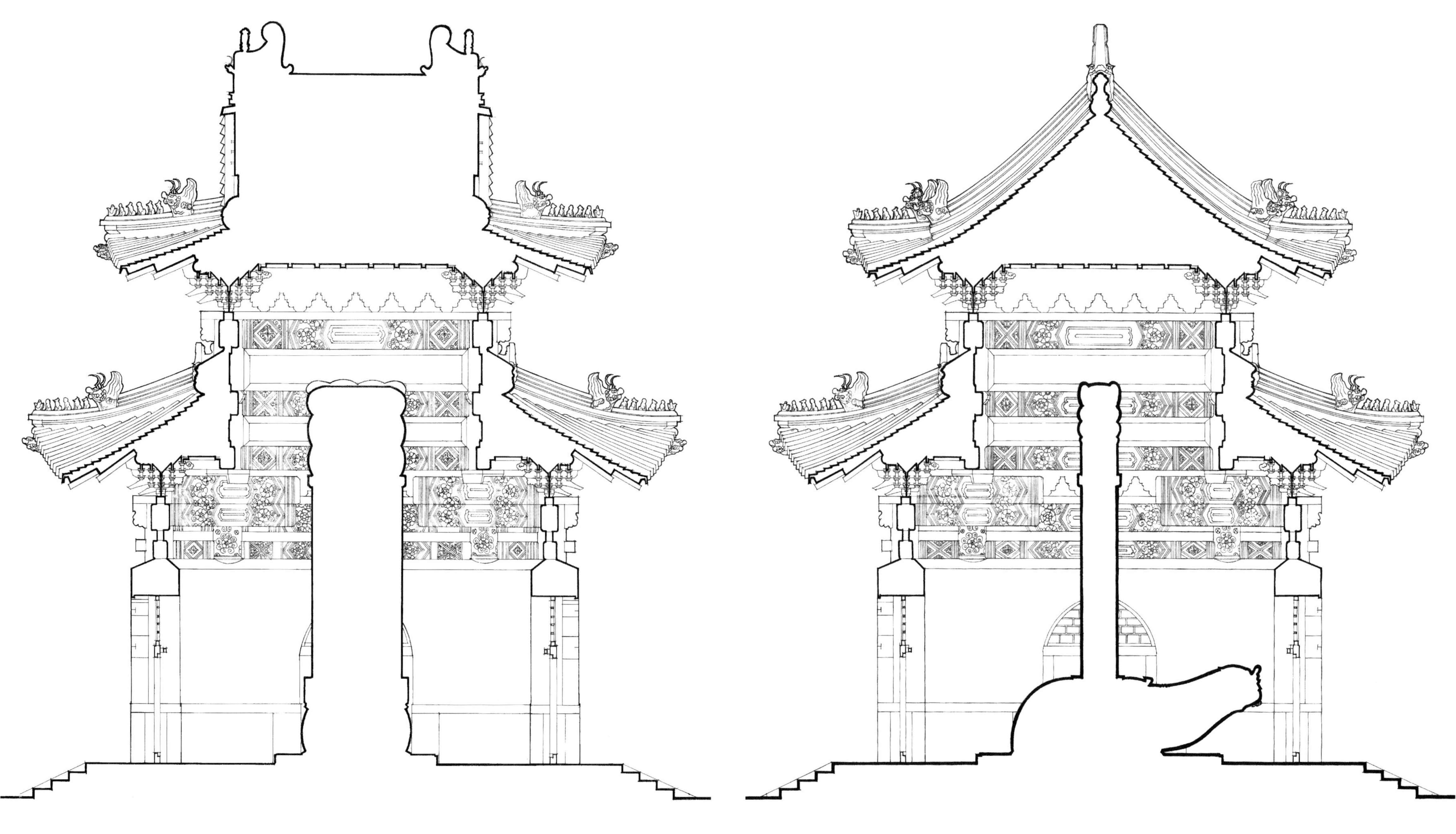

崇陵神道碑亭纵剖面图

Longitudinal section of the Pavilion for the Stela on the Spirit Way,Tomb of Serendipity

崇陵神道碑亭明间横剖面图

Cross section of the Central Chamber within the Pavilion for the Stela on the Spirit Way,Tomb of Serendipity

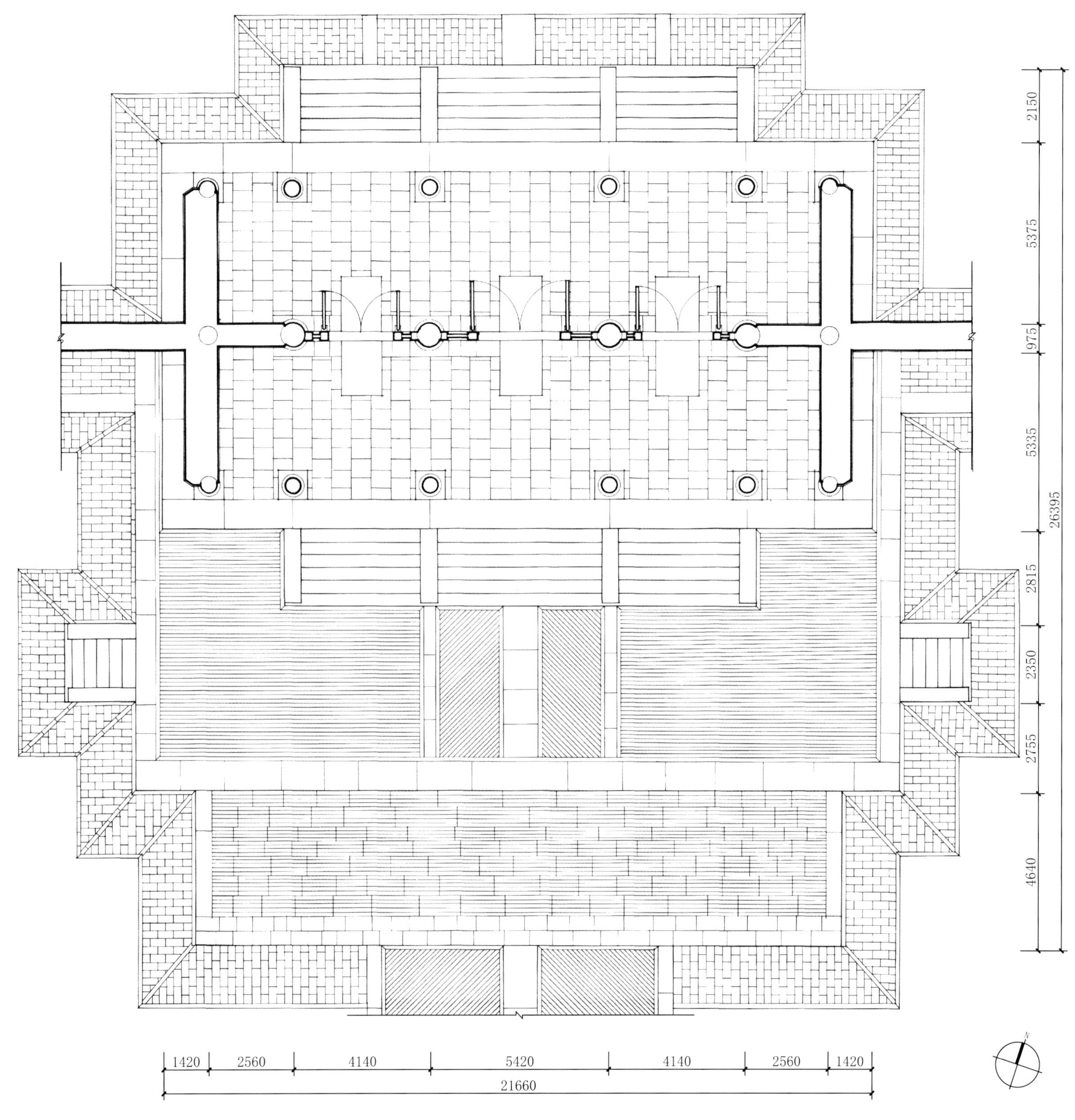

崇陵隆恩门平面图
Plan of the Gate of Monumental Grace,Tomb of Serendipity

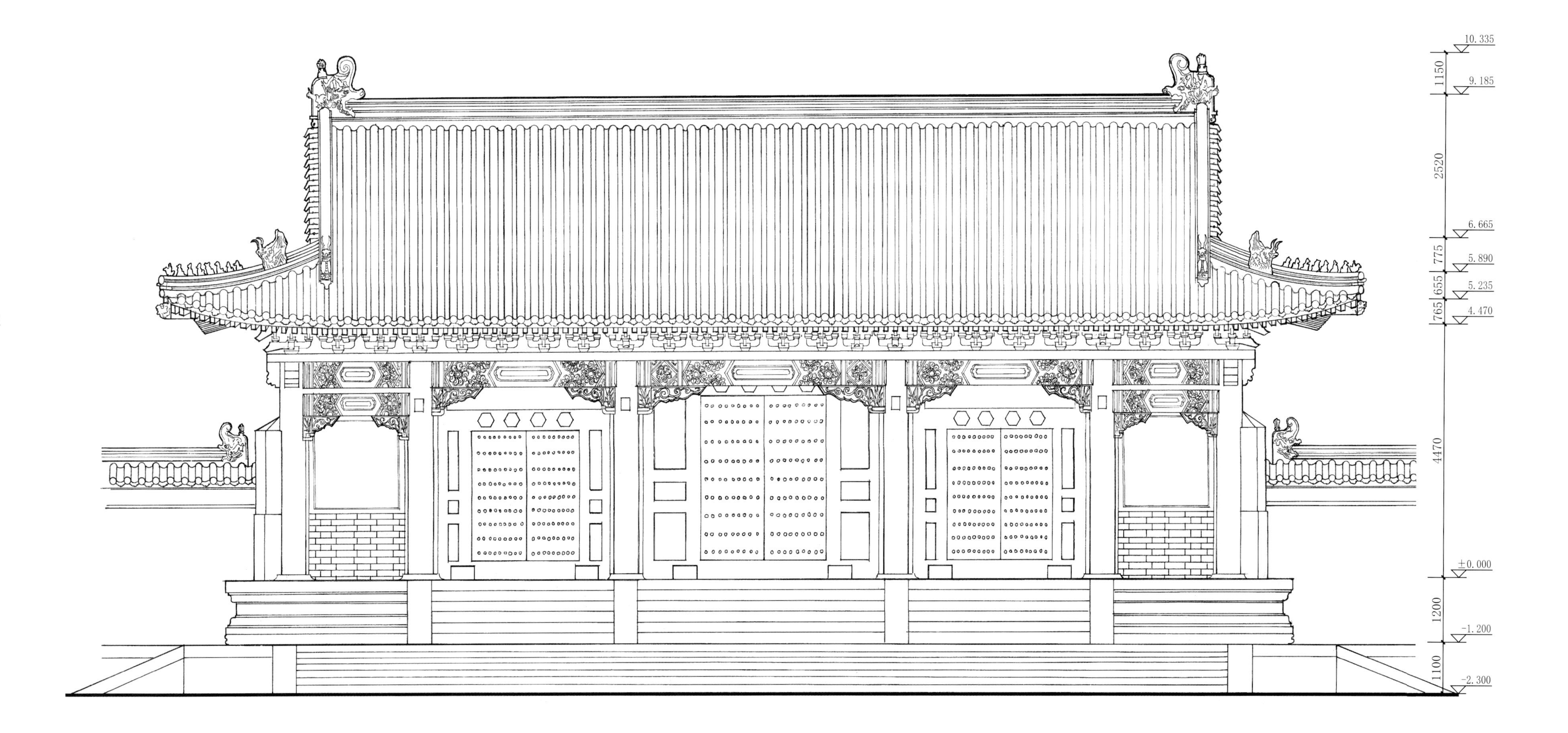

崇陵隆恩门正立面图

Front elevation of the Gate of Monumental Grace,Tomb of Serendipity

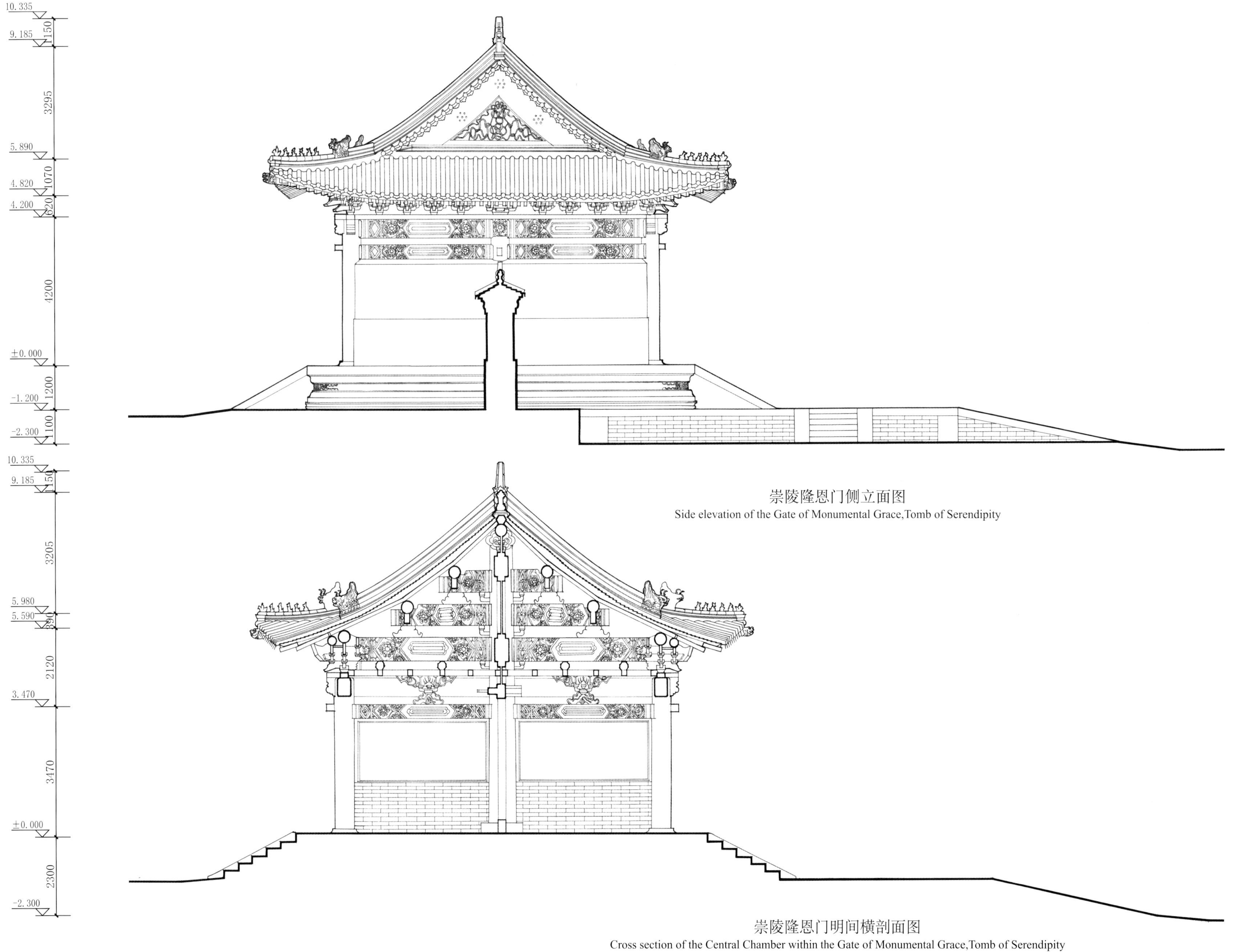

崇陵隆恩门侧立面图

Side elevation of the Gate of Monumental Grace,Tomb of Serendipity

崇陵隆恩门明间横剖面图

Cross section of the Central Chamber within the Gate of Monumental Grace,Tomb of Serendipity

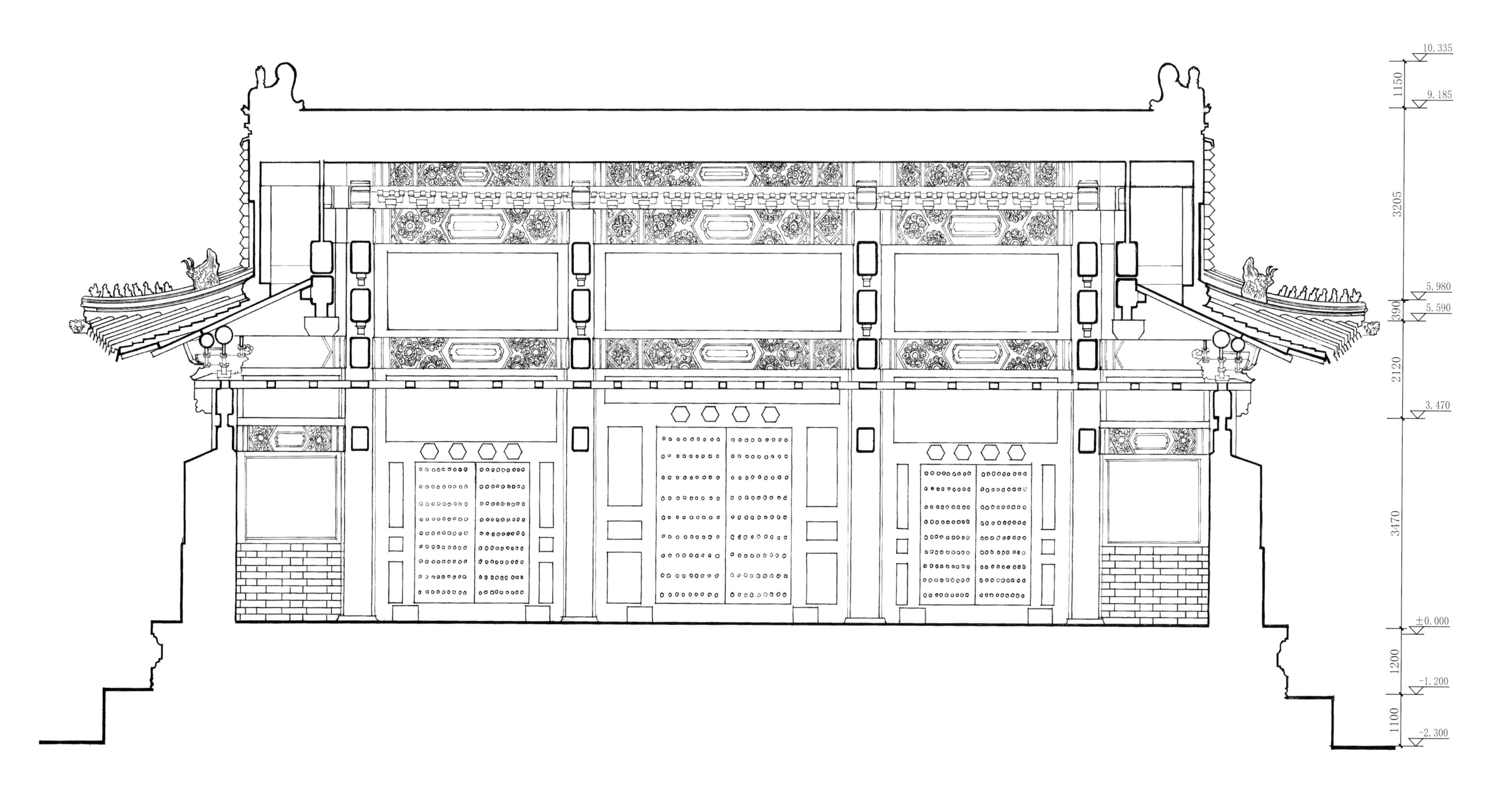

崇陵隆恩门纵剖面图
Longitudinal section of the Gate of Monumental Grace,Tomb of Serendipity

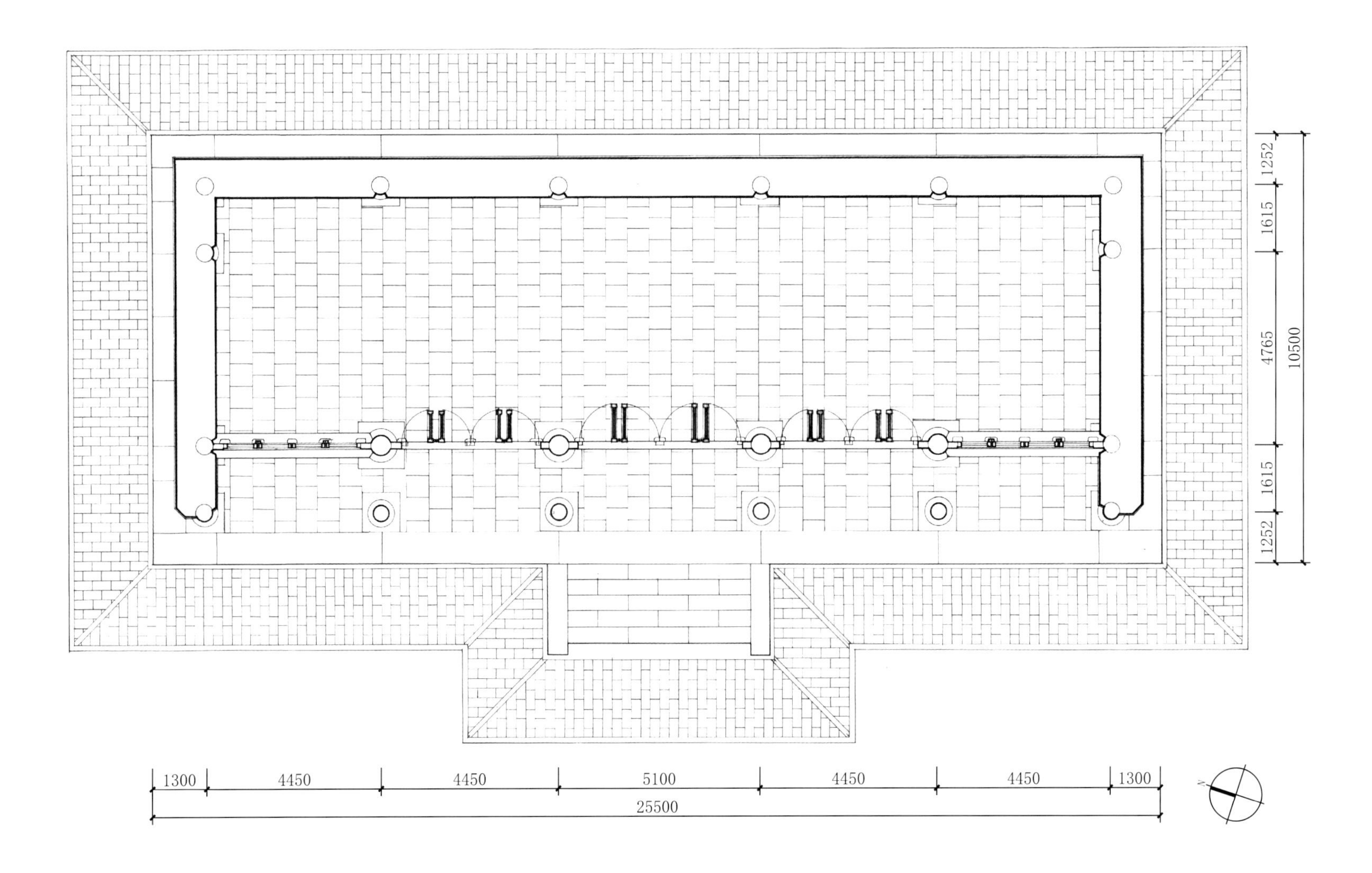

崇陵配殿平面图
Plan of the Side Hall,Tomb of Serendipity

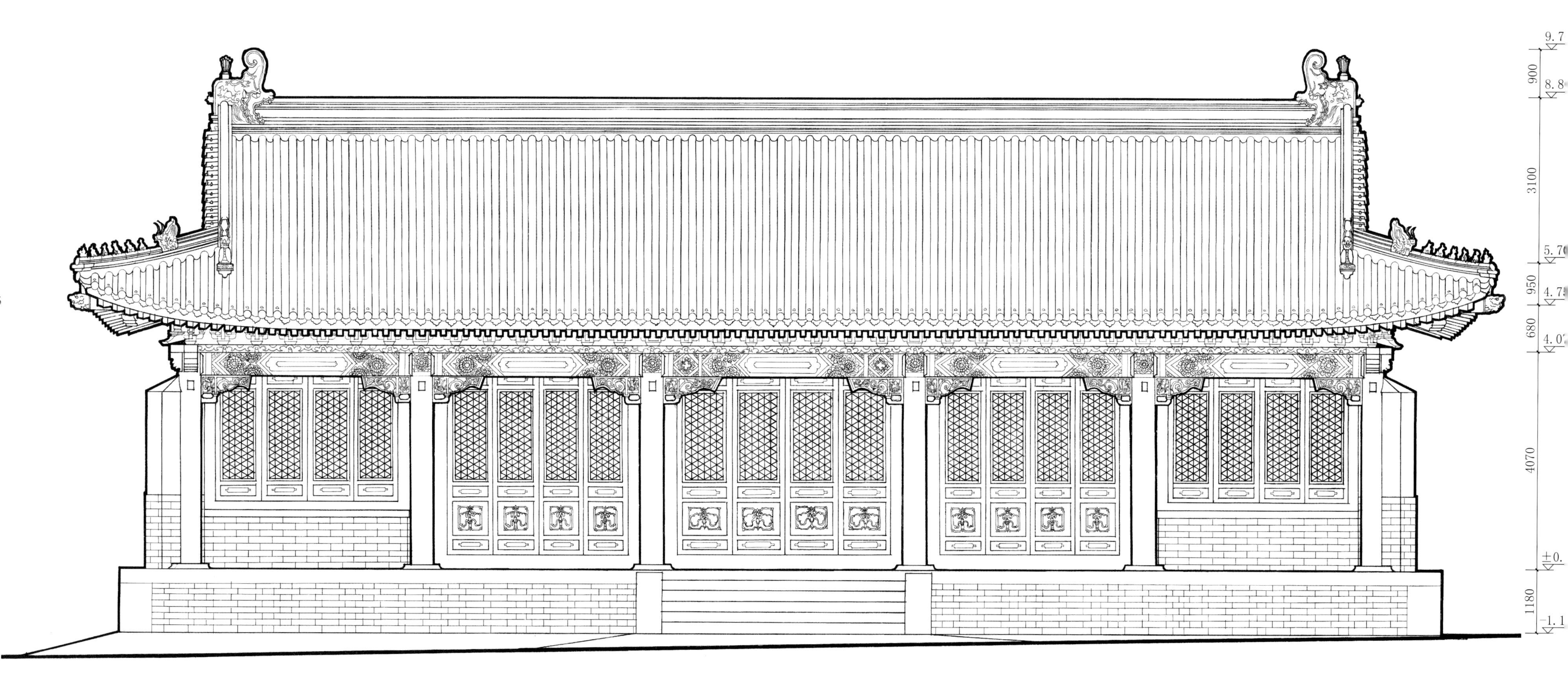

崇陵配殿正立面图
Front elevation of the Side Hall,Tomb of Serendipity

0 0.5 1 2m

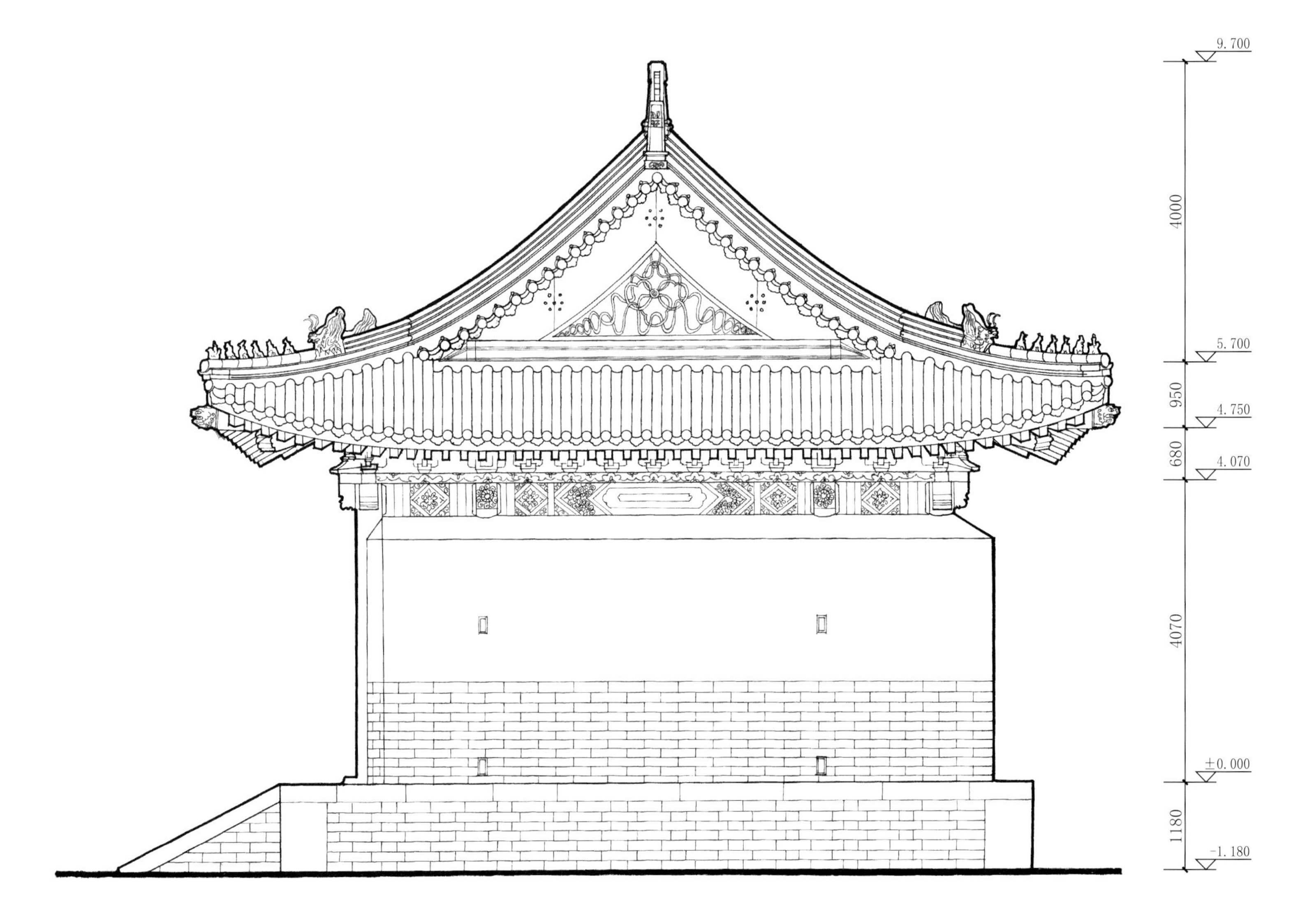

崇陵配殿侧立面图

Side elevation of the Side Hall,Tomb of Serendipity

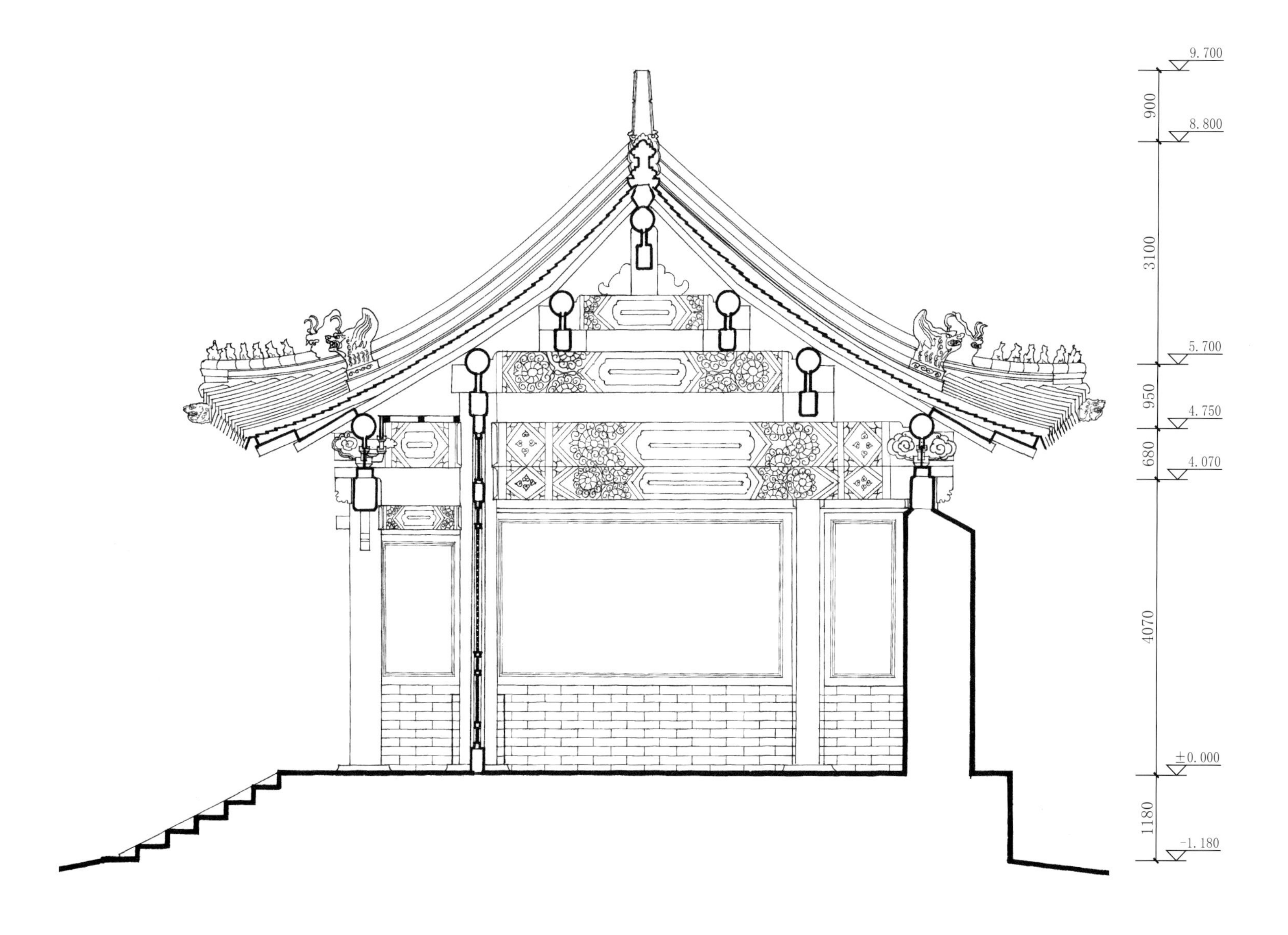

崇陵配殿明间横剖面图

Cross section of the Central Chamber within the Side Hall,Tomb of Serendipity

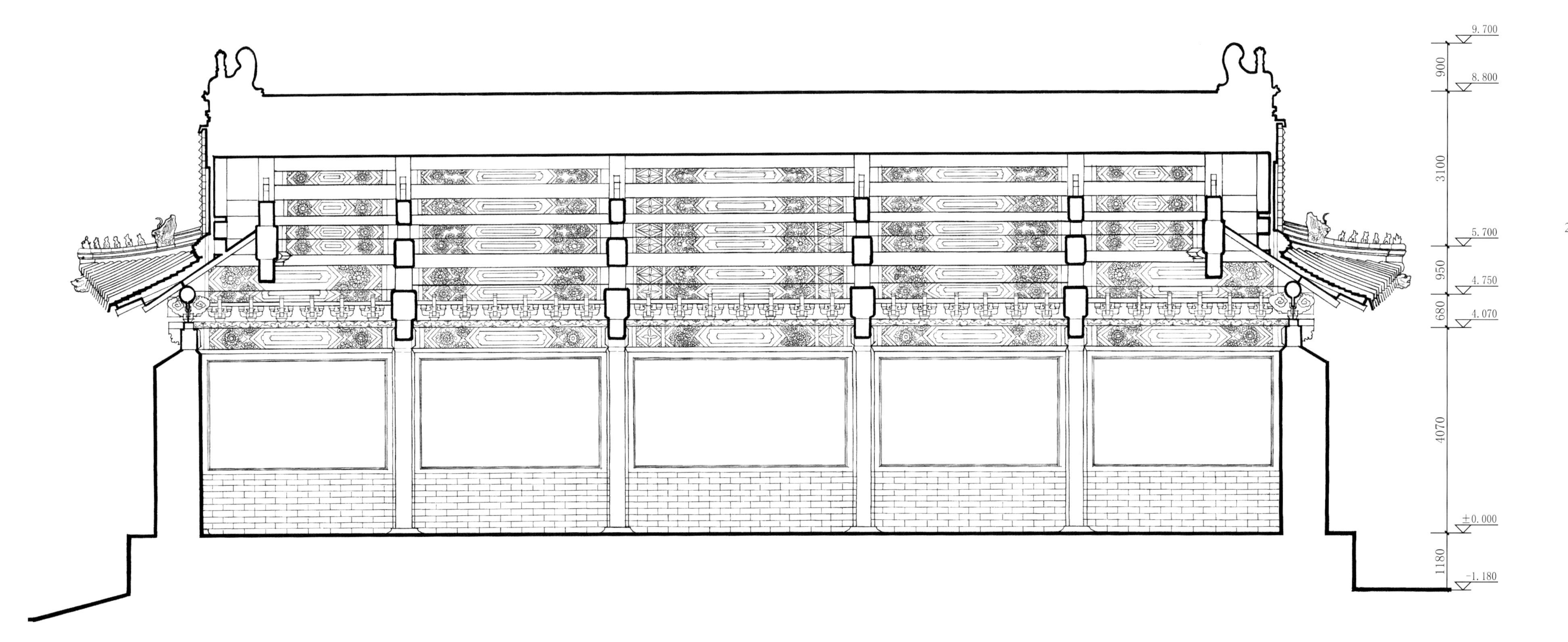

崇陵配殿纵剖面图
Longitudinal section of the Side Hall,Tomb of Serendipity

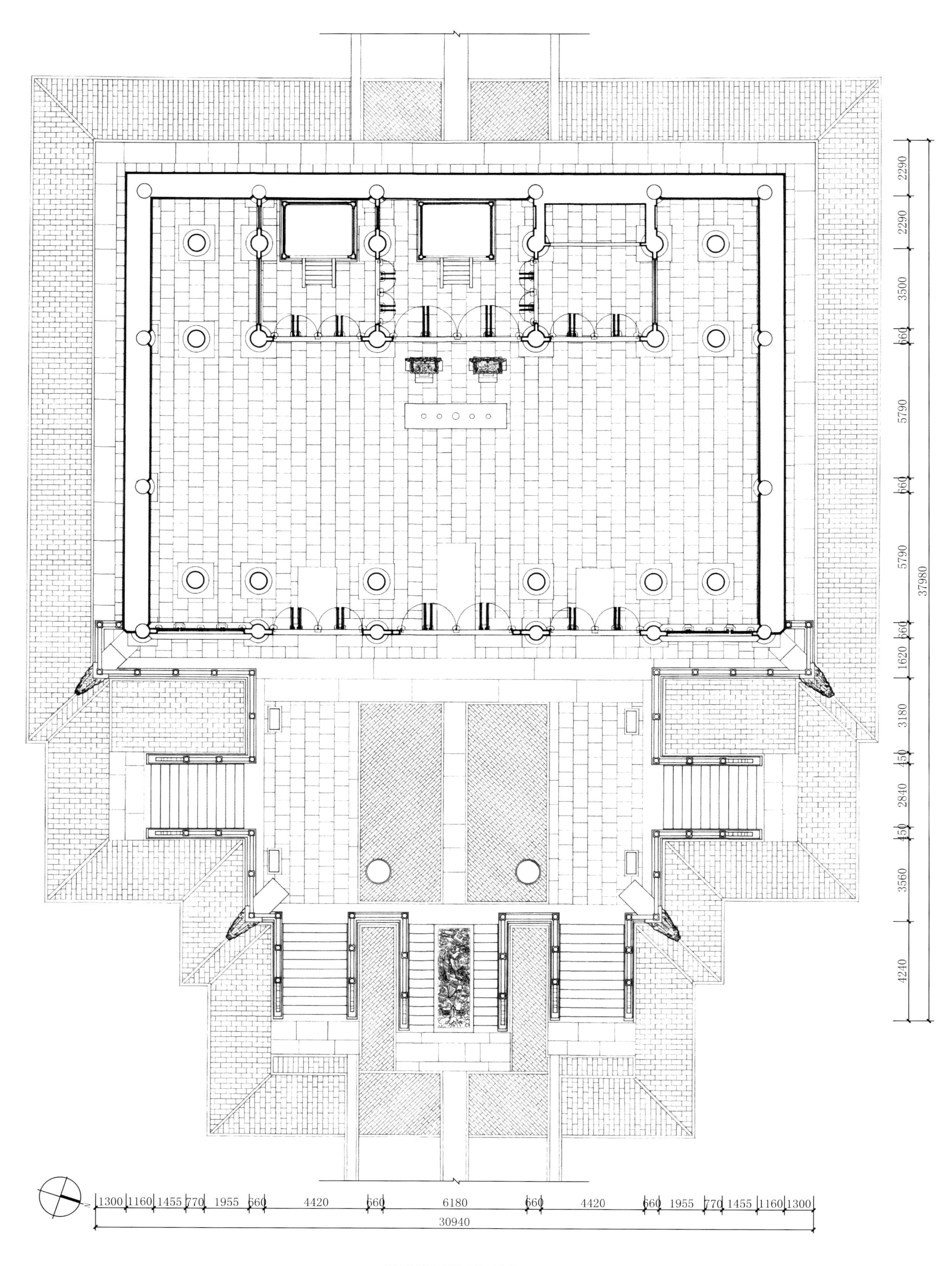

崇陵隆恩殿平面图
Plan of the Hall of Monumental Grace,Tomb of Serendipity

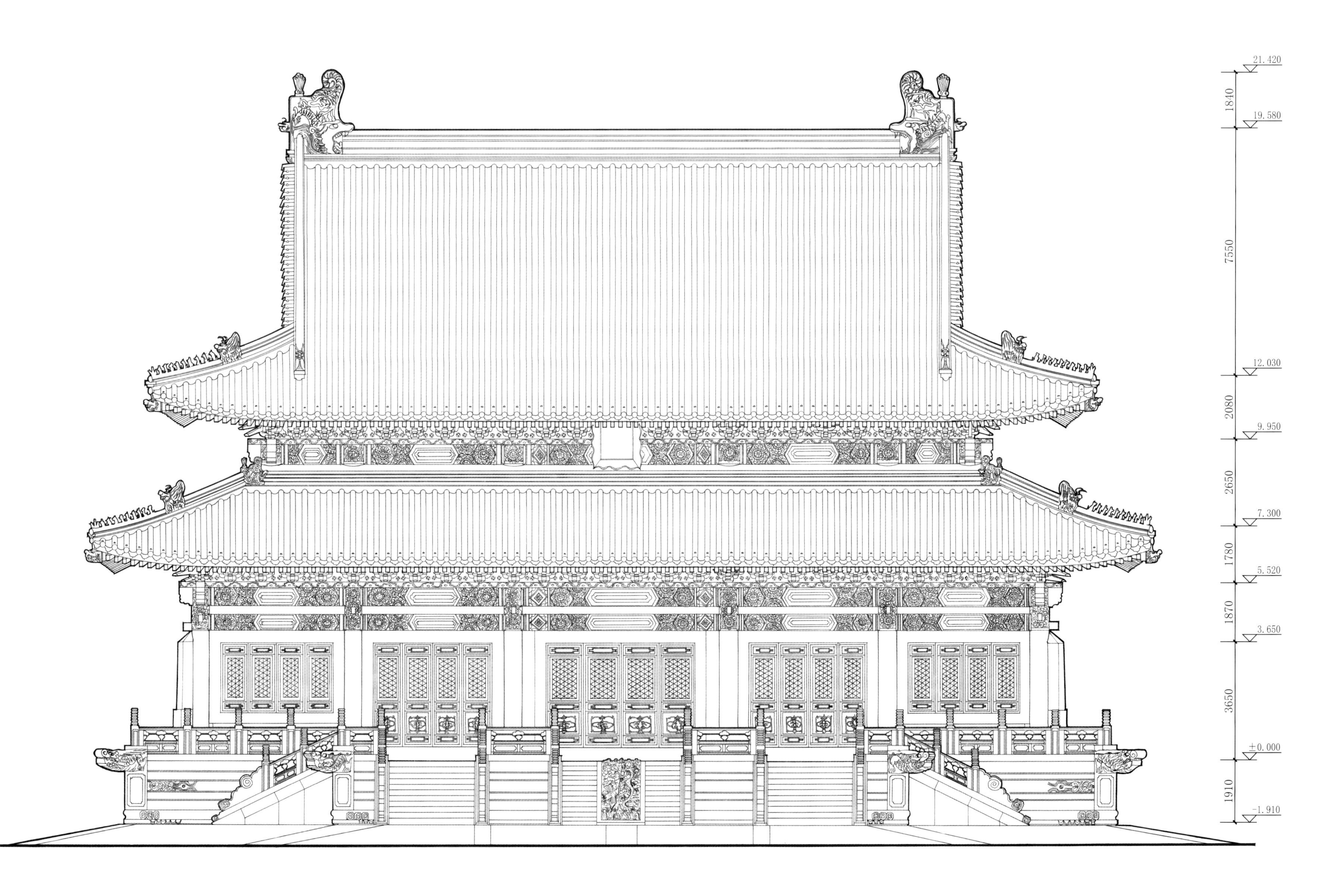

崇陵隆恩殿正立面图

Front elevation of the Hall of Monumental Grace,Tomb of Serendipity

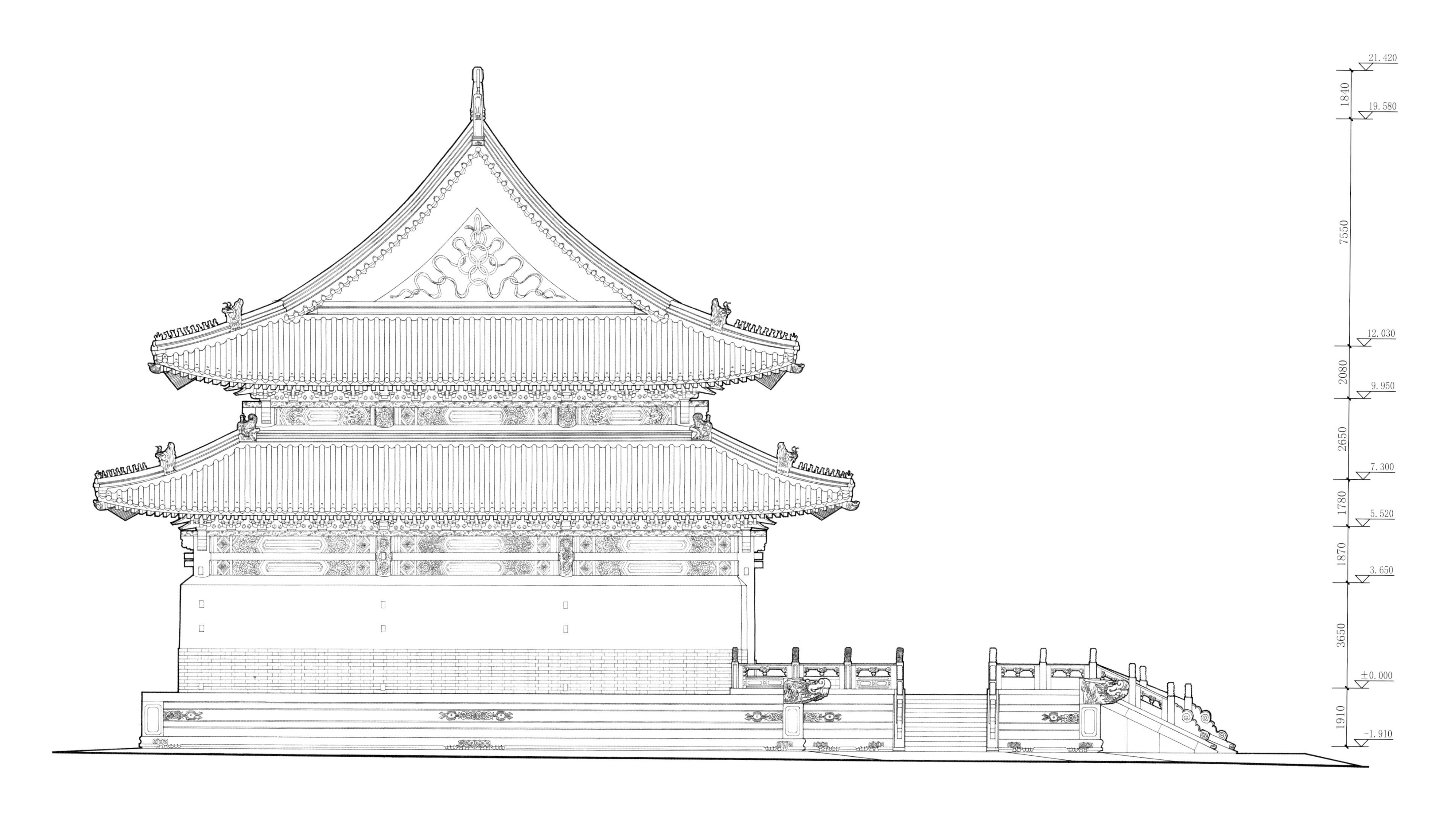

崇陵隆恩殿侧立面图

Side elevation of the Hall of Monumental Grace,Tomb of Serendipity

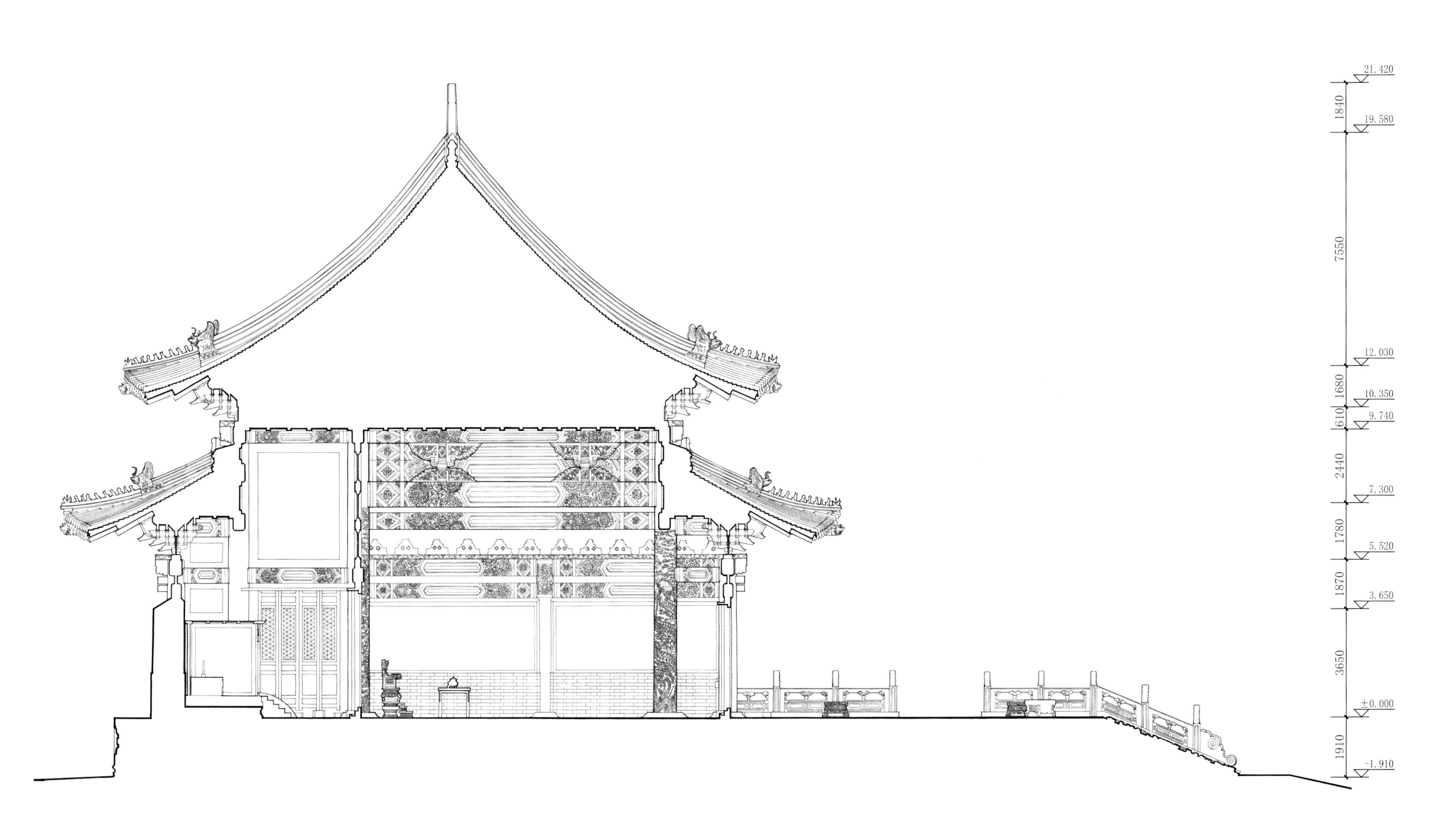

崇陵隆恩殿明间横剖面图
Cross section of the Central Chamber within the Hall of Monumental Grace,Tomb of Serendipity

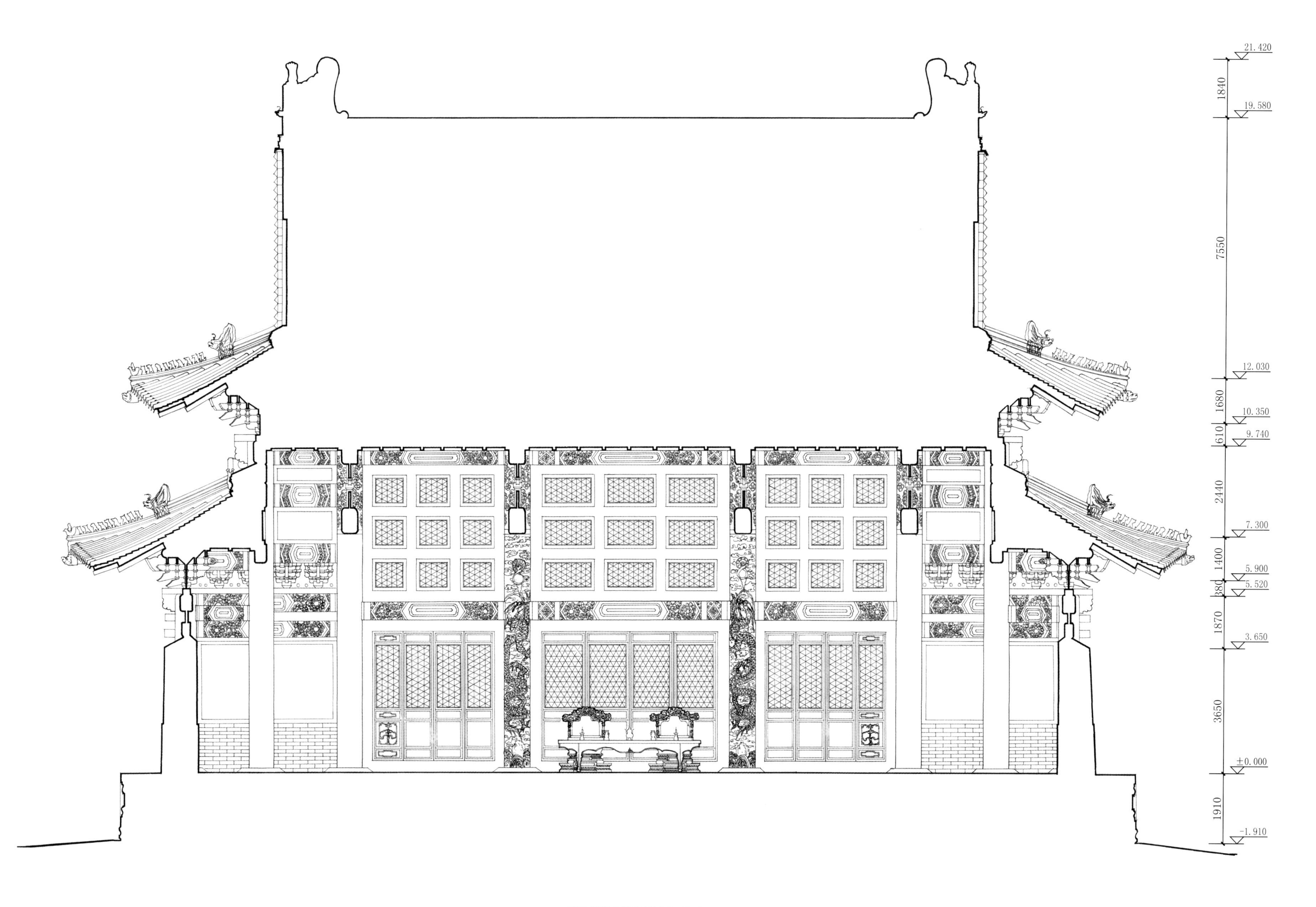

崇陵隆恩殿纵剖面图
Longitudinal section of the Hall of Monumental Grace,Tomb of Serendipity

崇陵隆恩殿宝床正立面图
Front elevation of the Precious bed of the Hall of Monumental Grace,Tomb of Serendipity

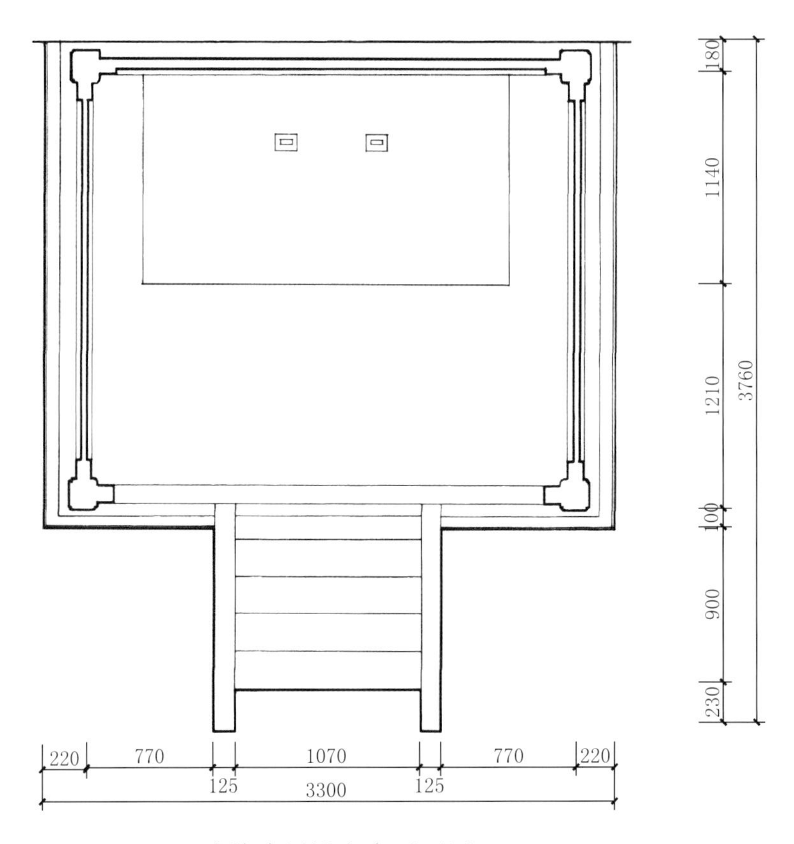

崇陵隆恩殿宝床平面图
Plan of the Precious bed of the Monumental Grace,Tomb of Serendipity

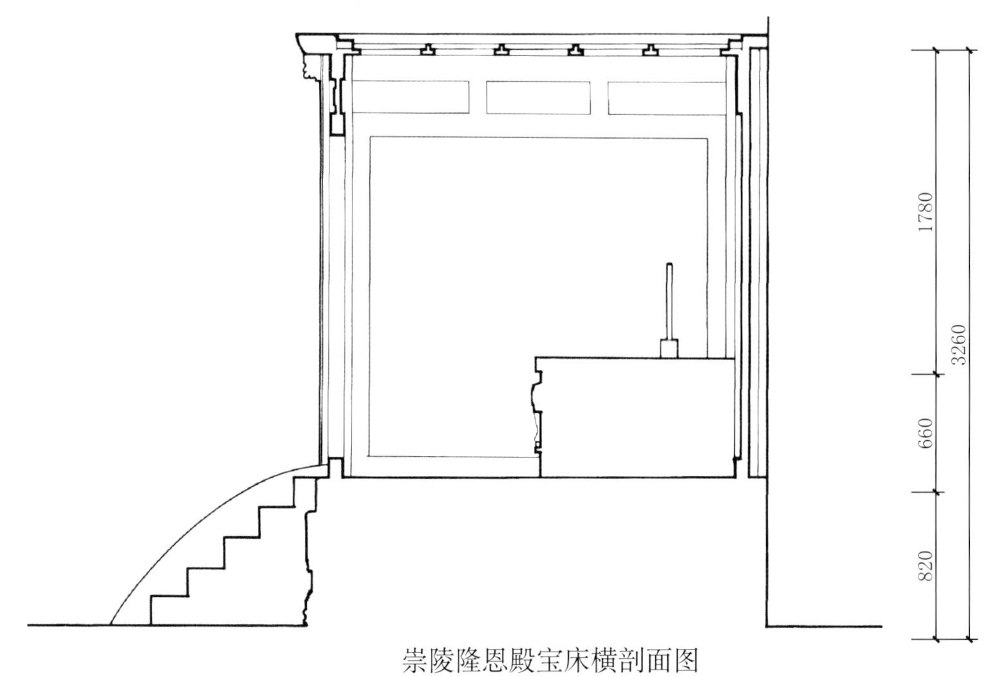

崇陵隆恩殿宝床横剖面图
Cross section of the Precious bed of the Hall of Monumental Grace,Tomb of Serendipity

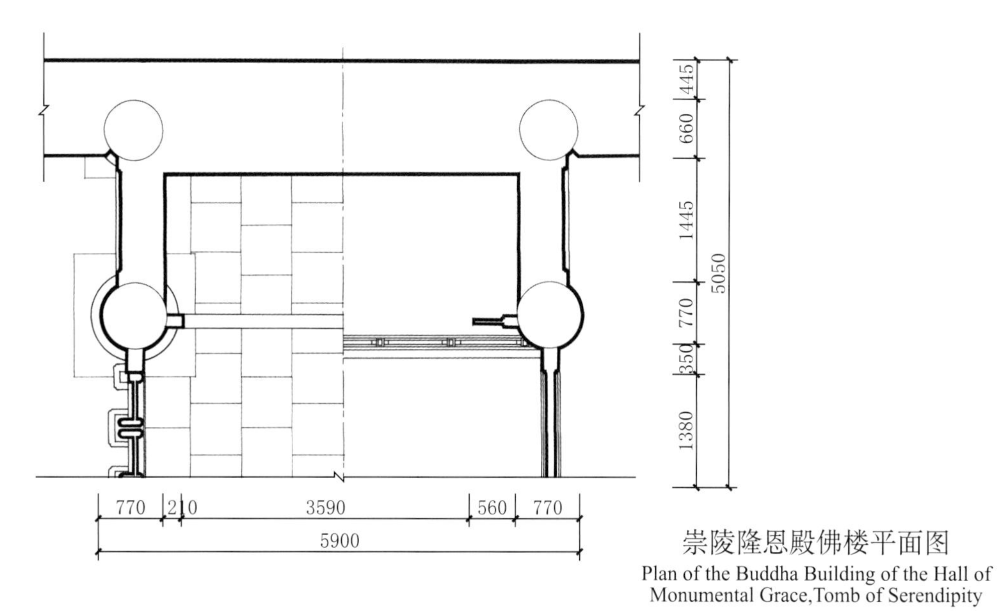

崇陵隆恩殿佛楼平面图
Plan of the Buddha Building of the Hall of Monumental Grace,Tomb of Serendipity

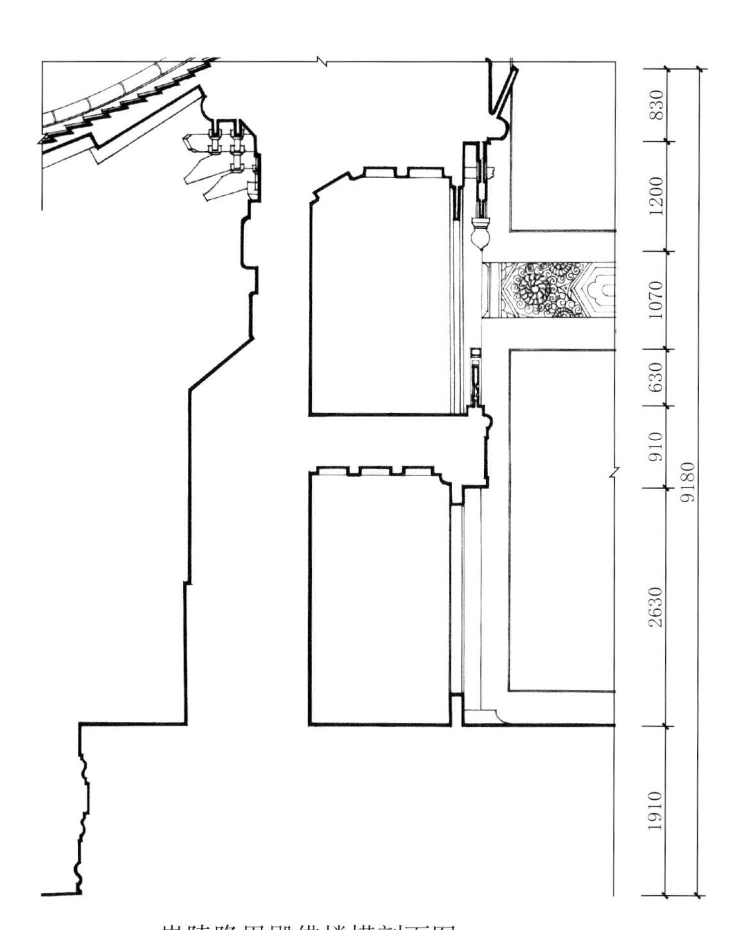

崇陵隆恩殿佛楼横剖面图
Cross section of the Buddha Building of the Hall of Monumental Grace,Tomb of Serendipity

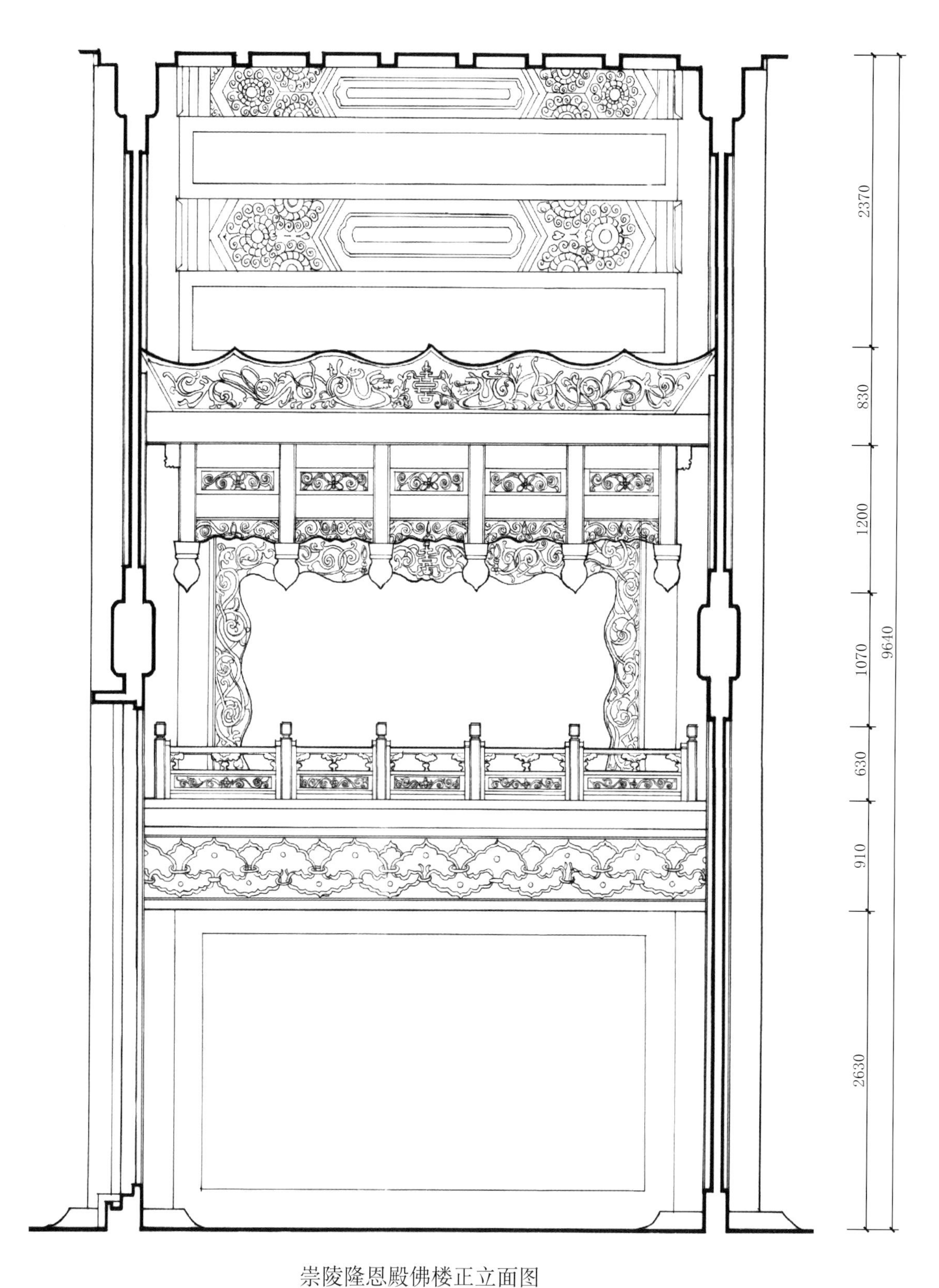

崇陵隆恩殿佛楼正立面图
Front elevation of the Buddha Building of the Hall of Monumental Grace,Tomb of Serendipity

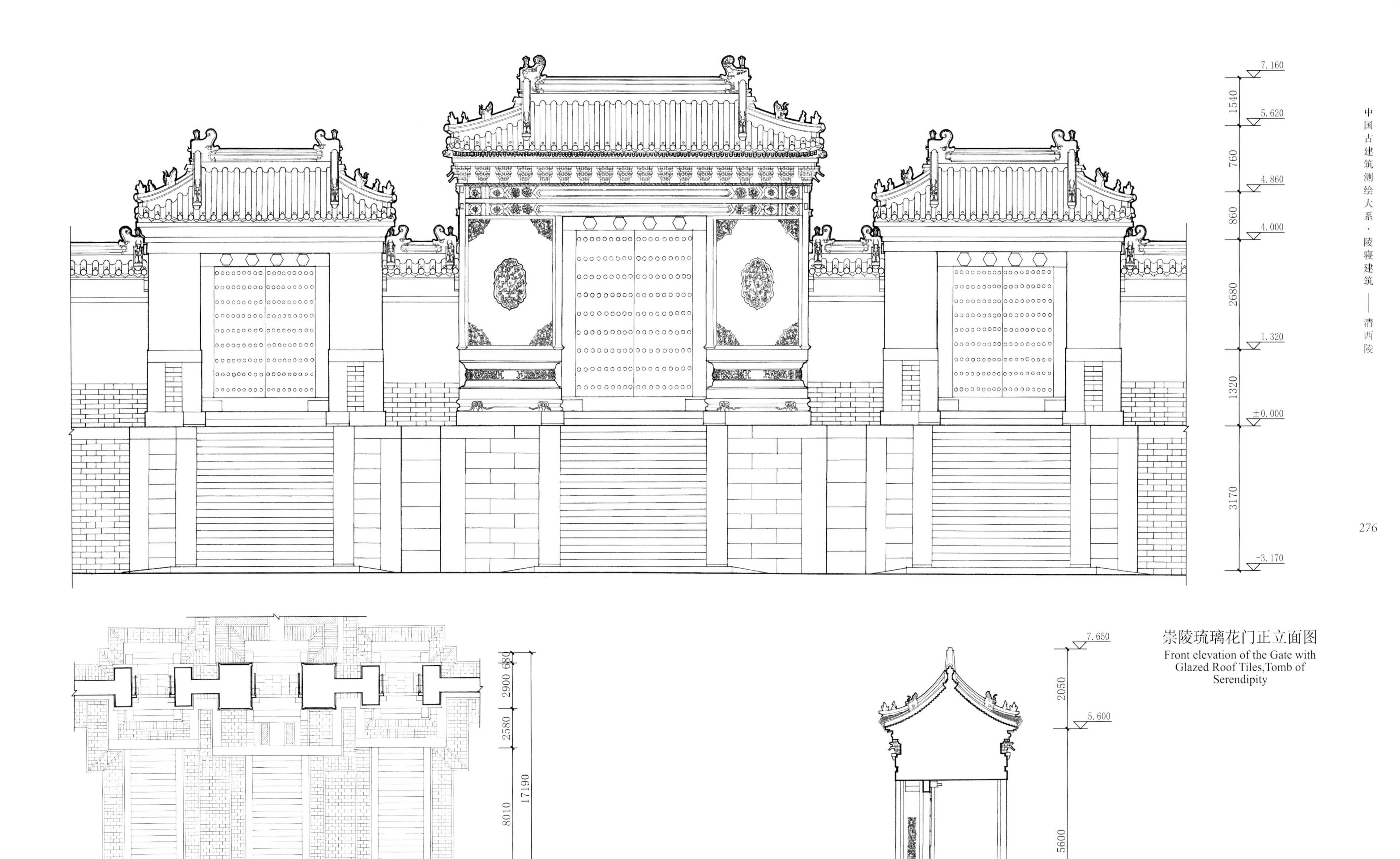

崇陵琉璃花门正立面图

Front elevation of the Gate with Glazed Roof Tiles,Tomb of Serendipity

崇陵琉璃花门平面图

Plan of the Gate with Glazed Roof Tiles,Tomb of Serendipity

崇陵琉璃花门横剖面图

Cross section of the Gate with Glazed Roof Tiles,Tomb of Serendipity

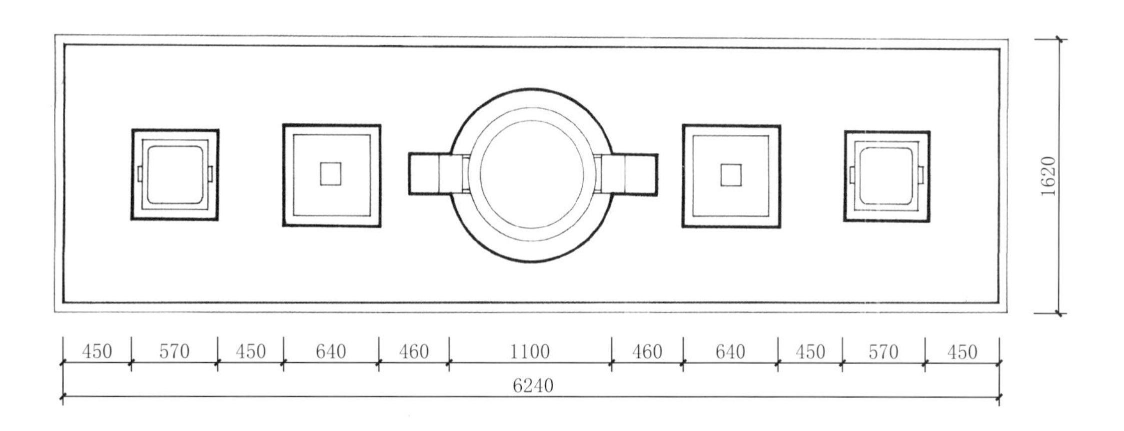

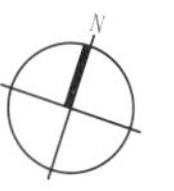

崇陵石台五供平面图
Plan of the Five Stone Ritual Vessels,Tomb of Serendipity

崇陵石台五供正立面图
Front elevation of the Five Stone Ritual Vessels,Tomb of Serendipity

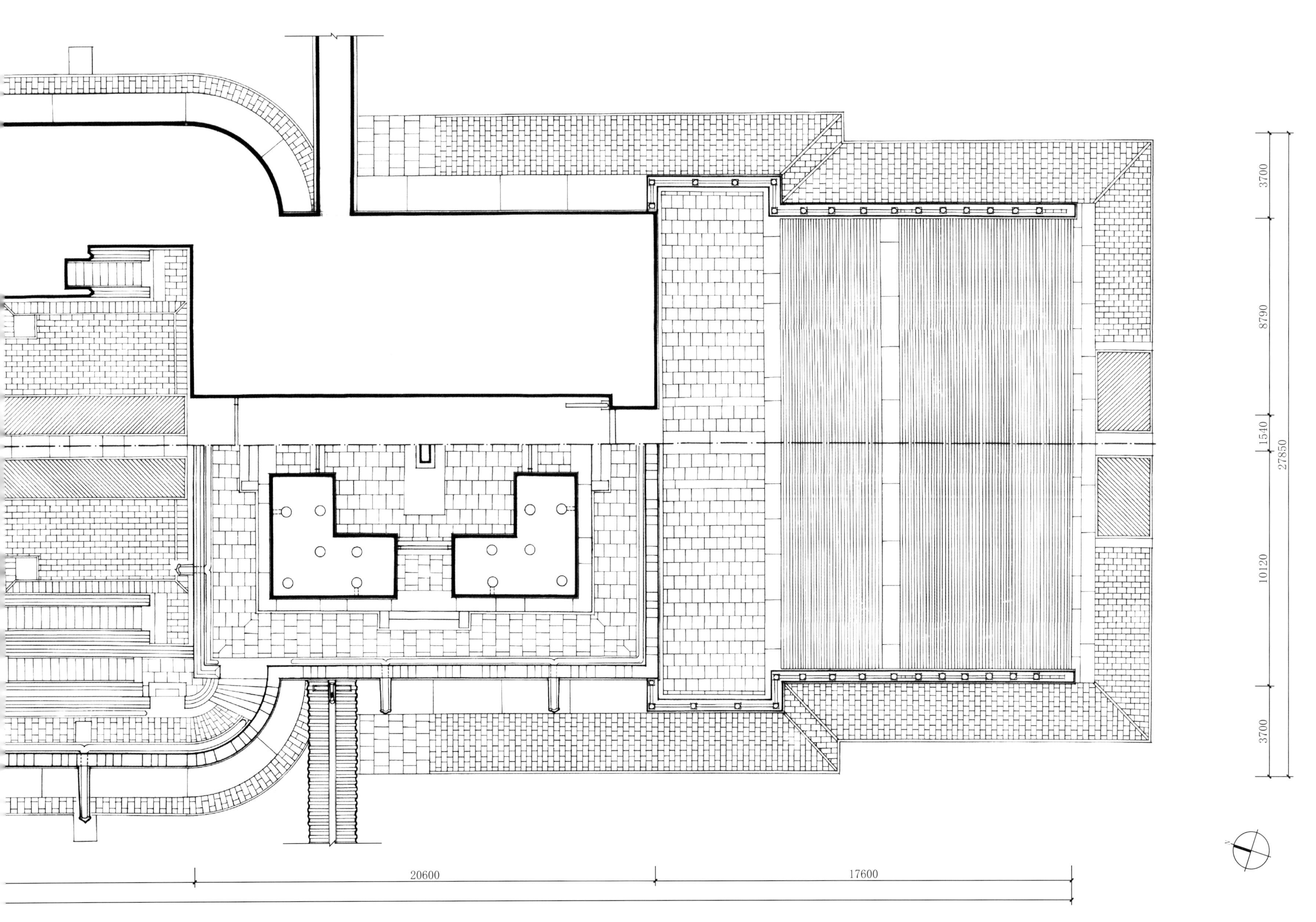

崇陵宝城及方城明楼宝城地宫平面图

Plan of the Encircled Realm of Treasure and the Underground Palace,the Square Walled Terrace and the Memorial Tower,Tomb of Serendipity

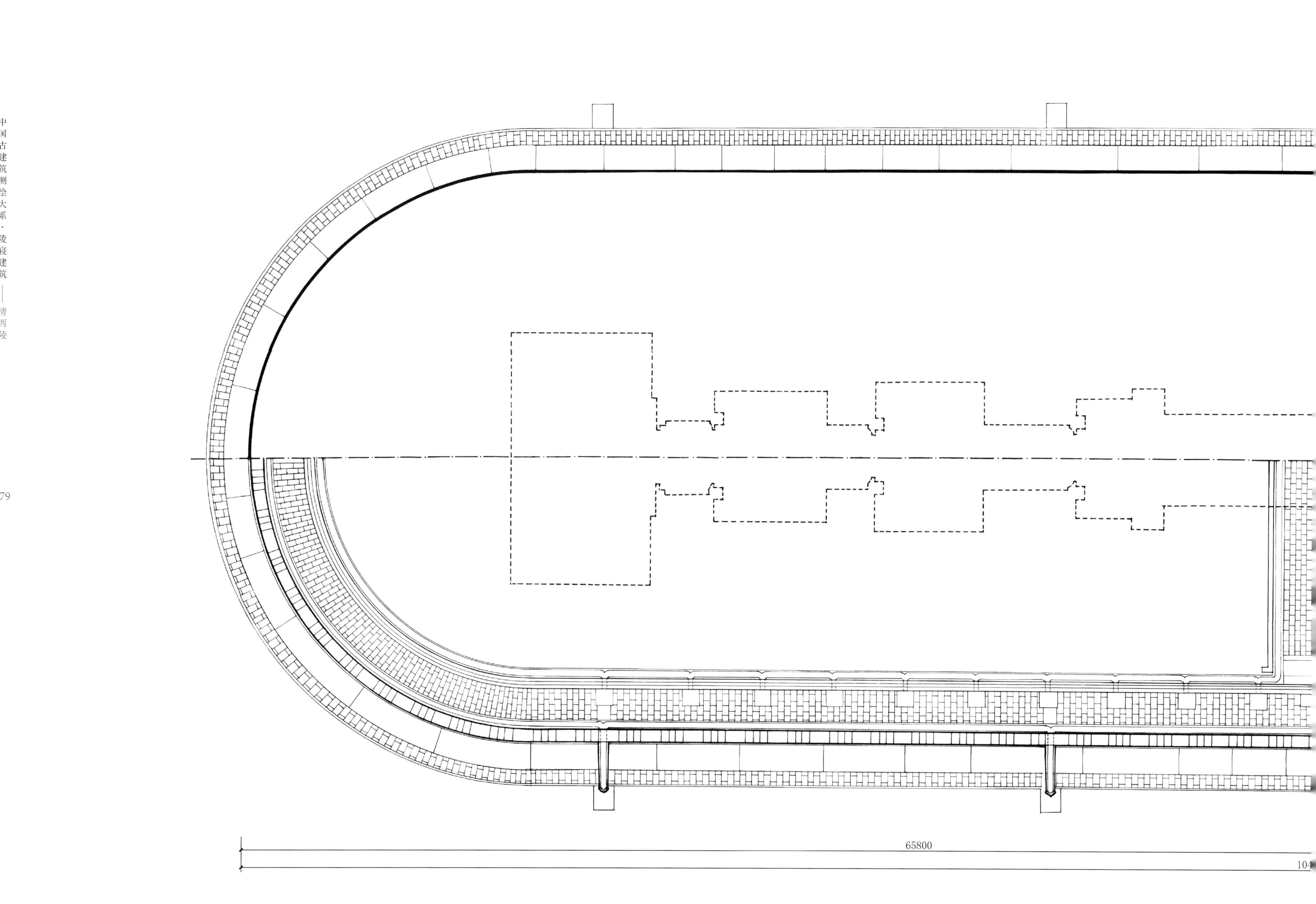
65800
104

崇陵方城明楼纵剖面图

Longitudinal section of the Square Walled Terrace and the Memorial Tower,Tomb of Serendipity

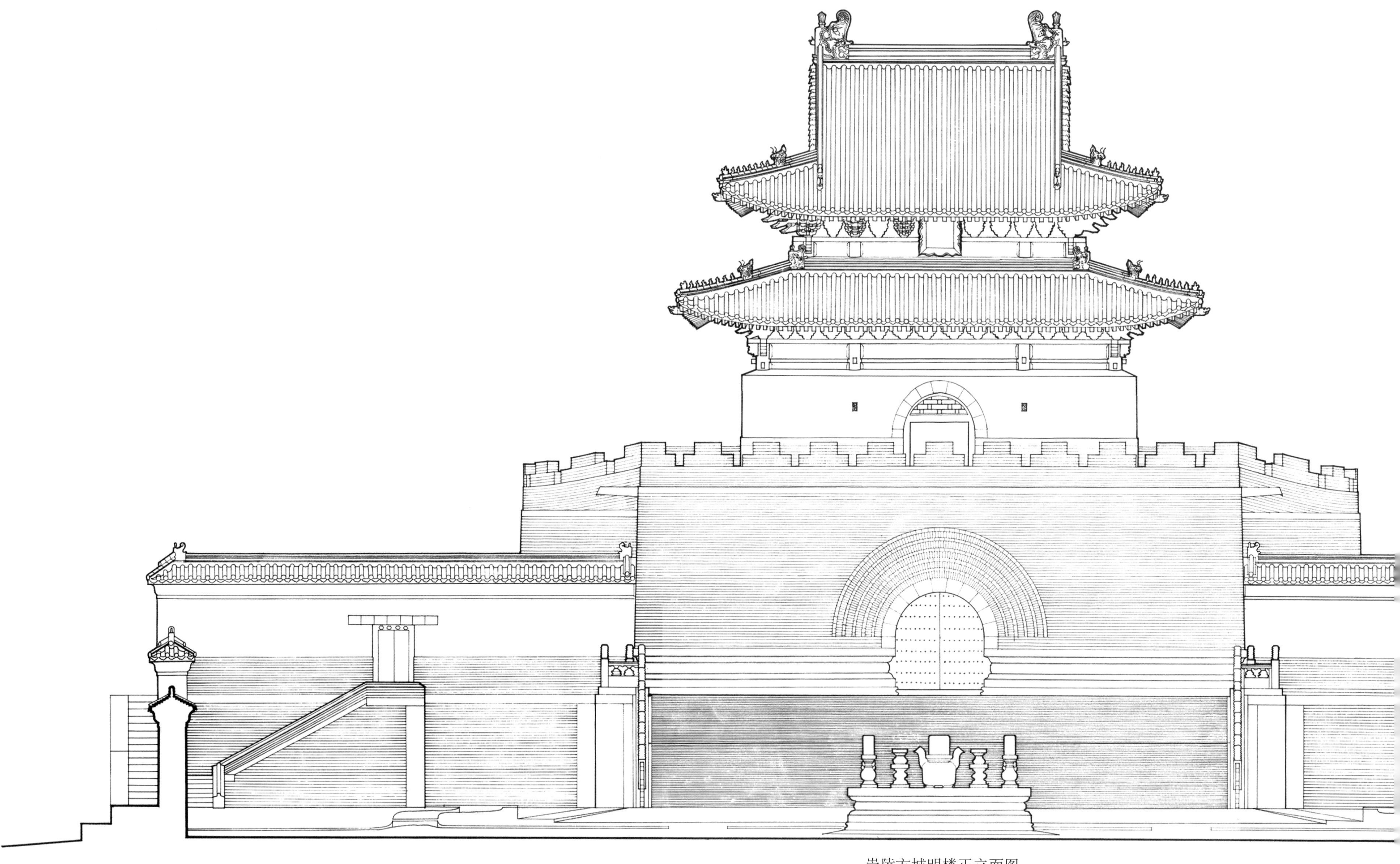

崇陵方城明楼正立面图

Front elevation of the Square Walled Terrace and the Memorial Tower,Tomb of Serendipity

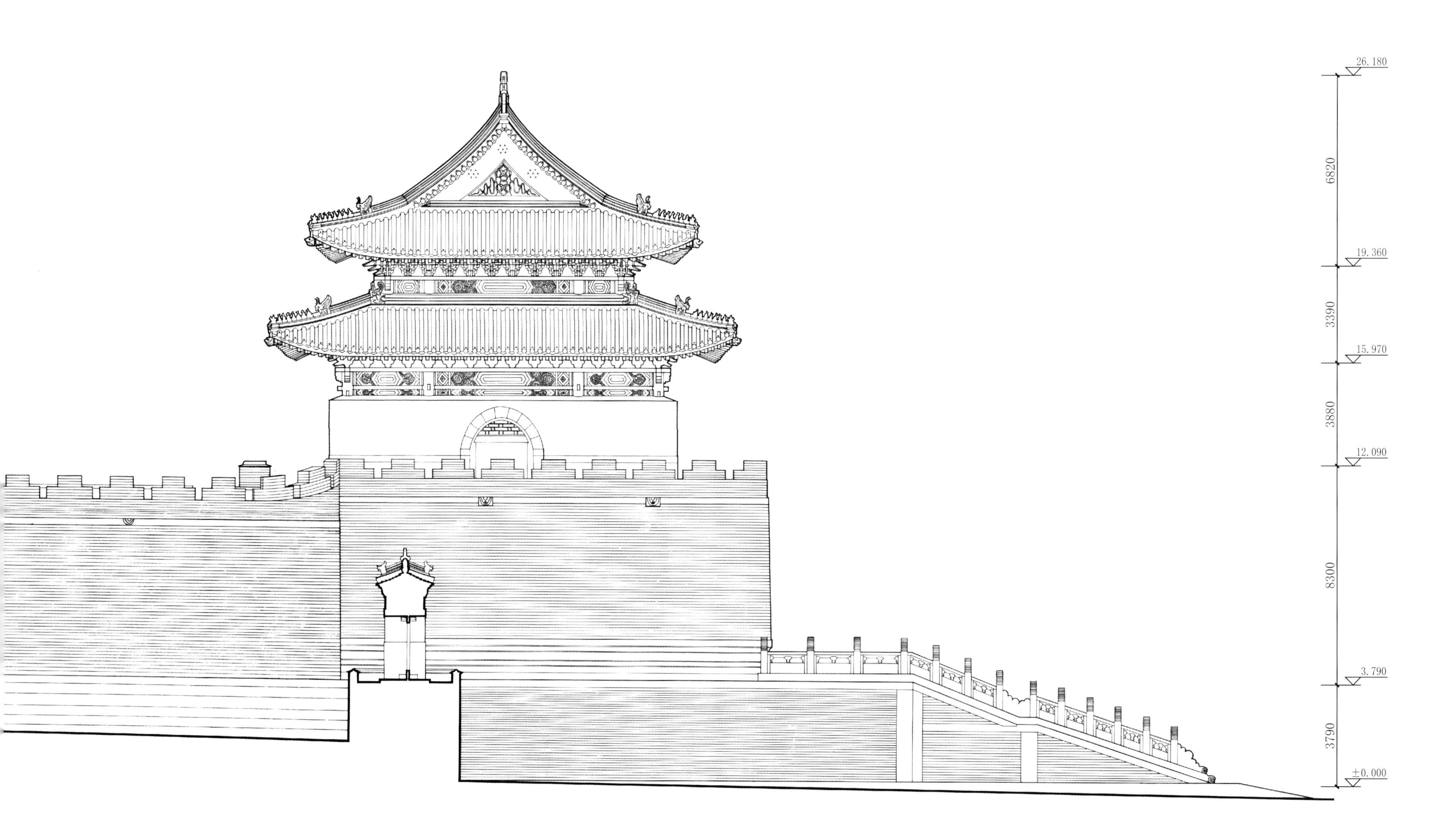

崇陵方城明楼宝城宝顶侧立面图

Side elevation of the Square Walled Terrace and the Memorial Tower as well as the Encircled Realm of Treasure and the Tumulus,Tomb of Serendipity

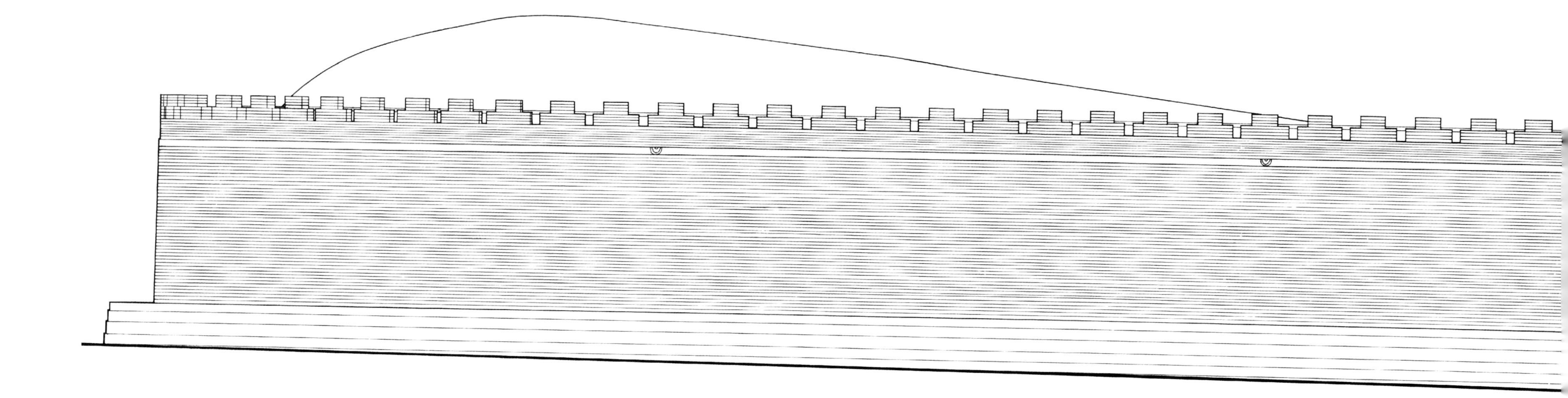

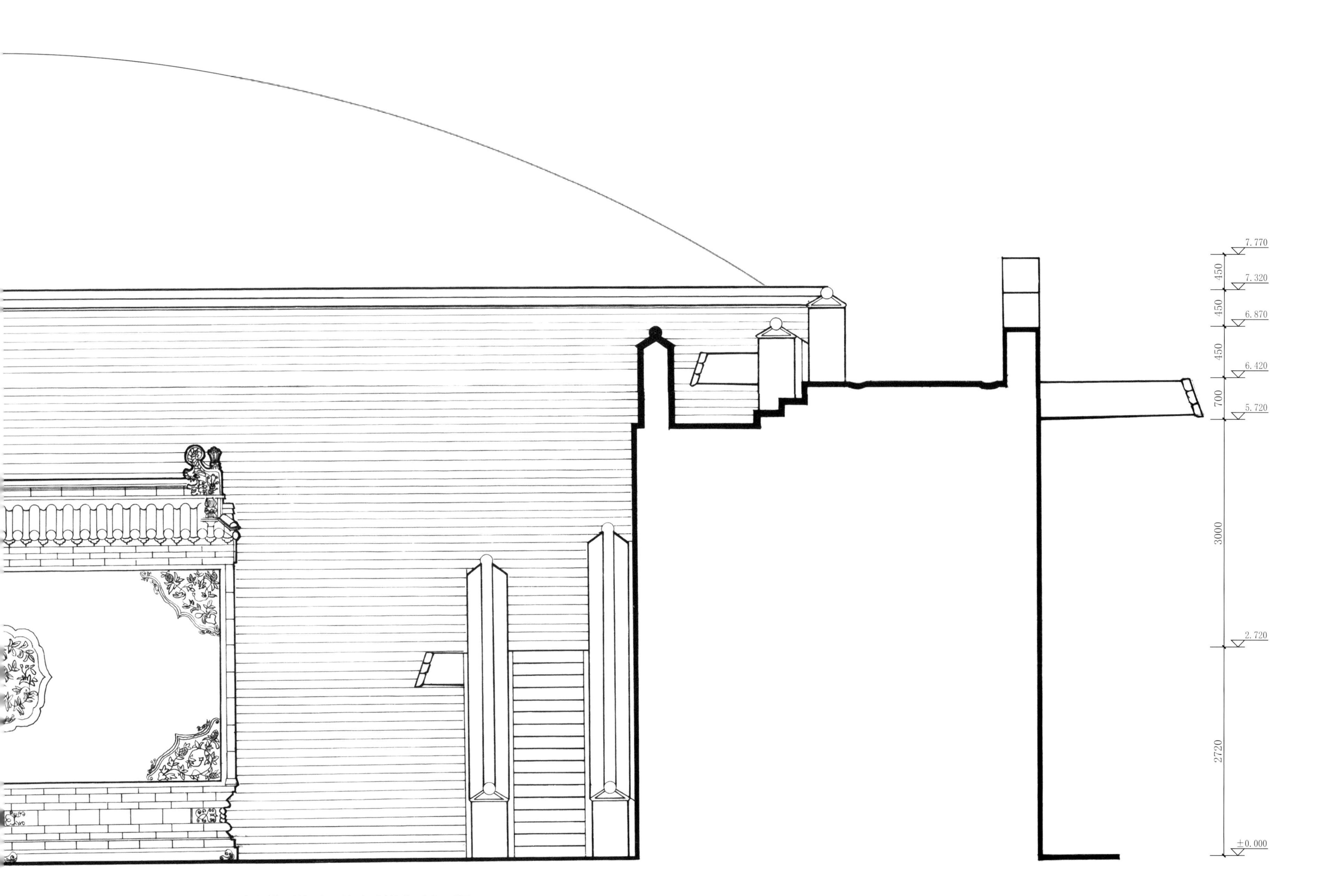

崇陵方城明楼哑巴院琉璃影壁正立面图

Front elevation of the Screen Wall of Glazed Tiles in Yaba Court,the Square Walled Terrace and the Memorial Tower,Tomb of Serendipity

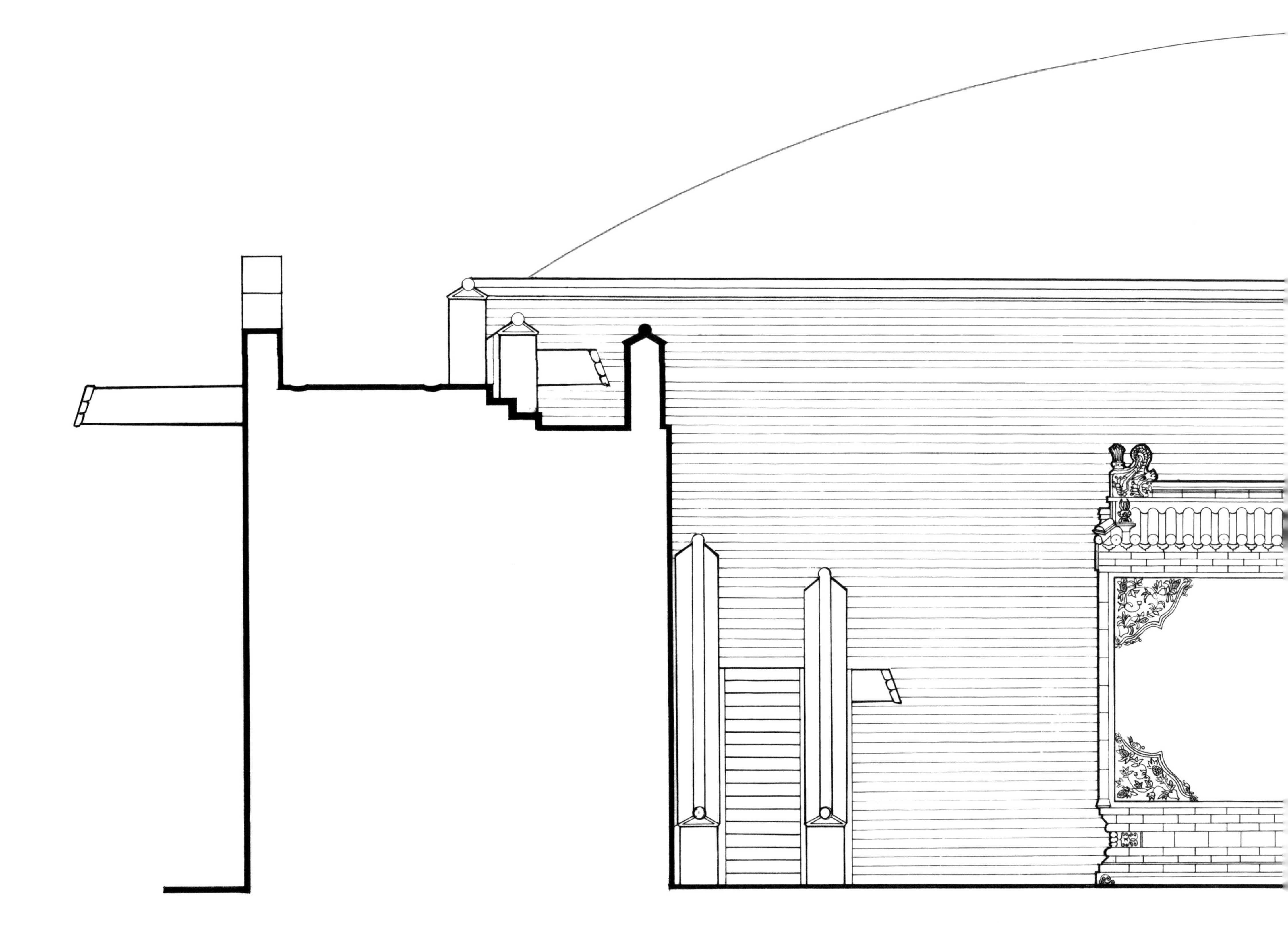

崇陵妃园寝

Imperial Consorts' Tombs affiliated with the Tomb of Serendipity (Chongling Feiyuanqin)

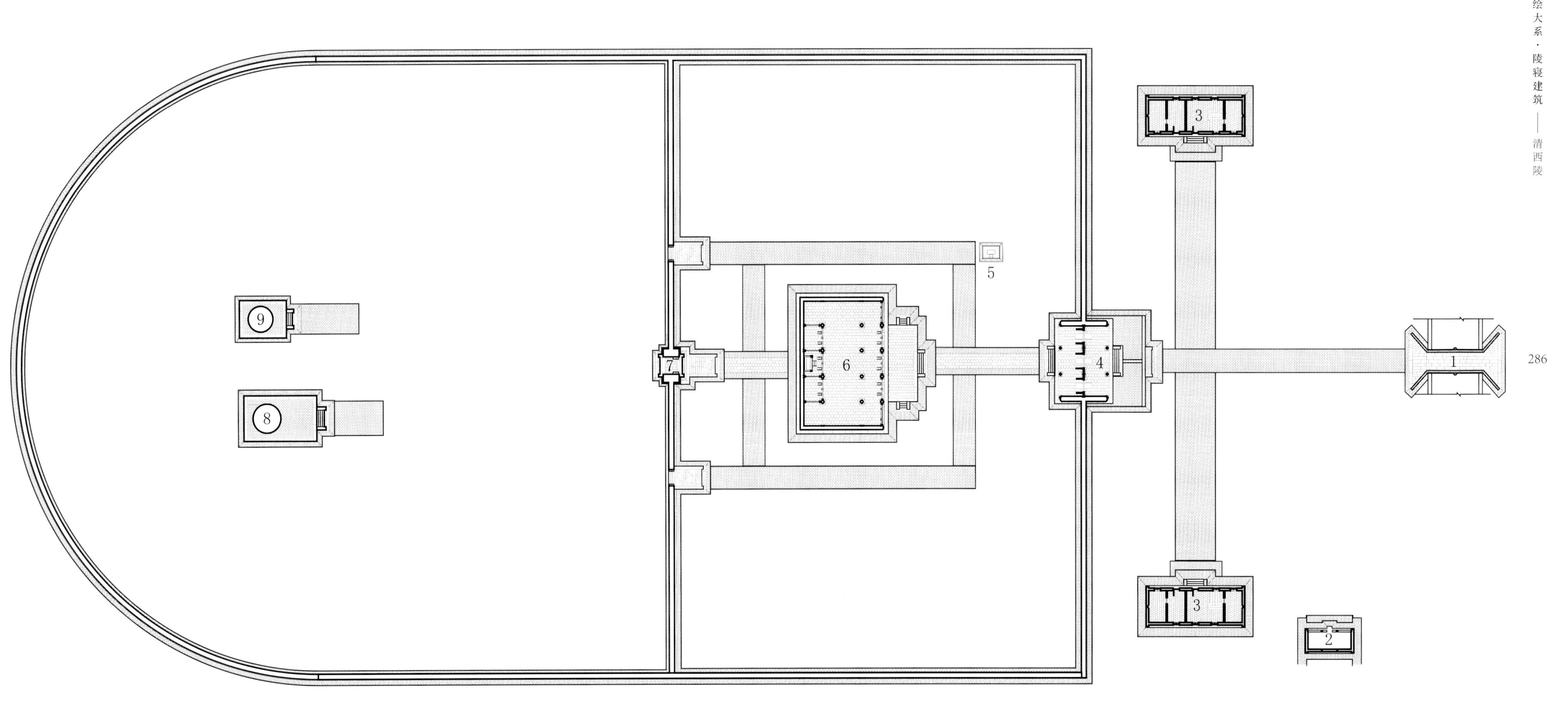

1 单孔石券桥 Single-Arched Stone Bridge
2 值房 Guard House
3 朝房 Reception Hall for Court Officials
4 宫门 Main Gate
5 焚帛炉 Sacrificial Burner
6 享殿 Hall of Ritual Sacrifice
7 琉璃花门 Gate with Glazed Roof Tiles
8 瑾妃宝顶 Tumulus of the Imperial Consort Jin
9 珍妃宝顶 Tumulus of the Imperial Consort Zhen

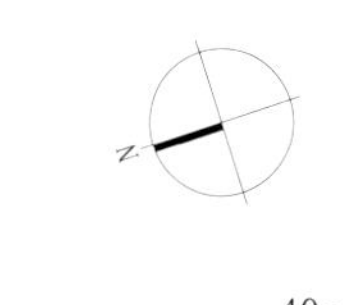

0 10 20 40m

崇陵妃园寝组群平面图

Site plan of the building complex of the Imperial Consorts'Tombs affiliated with the Tomb of Serendipity

端亲王园寝
Tomb for Prince Duan (Duanqinwang Yuanqin)

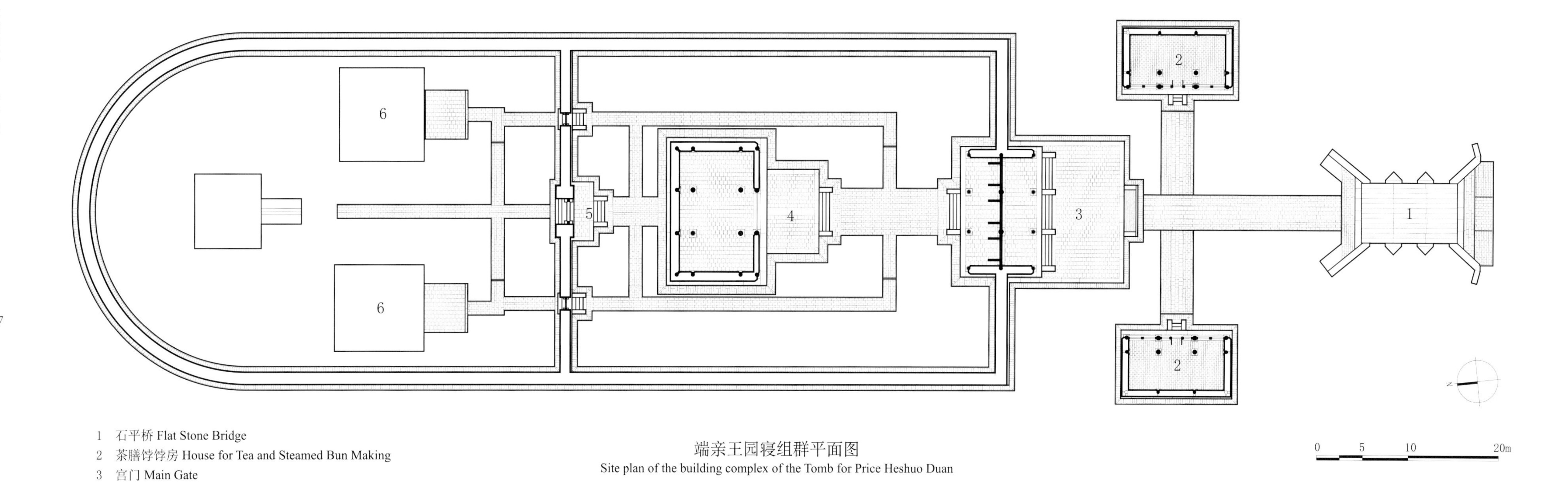

1 石平桥 Flat Stone Bridge
2 茶膳饽饽房 House for Tea and Steamed Bun Making
3 宫门 Main Gate
4 享殿 Hall of Ritual Sacrifice
5 砖门 Brick Gate
6 宝顶 Tumulus

端亲王园寝组群平面图
Site plan of the building complex of the Tomb for Price Heshuo Duan

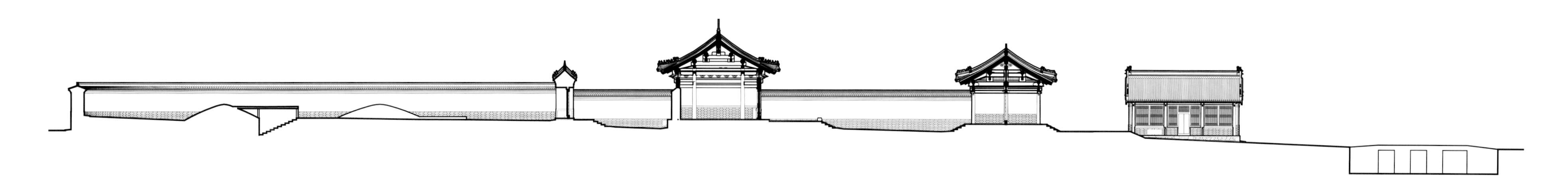

端亲王园寝组群剖面图
Site section of the building complex of the Tomb for Price Heshuo Duan

怀亲王园寝

Tomb for Prince Huai (Huaiqinwang Yuanqin)

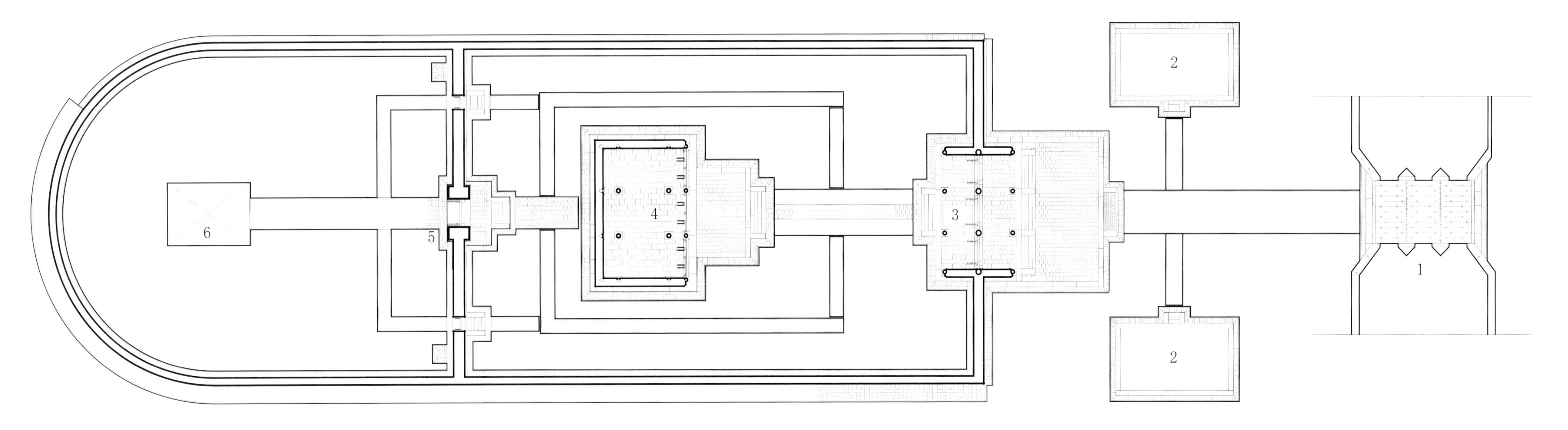

1 石平桥 Flat Stone Bridge
2 茶膳饽饽房 House for Tea and Steamed Bun Making
3 宫门 Main Gate
4 享殿 Hall of Ritual Sacrifice
5 琉璃花门 Gate with Glazed Roof Tiles
6 宝顶 Tumulus

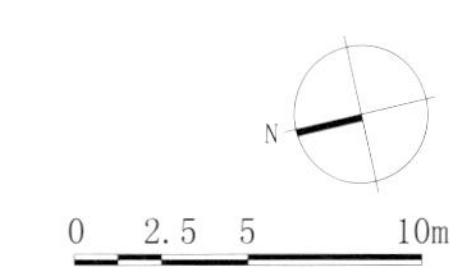

怀亲王园寝组群平面图

Site plan of the building complex of the Tomb for Prince Heshuo Huai

阿哥园寝
Tomb for a Son of Emperor Yongzheng (A' ge Yuanqin)

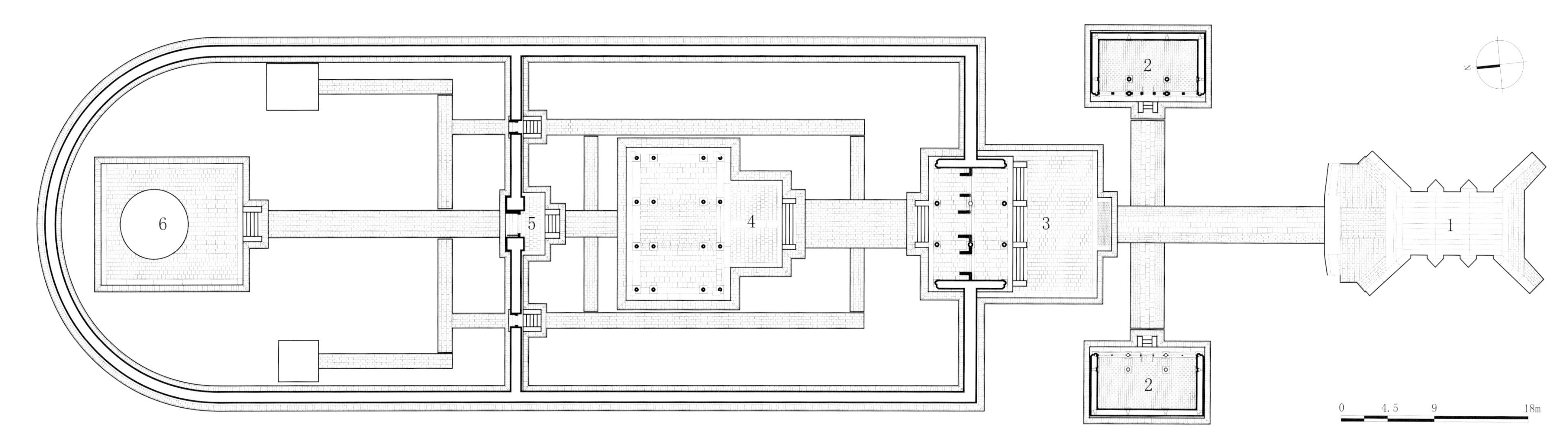

1 石平桥 Flat Stone Bridge
2 茶膳饽饽房 House for Tea and Steamed Bun Making
3 宫门 Main Gate
4 大殿遗址 Relics of the Main Hall
5 砖门 Brick Gate
6 宝顶 Tumulus

阿哥园寝总平面图
Site plan of the Tomb for a Son of Emperor Yongzheng

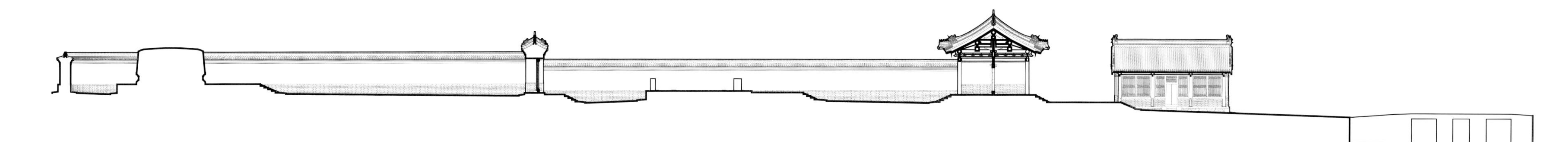

阿哥园寝组群剖面图
Site section of the building complex of the Tomb for a Son of Emperor Yongzheng

公主园寝
Tomb for Two Princesses (Gongzhu Yuanqin)

1 茶膳饽饽房 House for Tea and Steamed Bun Making
2 宫门 Main Gate
3 大殿遗址 Relics of the Main Hall
4 砖门 Brick Gate
5 宝顶 Tumulus

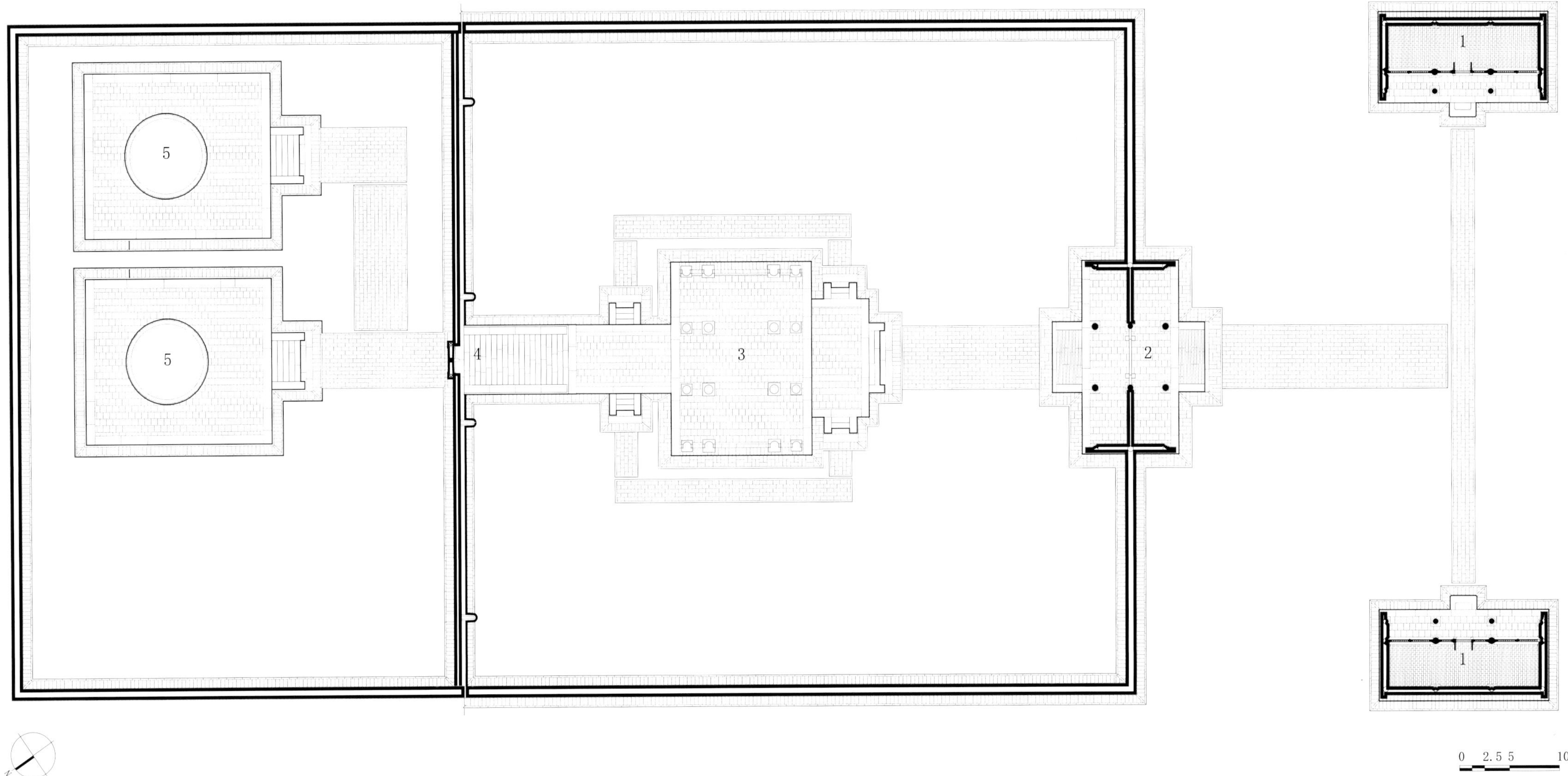

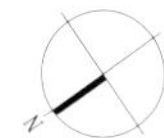

公主园寝组群平面图
Site plan of the building complex of the Tomb for Two Princesses

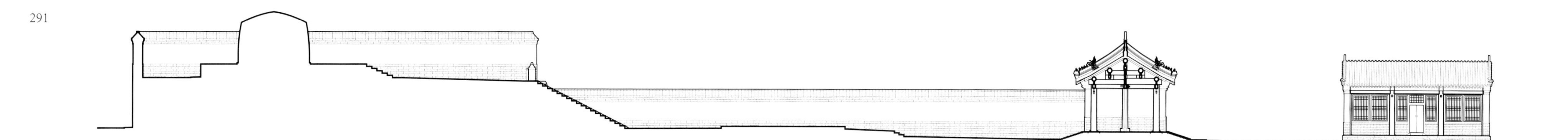

公主园寝组群剖面图

Site section of the building complex of the Tomb for Two Princesses

永福寺

Temple of Perpetual Happiness (Yongfu Si)

1

1 三孔桥 Three-Arch Bridge
2 山门 Main Gate
3 钟楼 Bell Tower
4 大雄宝殿 Great Buddha's Hall
5 牌坊 Memorial Gateway
6 碑亭 Pavilion for the Stela
7 配殿 Side Hall
8 普光明殿 Hall of Holy Light Illuminating All Things

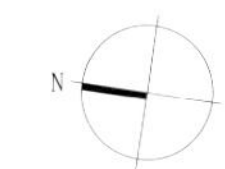

永福寺组群平面图
Site plan of the building complex of the Temple of Perpetual Happiness

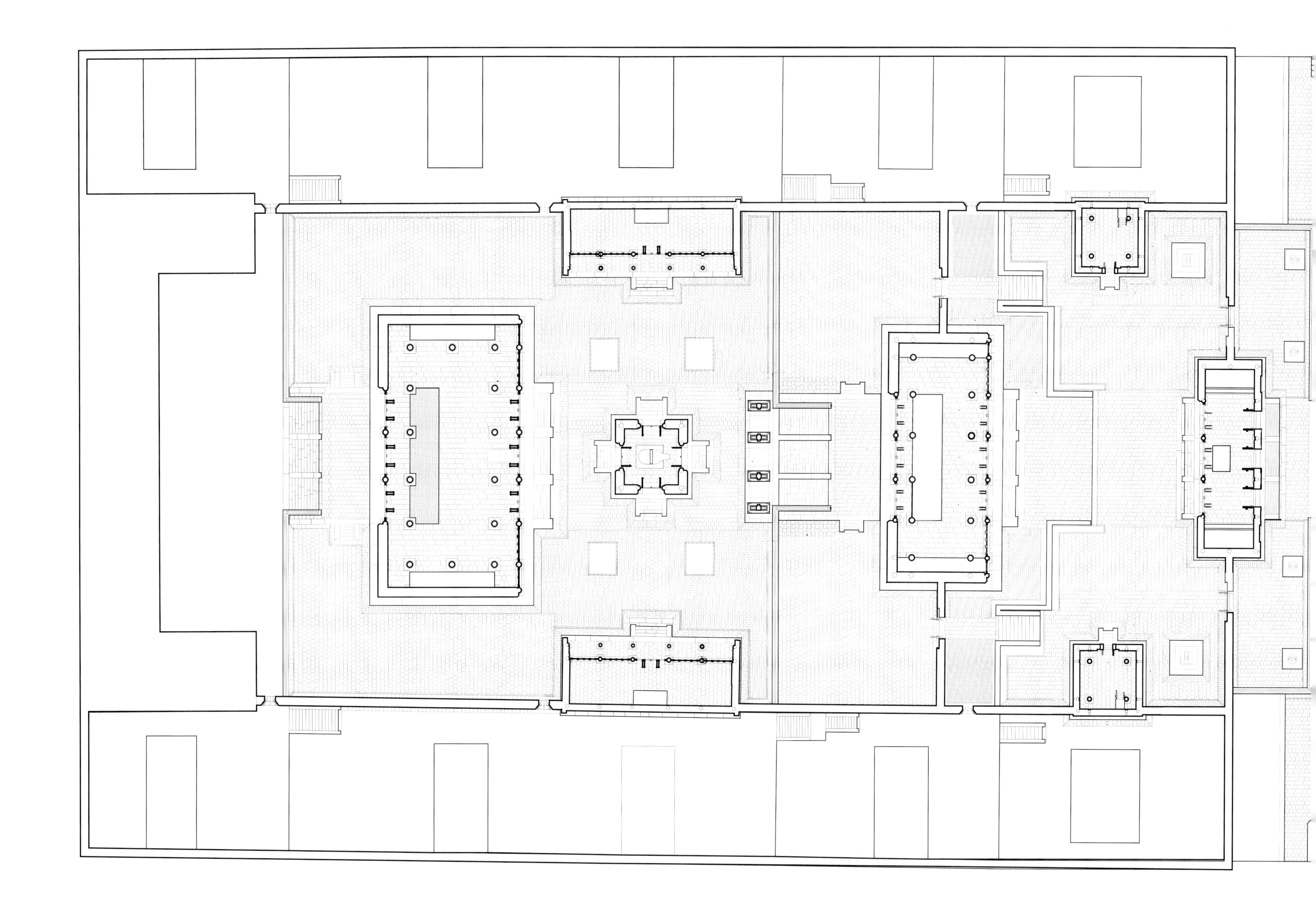

0 2 4 8m

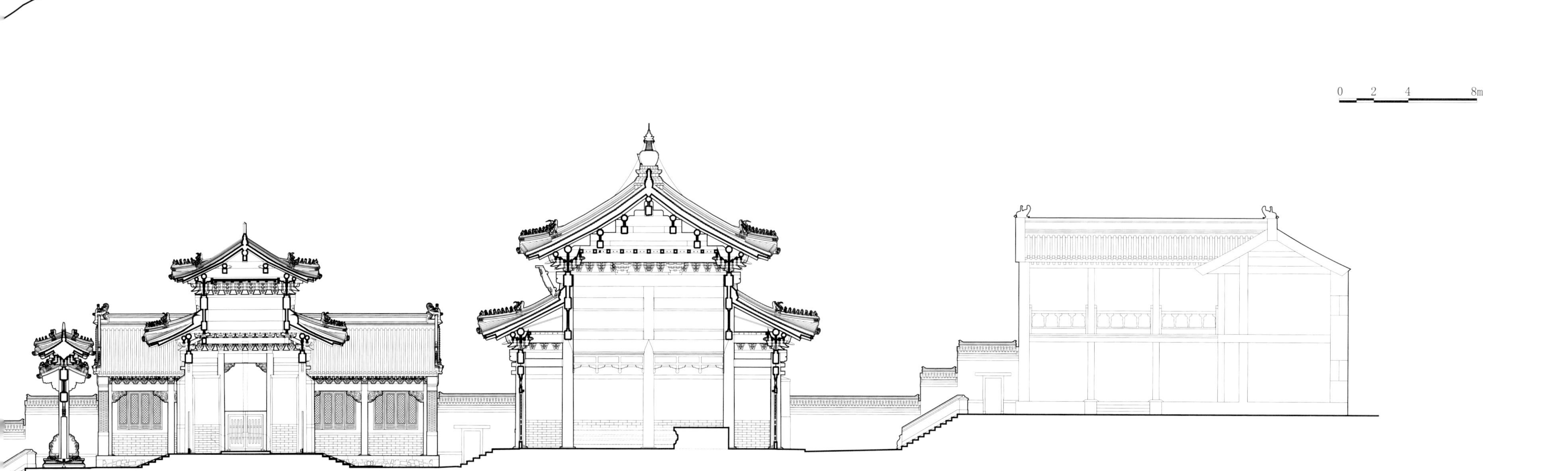

0 2 4 8m

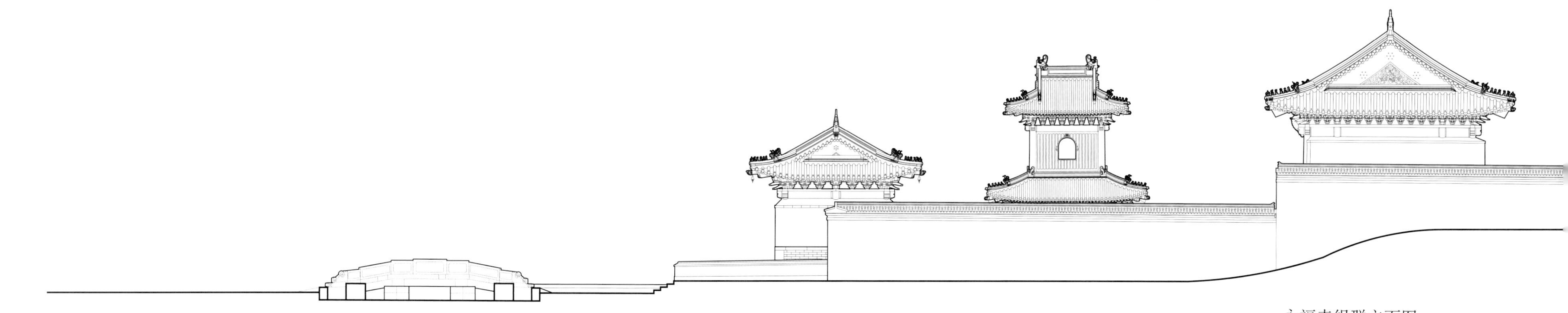

永福寺组群立面图

Site elevation of the building complex of the Temple of Perpetual Happiness

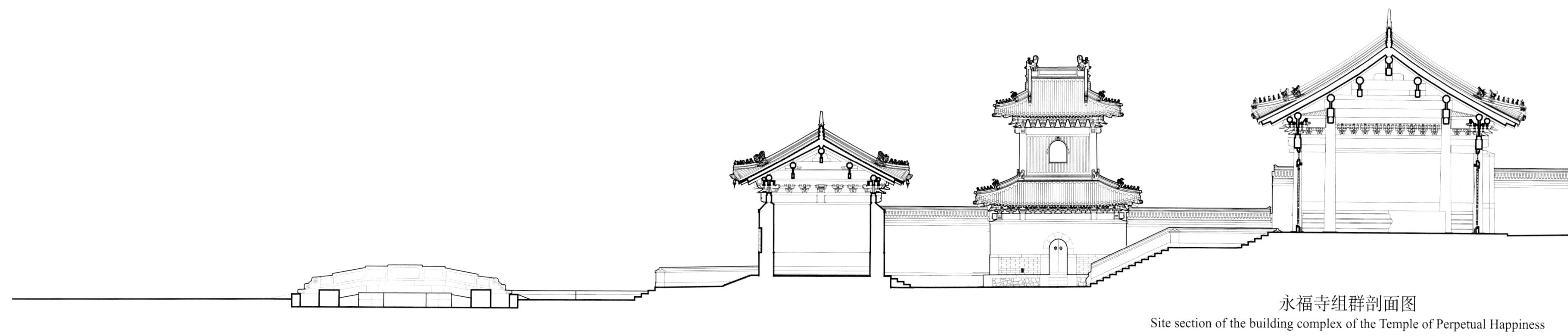

永福寺组群剖面图

Site section of the building complex of the Temple of Perpetual Happiness

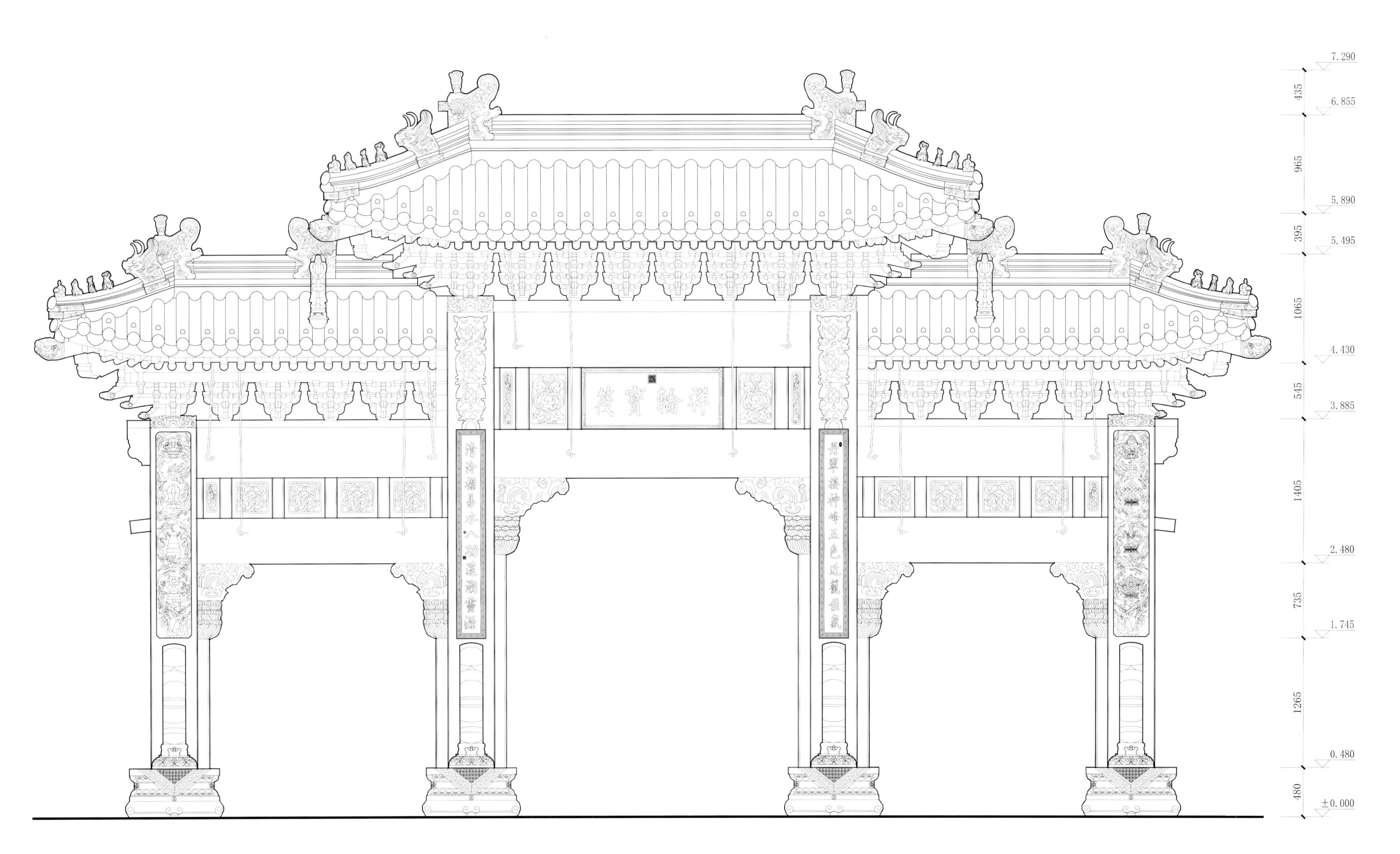

永福寺牌坊正立面

Front elevation of the Memorial Gateway,Temple of Perpetual Happiness

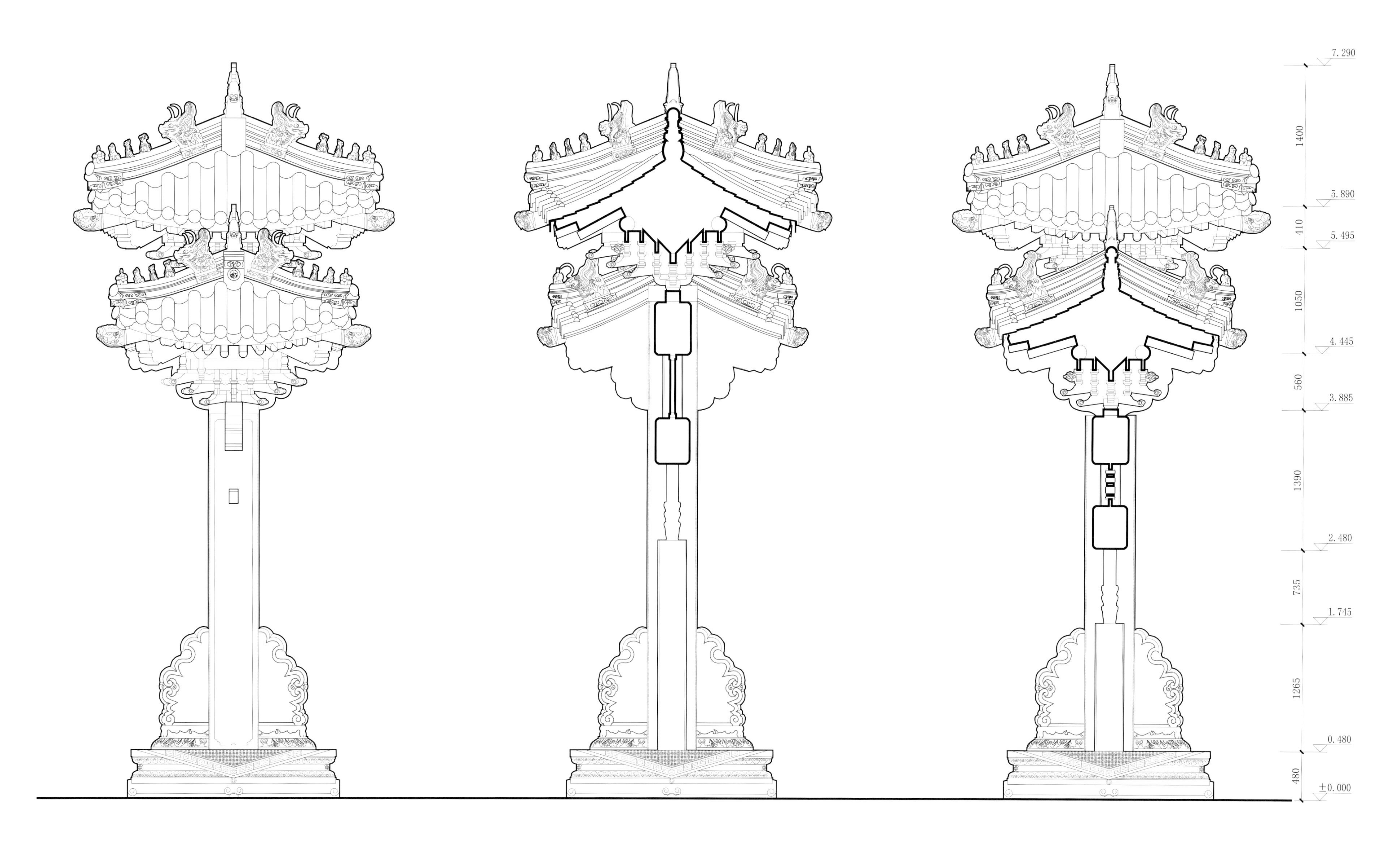

永福寺牌坊侧立面图
Side elevation of the Memorial Gateway,Temple of Perpetual Happiness

永福寺牌坊明间剖面图
Section of the Central Chamber within the Memorial Gateway,Temple of Perpetual Happiness

永福寺牌坊次间剖面图
Section of the Side Chamber within the Memorial Gateway,Temple of Perpetual Happiness

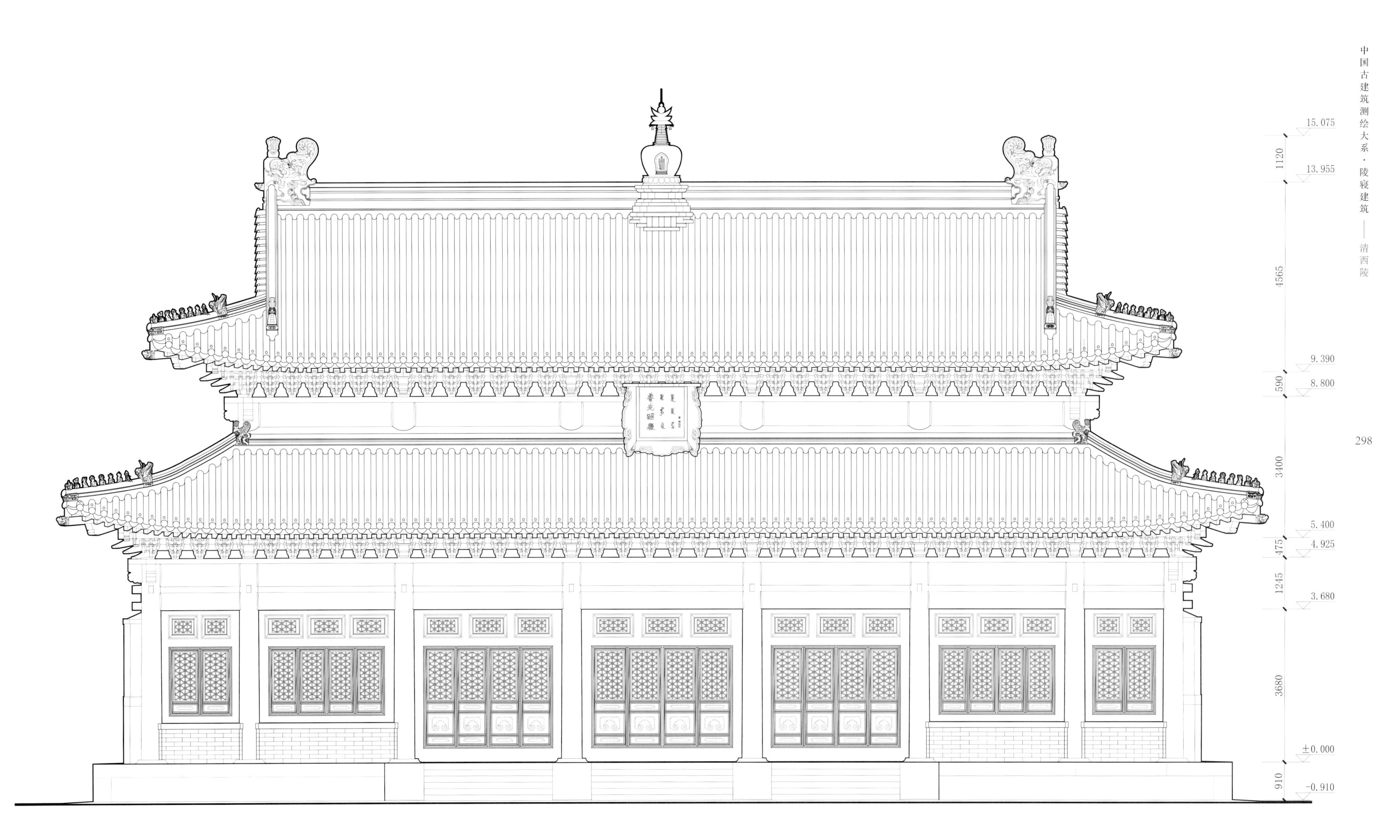

永福寺普光明殿正立面图
Front elevation of the Pu'guangming Hall,Temple of Perpetual Happiness

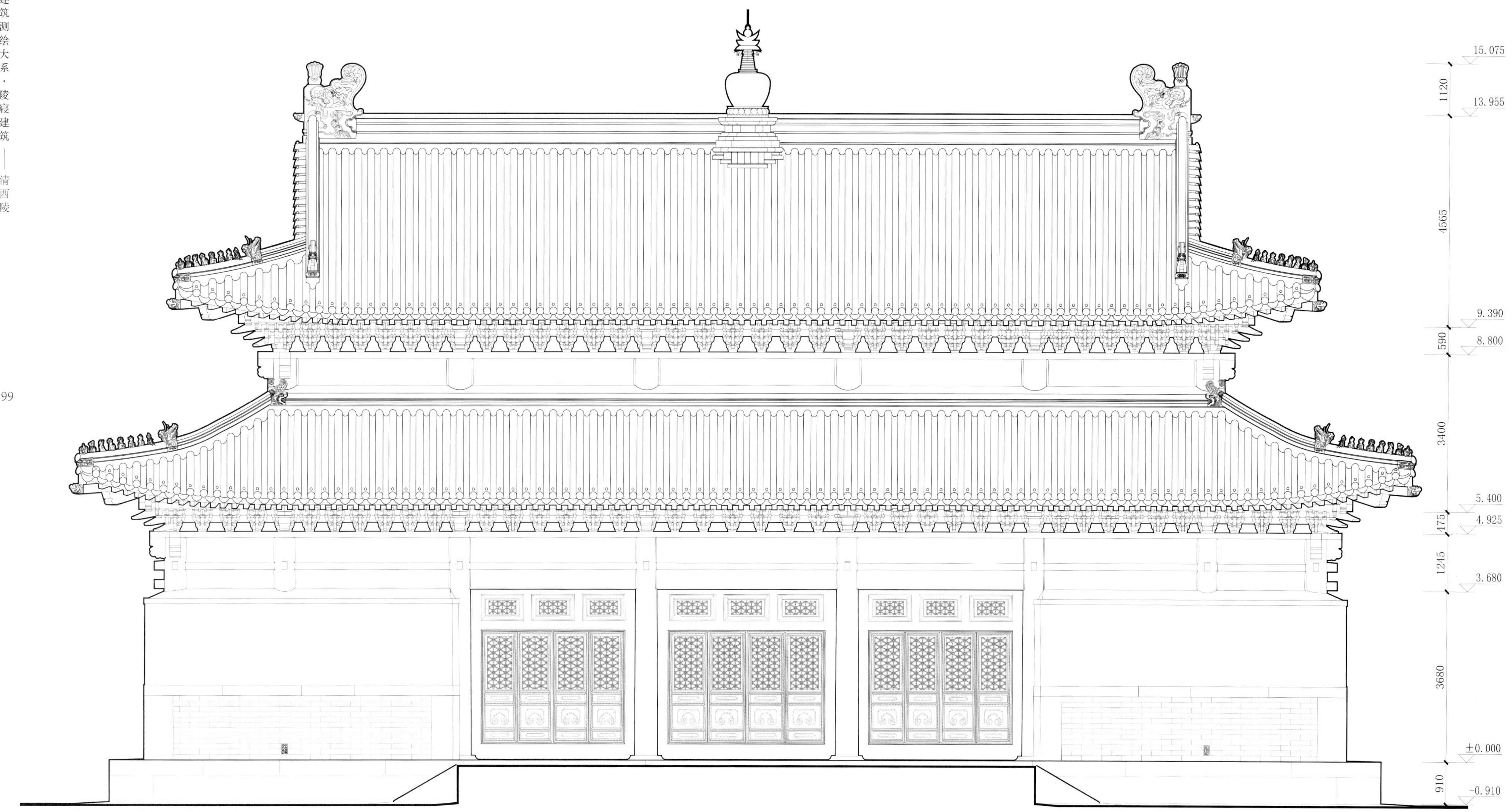

永福寺普光明殿背立面图
Back elevation of the Pu'guangming Hall,Temple of Perpetual Happiness

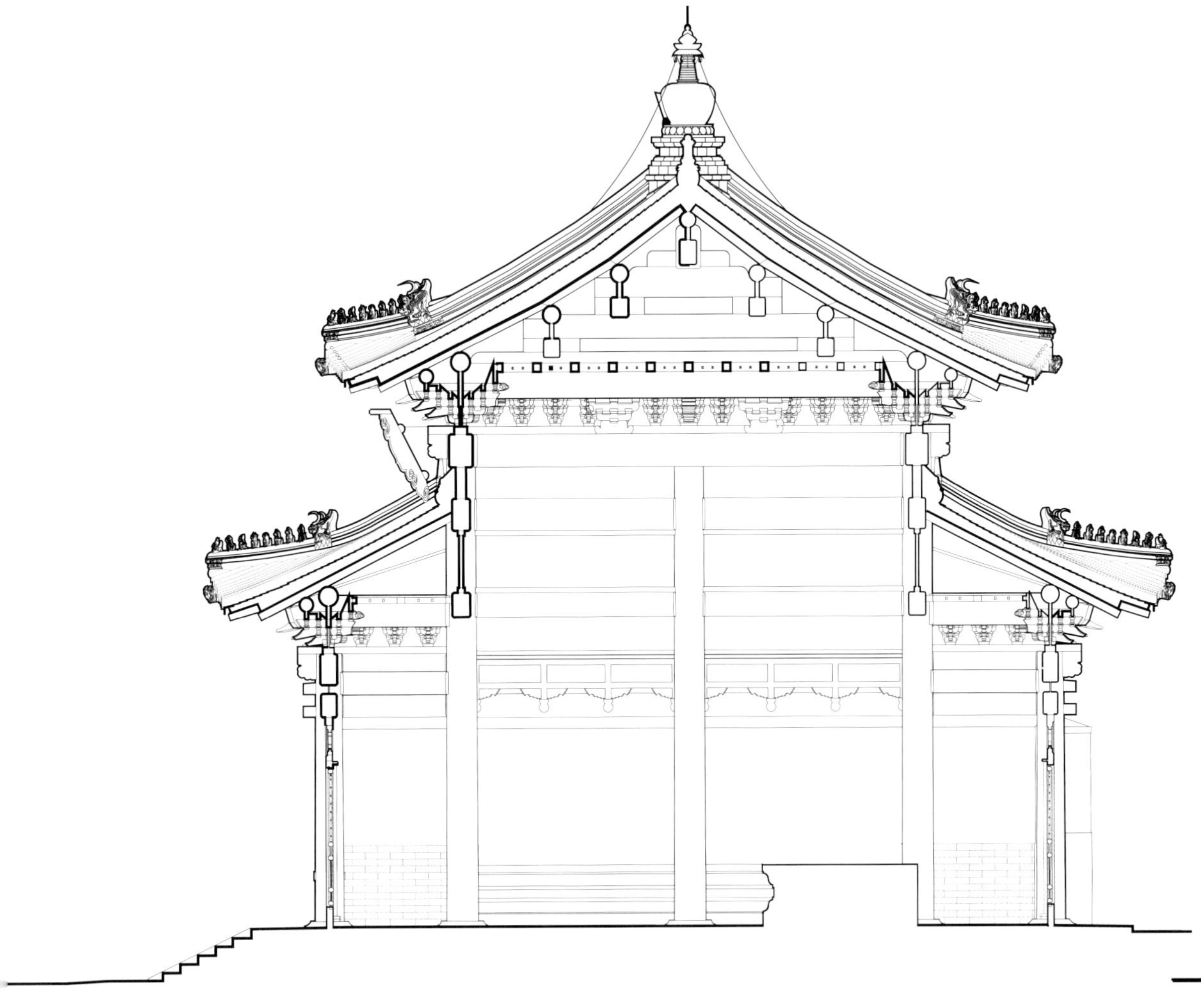

永福寺普光明殿明间剖面图

Section of the Central Chamber within the Pu'guangming Hall,Temple of Perpetual Happiness

永福寺普光明殿梢间剖面图

Section of the Third Chamber within the Pu'guangming Hall,Temple of Perpetual Happiness

梁格庄行宫
Lianggezhuang Palace (Lianggezhuang Xinggong)

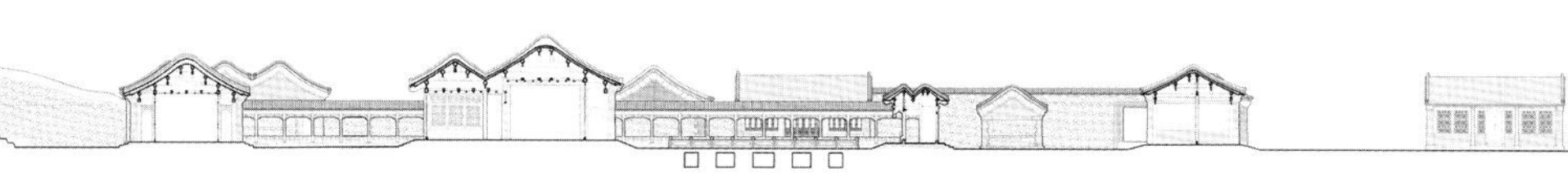

梁格庄行宫组群剖面图
Site section of the building complex of the Lianggezhuang Palace

1 石桥 Stone Bridge
2 朝房 Reception Hall for Court Officials
3 大宫门 Main Gate
4 值房 Guard House
5 膳房 Imperial Kitchen
6 值房 Guard House
7 垂花门 Floral-pendant Gate
8 厢房 Side Building
9 八方池 Octagonal Pool
10 书房 Reading Room
11 永慕斋 Perpetual Veneration Hall
12 东穿堂房 East Hall of Passage
13 三捲殿 Three-wavy-roof Palace
14 后照殿 Back Hall
15 正殿 Main Hall
16 后敞厅 Back Pavilion

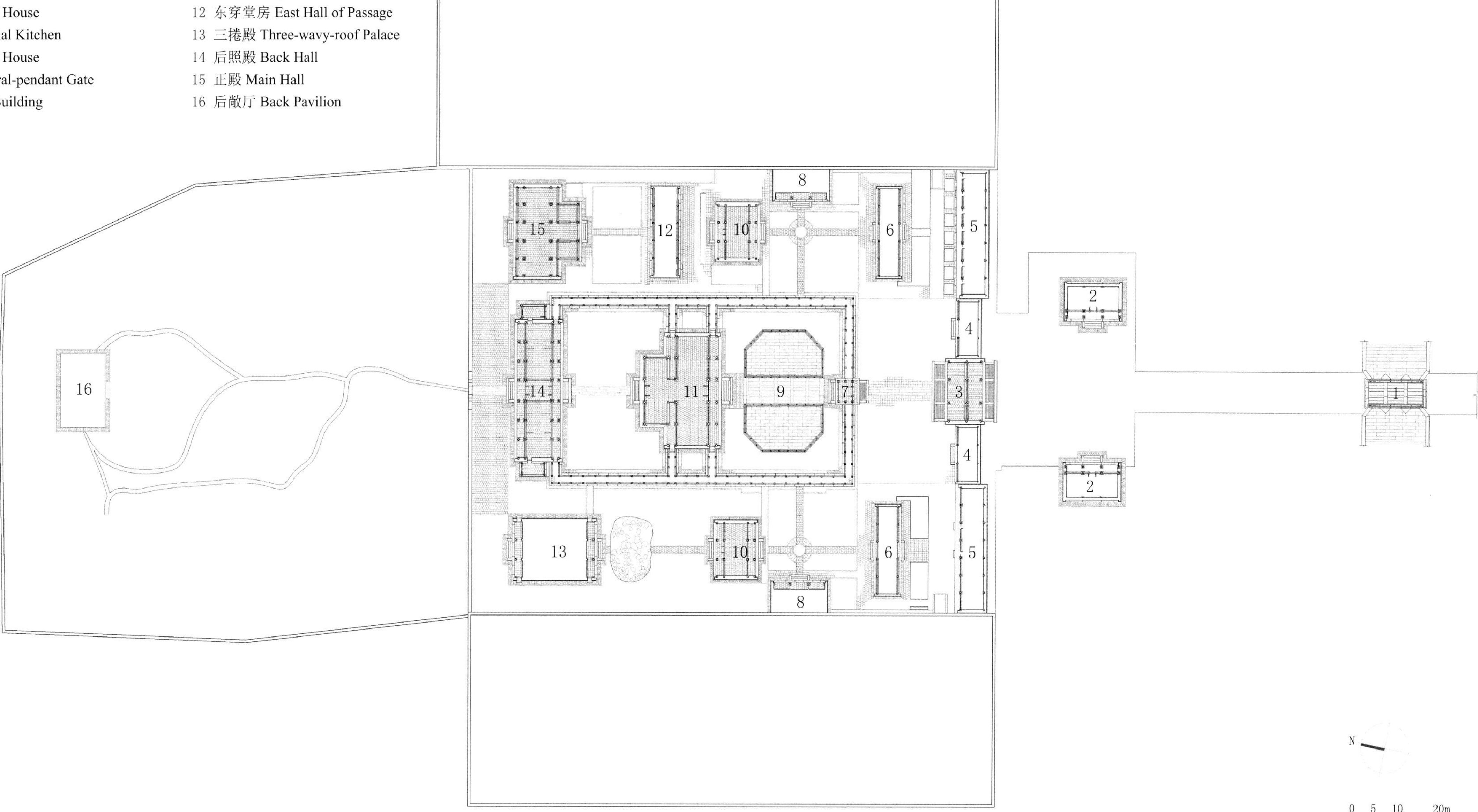

梁格庄行宫组群平面图
Site plan of the building complex of the Lianggezhuang Palace

火焰牌坊
Flame Memorial Gateway (Huoyan Paifang)

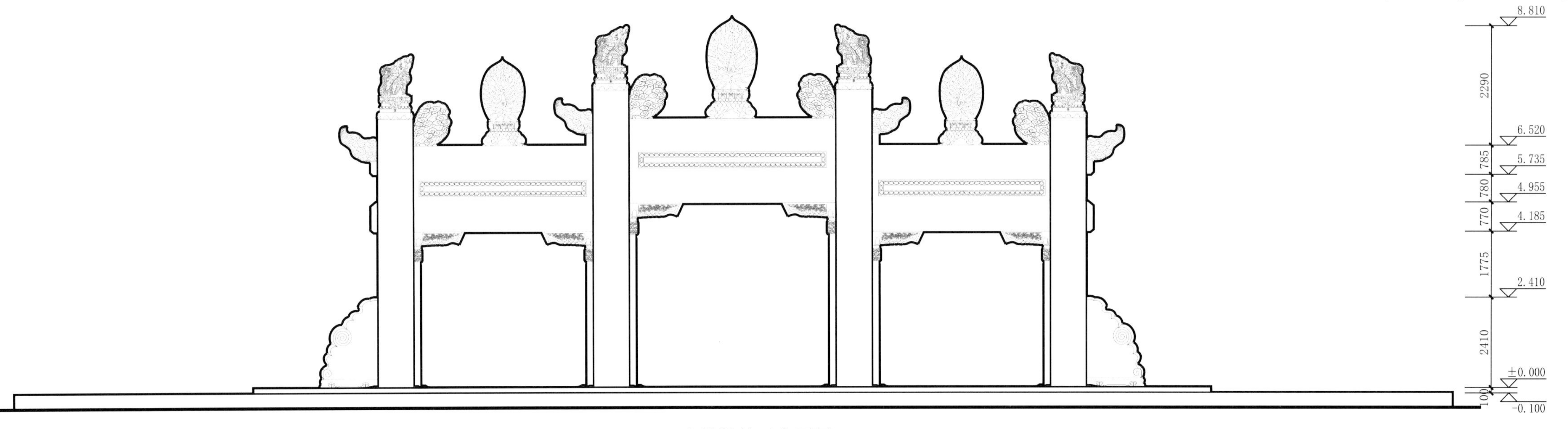

火焰牌坊正立面图
Front elevation of the Flame Memorial Gateway

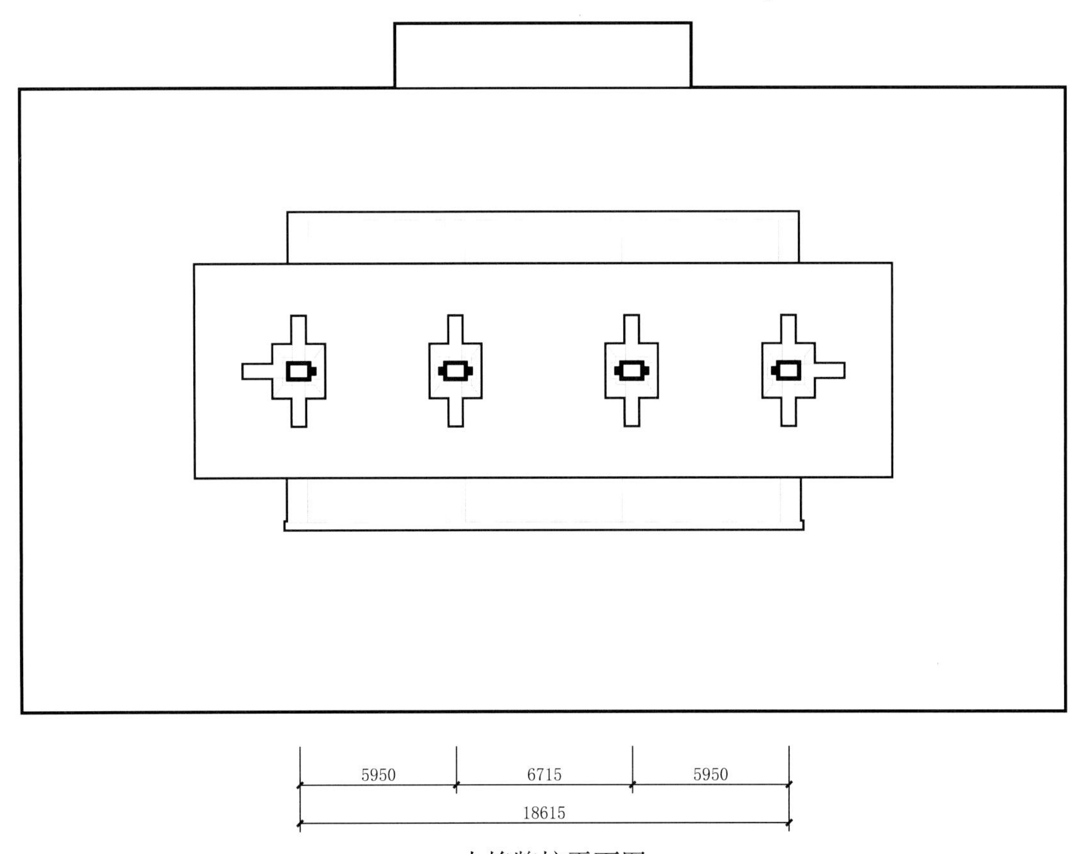

火焰牌坊平面图
Plan of the Flame Memorial Gateway

附录：建筑术语对照表

Appendix：Glossary of Architectural Terms

拼音 （**Pinyin**）	建筑术语 （**Architectural Terms**）	英文释义 （**Defining in English**）
banfang	班房	Duty House for the Guards
baocheng	宝城	Encircled Realm of Treasure
baoding	宝顶	Tumulus
beilou	碑楼	Large Pavilion for the Stela
beiting	碑亭	Pavilion for the Stela
Changling	昌陵	Tomb of Good Fortune
chaofang	朝房	Reception Hall for Court Officials
chishou	螭首	Dragon Heads
Chongling	崇陵	Tomb of Serendipity
chongtian paifang	冲天牌坊	Gateway with Columns Extending into the Sky
dahongmen	大红门	Great Red Gate Building
danbi	丹陛石	Marble Carving on the Steps
digong	地宫	Underground Palace
doubian	斗匾	Wooden Plaque
huabanshi	花斑石	Piebald Stone
erzhumen	二柱门	Gate with Two Columns
fangcheng	方城	Square Walled Terrace
fenbolu	焚帛炉	Sacrificial Burner
fengshuiqiang	风水墙	Geomantic Wall
geshanmen	槅扇门	Two-Panel Lattice Door
gongmen	宫门	Main Gate
huabiao	华表	Ceremonial Column
hundun jietiao	混沌阶条	Huge Cuboid Stones with a Height Equal to that of the Front Platform
huoyan paifang	火焰牌坊	Flame Memorial Gateway
jingting	井亭	Pavilion for the Well
jufudian	具服殿	Hall for Court Robes
kaziqiang	卡子墙	Division Wall
ling'endian	祾恩殿	Hall of Bliss Grace
ling'enmen	祾恩门	Gate of Bliss Grace
linggong	陵宫	Tomb Palace
liuli huamen	琉璃花门	Gate with Glazed Roof Tiles
liuli yingbi	琉璃影壁	Screen Wall of Glazed Tiles
long'endian	隆恩殿	Hall of Monumental Grace
long'enmen	隆恩门	Gate of Monumental Grace
longfengmen	龙凤门	Dragon and Phoenix Gate
longxugou	龙须沟	Covered Drainage Channels
minglou	明楼	Memorial Tower
Muling	慕陵	Tomb of Veneration
paifang	牌坊	Memorial Gateway
peidian	配殿	Side Hall
qilin	麒麟	Legendary Auspicious Animal
queti	雀替	Decorated Bracket
shenchuku	神厨库	Culinary Courtyard for Sacrifices
shendao	神道	Spirit Way
shengdeshengong bei	圣德神功碑	Stela of Sage Virtue and Divine Merit
shengdeshengong beiting	圣德神功碑亭	Pavilion for the Stela of Sage Virtue and Divine Merit
shipaifang	石牌坊	Marble Memorial Gateway
shiwugong	石台五供	Culinary Courtyard for Sacrifices
suiqiang jiaomen	随墙角门	Division Wall with Side Doors
taiji	台基	Tall Foundation Platform
Tailing	泰陵	Tomb of Peace
taiming	台明	Podium
wangzhu	望柱	Ornamental Column
wannian jidi	万年吉地	Burial Site
wujianliuzhu chongtianshi pailoumen	五间六柱冲天式牌楼门	Five-bay-six-into-the-sky-extending-column Memorial Gateway
wukongshiqiao	五孔石桥	Five-Arched Stone Bridge
xiamapai	下马牌	Stela Marking the Place for Dismounting from One's Horse
xiangdian	享殿	Main Sacrificial Hall
xiangfang	厢房	Side Building
xiangsheng	石像生	Stone Statue
xianlou	仙楼	Buddha Shrine
xinggong	行宫	Traveling Palace
xumizuo	须弥座	Mount-Sumeru-shaped Podium
xunfang	汛房	Guard Houses
yabayuan	哑巴院	Mute Courtyard
yashu	衙署	Government Offices
yingfang	营房	Camps
Yongfusi	永福寺	Temple of Perpetual Happiness
yuetai	月台	Front Platform
yueyacheng	月牙城	Crescent Wall

主要参考文献

References

[一] Eduard Kogel. [9783110401349 – The Grand Documentation] Field trips in China (19061909)[M]. Berlin/Boston: Walter de Gruyter GmbH, 2015.

[二] ERNST BOERSCHMANN. Chinesische Architektur[M]. Berlin: Ernst Wasmuth A.G., 1925.

[三] ERNST BOERSCHMANN. Baukunst und Landschaft in China[M]. Berlin: Ernst Wasmuth A.G., 1926.

[四] 刘敦桢. 易县清西陵[J]. 营造学社汇刊. 1941. 5(3). 68-109.

[五] 卢绳. 卢绳与中国古建筑研究[M]. 北京: 知识产权出版社. 2007.

[六] 天津大学建筑学院建筑历史与理论研究所. 天津大学古建筑测绘历程[M]. 天津: 天津大学出版社. 2017.

[七] 王璧文. 清官式石桥做法[J]. 营造学社汇刊. 1935. 5(4). 56-136.

[八] 王其亨. 风水理论研究[M]. 第二版. 天津: 天津大学出版社. 2005.

[九] 王其亨. 风水理论研究[M]. 天津: 天津大学出版社. 1986.

[十] 王其亨. 清代帝陵建筑制度沿革[M]// 清代宫史研究会. 清代皇宫陵寝. 北京: 紫禁城出版社. 1995: 3-22.

[十一] 王其亨. 清代妃园寝建筑制度沿革[M]// 清代宫史研究会. 清代皇宫陵寝. 北京: 紫禁城出版社. 1995: 38-50.

[十二] 王其亨. 清代后陵建筑制度沿革[M]// 清代宫史研究会. 清代皇宫陵寝. 北京: 紫禁城出版社. 1995: 23-37.

[十三] 王其亨. 清代陵寝工程的施工次序和礼仪[M]// 王树卿. 清代宫室从谈. 北京: 紫禁城出版社. 1996: 472-483.

[十四] 王其亨. 中国传统建筑外部空间设计理论探析[M]// 中国艺术研究院中国建筑艺术史编写组. 中国建筑艺术史. 北京: 文物出版社. 1999.

[十五] 王其亨. 中国建筑艺术全集（8）: 清代陵墓建筑[M]. 北京: 中国建筑工业出版社. 2002.

[十六] 徐广源. 清西陵史话[M]. 北京: 新世界出版社. 2004.

参与测绘及相关工作的人员名单

一、1983年测绘人员

指导教师：杨道明　王其亨　刘燕辉　范　挺　覃　力　杨昌鸣
　　　　　崔　愷　张　萍

测绘学生

本科生（1951级）：卜一秋　陈　波　陈根虎　陈向丽　陈新军　程波文
崔鸿麟　方　宏　龚宝光　何建清　贺　薇　江　宏
金丽昌　金卫钧　荆子洋　雷　建　李　定　李家祥
李亚莉　李镇龙　廉树欣　林　立　林　宁　刘恒谦
刘　晔　邱　康　盛　开　孙　通　孙　银　田川平
田　培　王　波　王　洪　王宏宇　王　辉　王健麟
王西京　王晓彤　吴丽娜　肖宇澄　徐江才　许晓春
杨　颖　于一平　于有为　余茂琳　张刚强　张　杰
张兆胜　张　铮　张中增　章　炜　赵建英　赵　齐
赵晓征　赵学军　周　恺　朱剑飞

二、1984年测绘人员

指导教师：冯建逵　王其亨　梁　雪　王　蔚　张玉坤　司小虎
　　　　　孙　刚

测绘学生

本科生（1982级）：常钟隽　陈　兵　陈向东　董静茹　董世海　杜海鹰
樊新和　范　为　韩吉明　何海清　贾锡江　靳小琳
靳元峰　李春舫　李建辉　刘　红　刘　杰　刘恩芳
刘　锋　刘冠华　刘自立　刘作军　马景忠　毛振强
潘　浩　庞志辉　曲　雷　盛海涛　石英波　苏　晨
苏蕴山　孙英波　唐大峰　田　华　涂慧斌　王　臣
王晨雨　王明浩　王太锋　王战洪　王志伟　吴　彬
吴振东　肖雅玲　邢　华　邢金利　薛晓东　郁林元
张　波　张　天（羽　一）　赵　安　祖万安

三、2008年测绘人员

指导教师：王其亨　丁　垚　常清华（研究生）　陈书砚（研究生）

季 宏（研究生） 刘 瑜（研究生） 杨 菁（研究生）
袁守愚（研究生） 赵向东（研究生） 耿 威（研究生）
林 佳（研究生） 朱 蕾（研究生）

测绘学生：

本科生（2006级）：丁 拓 杜 月 杜戎文 杜翛然 冯旭臣 高 梁
郭 俊 郭 珺 侯 凯 黄 墨 郎晓宇 李晓煜
李益帆 刘 畅 刘 晗 刘 寰 刘津津 刘文杰
吕思吉 马洁芳 梅振斌 孙煦暄 孙致伟 陶 昶
王 刚 王 剑 王 康 王 萌 王 倩 王 铮
王宇喆 武靓杰 向雪琪 徐 瑾 杨惠芳 袁小棠
翟建宇 张 驰 张 翀 张 舒 张思锐 张英伦
张英琦 赵 迪 周凤仪

四、2019年测绘人员

指导教师：曹 鹏
辅导员：王 伟
博士研究生：谢怡明 周俊良 李东祖 张静妮
硕士研究生：王 成 冯亚欣 刘欣佳 崔少斌 胡义夫 王艺璇
测绘学生：
本科生（2017级）：聂 月 王铭泽 周 铃 朱秋庚 李恒康 王芮蔓
毛 露 杨雯琪 隋 琦 崔佳星 王星凯 宋 昕
王怡雯 翁童曦 陈鑫怡 李 杰 梁季伟 田培培
席坤杨 刘静娴 闫炳彤 徐彤伟 李沛东 林伟涵
穆荣轩 王雨萌 陈凯昕 常涵宇 江柏霆 秦浩宸
张思远 钱广宇 张亚楠 林雨燕 张琪明 楚雪楠
陈世昭 许佳时 武 璇 於昊臻 翟月阳 路 易
王 越 王舒海 党捷然 王筱琪 郑明远 付天怡
桂溟悦 杨钦惠 刘浩月 刘碧静 申 烁 王雨佳
伍叶子 未爱霖 寇新卓 朱利民 张逸君 索 曼
任英辉 楚田竹 王国政

五、保护规划编制人员：

王其亨 朱 蕾 王茹茹 陈书砚 林 佳 王方捷
李 茜 徐龙龙 张 骏 吕 双 常清华 赵向东

六、清西陵测绘图出版整理

图纸审阅：王其亨 王 蔚 朱 蕾 陈书砚
图纸整理修改：彭思博 朱 琨 张恒媛 李文迪 王 晨 王 越
张 涵 李 雪 何成军 张庆宏 蔡燕丹 陶禹竹
英文统筹：[奥]荷雅丽 吴 葱 朱 蕾 李文迪 张恒媛
英文翻译：[奥]荷雅丽 Michael Norton 周彦邦 庄 岳

List of Participants Involved in Architectural Survey, Drawing, and Work related to the Publication

1. Survey and Drawing in 1983

Supervising Instructor: YANG Daoming, WANG Qiheng, LIU Yanhui, FAN Ting, QIN Li, YANG Changming, CUI Kai, ZHANG Ping

Team Members (Bachelor Students in 1981): BU Yiqiu, CHEN Bo, CHEN Genhu, CHEN Xiangli, CHEN Xinjun, CHENG Bowen, CUI Honglin, FANG Hong, GONG Baoguang, HE Jianqing, HE Wei, JIANG hong, JIN Lichang, JIN Weijun, JING Ziyang, LEI Jian, LI Ding, LI Jiaxiang, LI Yali, LI Zhenlong, LIAN Shuxin, LIN Li, LIN Ning, LIU Hengqian, LIU Ye, QIU Kang, SHENG Kai, SUN Tong, SUN Yin, TIAN Chuanping, TIAN Pei, WANG Bo, WANG Hong, WANG Hongyu, WANG Hui, WANG Jianlin, WANG Xijing, WANG Xiaotong, WU Lina, XIAO Yucheng, XU Jiangcai, XU Xiaochun, YANG Ying, YU YIping, YU Youwei, YU Maolin, ZHANG Gangqiang, ZHANG Jie, ZHANG Zhaosheng, ZHANG Zheng, ZHANG Zhongzeng, ZHANG Wei, ZHAO Jianying, ZHAO Qi, ZHAO Xiaozheng, ZHAO Xuejun, ZHOU Kai, ZHU Jianfei

2. Survey and Drawing in 1984

Supervising Instructor: FENG Jiankui, WANG Qiheng, LIANG Xue, WANG Wei, ZHANG Yukun, SI Xiaohu, SUN Gang

Team Members (Bachelor Students in 1982): CHANG Zhongjuan, CHEN Bing, CHEN Xiangdong, DONG Jingru, DONG Shihai, DU Haiying, FAN Xinhe, FAN Wei, HAN Jiming, HE Haiqing, JIA Xijiang, JIN Xiaolin, JIN Yuanfeng, LI Chunfang, LI Jianhui, LIU Hong, LIU Jie, LIU Enfang, LIU Feng, LIU Guanhua, LIU Zili, LIU Zuojun, MA Jingzhong, MAO Zhenqiang, PAN Hao, PANG Zhihui, QU Lei, SHENG Haitao, SHI Yingbo, SU Chen, SU Yunshan, SUN Yingbo, TANG Dafeng, TIAN Hua, TU Huibin, WANG Chen, WANG Chenyu, WANG Minghao, WANG Taifeng, WANG Zhanhong, WANG Zhiwei, WU Bin, WU Zhendong, XIAO Yaling, XING Hua, XING Jinli, XUE Xiaodong, YU Linyuan, ZHANG Bo, ZHANG Tian(yuyi), ZHAO An, ZU Wan'an

3. Survey and Drawing in 2008

Supervising Instructor: WANG Qiheng, DING Yao, CHANG Qinghua (Postgraduate student), CHEN Shuyan(Postgraduate student), JI Hong (Postgraduate student), LIU Yu (Postgraduate student), YANG Jing (Postgraduate student), YUAN Shouyu (Postgraduate student), ZHAO Xiangdong (Postgraduate student), GENG Wei (Postgraduate student), LIN Jia (Postgraduate student), ZHU Lei (Postgraduate student)

Team Members (Bachelor Students in 2006): DING Tuo, DU Yue, DU Rongwen, DU Xiaoran, FENG Xuchen, GAO Liang, GUO Jun, GUO Jun, HOU Kai, HUANG Mo, LANG Xiaoyu, LI Xiaoyu, LI Yifan, LIU Chang, LIU Han, LIU huan, LIU Jinjin, LIU Wenjie, LYU Siji, MA Jiefang, MEI Zhenbin, SUN Xuxuan, SUN Zhiwei, TAO Chang, WANG Gang, WANG Jian, WANG Kang, WANG Meng, WANG Qian, WANG Zheng, WANG Yuzhe, WU Liangjie, XIANG Xueqi, XU Jin, YANG Huifang, YUAN Xiaotang, ZHAI Jianyu, ZHANG Chi, ZHANG Chong, ZHANG Shu, ZHANG Sirui, ZHANG Yinglun, ZHANG Yingqi, ZHAO Di, ZHOU Fengyi

4. Survey and Drawing in 2019

Supervising Instructor: CAO Peng

Student Administrator: WANG Wei

PHD Students: XIE Yiming, ZHOU Junliang, LI Dongzu, ZHANG Jingni

Master Students: WANG Cheng, FENG Yaxin, LIU Xinjia, CUI Shaobin, HU Yifu, WANG Yixuan

Team Members (Bachelor Students in 2019): NIE Yue, WANG Mingze, ZHOU Ling, ZHU Qiugeng, LI Hengkang, WANG Ruiman, MAO Lu, YANG Wenqi, SUI Qi, CUI Jiaxing, WANG Xingkai, SONG Xin, WANG Yiwen, WENG Tongxi, CHEN Xinyi, LI Jie, LIANG Jiwei, TIAN Peipei, XI Kunyang, LIU Jingxian, YAN Bingtong, XU Tongwei, LI Peidong, LIN Weihan, MU Rongxuan, WANG Yumeng, CHEN Kaixin, CHANG Hanying, JIANG Baiting, QIN Haochen, ZHANG Siyuan, QIAN Guangyu, ZHANG Yanan, LIN Yuyan, ZHANG Qiming, CHU Xuenan, CHEN Shizhao, XU Jiashi, WU Xuan, YU Haozhen, ZHAI Yueyang, LU Yi, WANG Yue, WANG Shuhai, DANG Jieran, WANG Xiaoqi, ZHENG Mingyuan, FU Tianyi, GUI Mingyue, YANG Qinhui, LIU Haoyue, LIU Bijing, SHEN Shuo, WANG Yujia, WU Yezi, WEI Ailin, KOU Xinzhuo, ZHU Limin, ZHANG Yijun, SUO Man, REN Yinghui, CHU Tianzhu, WANG Guozheng

5. Protection Planning

WANG Qiheng, ZHU Lei, WANG Ruru, CHEN Shuyan, LIN Jia, WANG Fangjie, LI Qian, XU Longlong, ZHANG Jun, LYU Shuang, CHANG Qinghua, ZHAO Xiangdong

6. Participant involved in the compiling of survey drawings of the Western Qing Tombs for publication

Drawings Review: WANG Qiheng, WANG Wei, ZHU Lei, CHEN Shuyan

Team Members: PENG Sibo, ZHU Kun, ZHANG Hengyuan, LI Wendi, WANG Chen, WANG Yue, ZHANG Han, LI Xue, HE Chengjun, ZHANG Qinghong, CAI Yandan, TAO Yuzhu

Translator in chief: Alexandra Harrer, WU Cong, ZHU Lei, LI Wendi, ZHANG Hengyuan

Team Member: Alexandra Harrer, Michael Norton, CHOU Yen Pang, ZHUANG Yue

图书在版编目（CIP）数据

清西陵 =WESTERN QING TOMBS / 王其亨主编；天津大学建筑学院，易县清西陵文物管理处合作编写；朱蕾，陈书砚，王其亨编著 . —北京：中国建筑工业出版社，2018.6

（中国古建筑测绘大系 . 陵寝建筑）

ISBN 978-7-112-22327-5

Ⅰ . ①清… Ⅱ . ①王… ②天… ③易… ④朱… ⑤陈… Ⅲ . ①陵墓—建筑艺术—河北—清代—图集 Ⅳ . ① TU251.2-64

中国版本图书馆CIP数据核字（2018）第123632号

从书策划 / 王莉慧
责任编辑 / 李　鸽　陈海娇
书籍设计 / 付金红
责任校对 / 王　烨

中国古建筑测绘大系・陵寝建筑

清西陵

天津大学建筑学院
易县清西陵文物管理处　合作编写

王其亨　主编

朱　蕾　陈书砚　王其亨　编著

Traditional Chinese Architecture Surveying and Mapping Series: Tomb Architecture

WESTERN QING TOMBS

Compiled by School of Architecture, Tianjin University & the Western Qing Tombs Cultural Relics Management Office, Yixian
Chief Edited by WANG Qiheng
Edited by ZHU Lei, CHEN Shuyan, WANG Qiheng

*

中国建筑工业出版社出版、发行（北京海淀三里河路9号）
各地新华书店、建筑书店经销
北京方舟正佳图文设计有限公司制版
北京雅昌艺术印刷有限公司印刷

*

开本：787 毫米 ×1092 毫米　横 1/8　印张：42½　字数：1126 千字
2022 年 2 月第一版　2022 年 2 月第一次印刷
定价：**328.00** 元
ISBN 978-7-112-22327-5
（32150）